By Al
James Acret ESQ.

CONSTRUCTION NIGHTMARES

Jobs From Hell And How To Avoid Them

BNi Building News

Los Angeles • New England • Anaheim • Washington D.C.

**Stories by Arthur F. O'Leary, Architect
Legal Analysis by James Acret, Lawyer**

BNi Building News

Editor-In-Chief
William D. Mahoney, P.E.

Technical Services
Rod A. Yabut

Design
Robert O. Wright

BNi PUBLICATIONS, INC.

LOS ANGELES
10801 National Blvd., Ste.100
Los Angeles, CA 90064

ANAHEIM
1612 S. Clementine St.
Anaheim, CA 92802

NEW ENGLAND
172 Taunton Ave. Ste.2
PO BOX 14527
East Providence, RI 02914

WASHINGTON, D.C.
502 Maple Ave. West
Vienna, VA 22180

1-800-873-6397

Copyright © 1999 by BNi Publications, Inc. All rights reserved. Printed in the United States of America. Except as permitted under the United States Copyright Act of 1976, no part of this publication may be reproduced or distributed in any form or by any means, or stored in a data base or retrieval system, without the prior written permission of the publisher.

While diligent effort is made to provide reliable, accurate and up-to-date information, neither BNI Publications Inc., nor its authors or editors, can place a guarantee on the correctness of the data or information contained in this book. BNI Publications Inc., and its authors and editors, do hereby disclaim any responsibility or liability in connection with the use of this book or of any data or other information contained therein.

ISBN 1-55701-298-9

About the Authors

James Acret, a member of the California State Bar, is a graduate of UCLA Law School and has practiced Construction Law over four decades. In his years of service to the construction industry, he has represented numerous contractors, architects, owners, developers, and sureties. He was a member of the committee that rewrote the California Mechanics Lien Law in 1969 and is considered the Dean of construction lawyers. He is the author of several legal publications and is the Editor of California Construction Law Reporter. He lives with his wife, Laiah Lee, in Pacific Palisades, California.

Arthur F. O'Leary, FAIA, MRIAI, co-founded the architectural firm of O'Leary Terasawa Partners now merged into Widom Wein Cohen O'Leary Terasawa, Santa Monica, California. He has served as arbitrator and consultant on more than 300 industry litigations and has written, lectured, and published extensively in over 40 years of professional activity. He received his Bachelor of Architecture from the University of Southern California, where he served on the faculty for 10 years. He also received a Certificate in Real Estate from the UCLA School of Management. He lives with his wife, Inny, near Dublin in Drogheda, County Louth, Ireland.

Contents

Acknowledgements .... *IX*

1
The Foolproof Construction Contract
An Attempt to Write a No-Extra Contract *1*
Points of Law
Who Pays for the Extra Work *9*

2
ABC Warehouse I
A Construction Defect *11*
Points of Law
Allen Gets a Cracked Slab *21*

3
ABC Warehouse II
Winding Up the Job *27*
Points of Law
Job Close-Out *37*

4
ABC Warehouse III
A Maintenance Failure *39*
Points of Law
Plugged Roof Drains *47*

5
The Tulare Job
A Costly Labor Relations Problem. A Disastrous Subcontract *49*
Points of Law
Problems With Labor Relations *59*

6
ABC Warehouse IV
The Mysterious Retaining Wall Collapse *61*
Points of Law
Retaining Wall Collapse *73*

7
ABC Warehouse V
Earthquake Damage or a Latent Defect? *75*
Points of Law
Glulam Beam Failure *87*

8
The Joint Venture
A Bonded Roofing Contract *89*
Points of Law
Choose Your Joint Venturer More Carefully Than Your Spouse *97*

9
Shangri-La Gardens
The Short Life of Eternal Waterproofing *101*
Points of Law
Intentional Misconduct *113*

10
ABC Warehouse VI
The Pressure Regulator Ordered by the Building Inspector *117*
Points of Law
Who Pays for the Pressure Regulator? *123*

11
Davren Effingwell dba Economy Home Improvement Company I
Who Protects the Owner? Conflicts of Interest *125*
Points of Law
The dangers of Contracting With a Friend *137*

12
ABC Warehouse VII
Faulty Contractor Selection and Contract Administration *141*
Points of Law
Allen Hires His High School Buddy *153*

13
Davren Effingwell dba Economy Home Improvement Company II
Cleaning Up the Mess *159*
Points of Law
The Aftermath *171*

14
ABC Warehouse VIII
Shop Drawings Procedure *173*
Points of Law
Shop Drawings, Changes, Claims, and Dispute Resolution *185*

15
Venture Tower
Responsibility for Window Wall Leaking *191*
Points of Law
Fraudulent Shop Drawings *207*

16
Psi Creativity Center I
The Design and Bidding Phases *211*
Points of Law
A Difficult Client for an Architect *223*

17
Psi Creativity Center II
The Construction Period *229*
Points of Law
Quagmire Employs His Patient as a Contractor *241*

18
ABC Warehouse IX
Owner-Nominated Subcontractor *247*
Points of Law
Allen Helps His Brother-in-Law *261*

19
ABC Warehouse X
Owner's Separate Contractor *265*
Points of Law
Allen Hires a Decorator *279*

20
Pacific Horizon I
Value Engineering – The Contract Negotiation Phase *283*
Points of Law
Selling the Job *295*

21
Pacific Horizon II
Value Engineering – The Construction Phase *299*
Points of Law
The Construction Period *311*

22
Pacific Horizon III
Value Engineering – After Completion of Construction *315*
Points of Law
Living With the Building *325*

23
ABC Warehouse XI
Substitutions in Lieu of Specified Materials *327*
Points of Law
Substitutions *335*

24
Parc L'Cockaigne
Discovering the Cause of Construction Defects *337*
Points of Law
Condominium Defects Caused By Contractor Ignorance *345*

25
ABC Warehouse XII
A corrupt Alliance – A Criminal Conspiracy *347*
Points of Law
Skullduggery and Corruption *355*

26
ABC Warehouse XIII
Home Improvements *359*
Points of Law
Allen's Deck – The Independent Contractor *371*

27
Utopian Villas
Keeping Costs Under Control *373*
Points of Law
Construction Documents for an Owner-Builder-Developer *387*

28
The Chief Estimator
Getting the Job by Hook or by Crook *389*
Points of Law
The Crooked Estimator *401*

Index –
Stories *405*

Index –
Points of Law *410*

Acknowledgements

The construction stories in this book illustrate in realistic context various of the architectural, construction, and legal principles that permeate the design, construction, and construction administration processes. The Points of Law following each story identify and explain these principles in clear and simple language.

Although the stories and characters are fictional, many of the situations are based on actual events encountered in a busy Southern California architectural practice spanning four decades.

Most of the stories are based on the use of the standard AIA document system and refer to the documents current in the relevant time frames. AIA's documents undergo a constant program of review and revision. Therefore, anyone contemplating administrative or legal action based on these fictional situations should check the latest versions of the AIA forms and, when in doubt, should seek professional advice.

The stories originally appeared in the California Construction Law Reporter, a monthly periodical read by construction lawyers. We thank the publisher, Bancroft Whitney, a division of West Group, for permission to reprint.

Technical information needed to make the stories realistic was obtained from these professional associates who submitted generously and willingly to shameless brain-picking: Paul Bennett, Mechanical Engineer; Sidney Berke, Electrical Engineer; Richard Hunter, AIA, RIBA; Lawrence O. Mackel, Architect and Structural Engineer; Robert M. Zigman, AIA.

James Acret, of the California State Bar

Arthur F. O'Leary, FAIA, MRIAI

The Foolproof Construction Contract
An Attempt to Write a No-Extra Contract

Baker/Doyle was a thriving construction company, highly experienced in the supermarket field. In addition to over 50 market projects they had completed for customers, they had built and still owned 15 under their investment subsidiary, Baker/Doyle Properties. All 15 were on long-term leases to Frugalmart Stores, Inc, one of the more stable retail operators. There was little that the partners, Abel Baker and Charles Doyle, didn't know about supermarket construction.

The Frugalmart buildings were typically 150 feet by 180 feet, 27,000 square feet, with a generous loading dock area and a 135 car landscaped parking area.

In early October 1992, Abel Baker had spotted a suitable piece of recently cleared land, about 100,000 square feet, in a district where Frugalmart was interested in locating. The real estate broker readily disclosed the existence of several feet of old fill soil over most of the site. This did not bother Baker/Doyle, as they would simply dig a test hole at each of the four corners of the proposed building location and allow for the estimated extra cost of deeper foundations.

Upon obtaining permission to dig the test holes, they asked Ivan Jenson, their excavation contractor, to send over a man and a backhoe to sink the four holes. The site was more or less level and they recorded the depths of fill at the four corners as 2 feet at the NW corner, 5 feet at the NE corner, 3 feet at the SW corner, and 6 feet at the SE corner. They logically assumed that it would be a straight line, more or less, between the four measured natural earth levels.

After allowing for their estimate of extra foundation expense, the land cost was still reasonable so they optioned it for 60 days to allow time to finalize a lease with Frugalmart.

Negotiations with the leasing executive at Frugalmart resulted in a satisfactory long term lease, so Baker/Doyle properties exercised their option to purchase the property on November 23, 1992. On the same day they visited and hired Flynn & Hill, Architects and Engineers, who had designed all their previous supermarkets.

The partnership of Eugene Flynn, Architect and Gerald Hill, Structural Engineer had a well-earned reputation for designing attractive and economical supermarkets and their work was acceptable to the major supermarket tenants. They were known among their peers for design innovation and competent documentation.

* * *

At their first meeting with Baker/Doyle, Gene and Gerry insisted on a boundary and topographic survey of the property and a foundation investigation. Charlie Doyle said okay on the land survey but was not keen on the soil testing. He rationalized, "We already know about the fill and its depth so why waste good money on soil testing?"

Gerry Hill advised, "If we don't have a soil test then we'll have to assume that the maximum allowable pressure on the natural soil is 1500 pounds per square foot. But if we obtain a soil test it might disclose a higher allowable value and you could save some money on concrete and reinforcing."

Doyle said they had built on a similar site nearby and the soil test substantiated a 2000 pounds per square foot value, "So let's use 2000 psf."

Hill replied, "No, Charlie, we can use only 1500 psf unless we see a soil test. The Building Department will take the same position. Although 1500 psf is fairly conservative, we'd be still happier with a soil test." Charlie Doyle authorized the land survey but not the foundation investigation.

* * *

So Flynn & Hill designed the building. Baker/Doyle was satisfied with the design sketches and submitted them to Frugalmart who promptly approved them. Flynn & Hill proceeded with the construction drawings and specifications on December 15, 1992. On the foundation plan and details they showed the wall and column footings extending down through the fill soil and 18 inches into the natural soil. The structural engineers used 1500 pounds per square foot soil pressure for designing the footings. On the foundation plan they placed a note saying: "All footings shall be founded 18 inches into natural undisturbed soil. Minimum bearing capacity shall be 1500 pounds per square foot."

Flynn & Hill was hired only for design, production of the construction documents, and obtaining the building permit but not for administration of the construction contract. Baker/Doyle felt they could handle the contract administration themselves. After all, they reasoned, they were experienced general contractors and it was their own building. Field inspection or consultation, if any is needed, would be on an as-called-for basis.

The architects and engineers worked diligently, finished the plans and specifications, and applied for the building permit on February 15, 1993. It was issued on April 2, 1993.

Baker/Doyle was quite busy at the time with three supermarkets and two warehouses then under construction, tying up their best superintendents. They didn't feel they could do their own job justice and complete it on time to satisfy their commitment to Frugalmart. So they decided to hire Lee-Noonan Constructors, experienced supermarket builders, to build it for them. Before going into the contracting business together, Kenneth Lee had been an estimator with Baker/Doyle, and Michael Noonan had been a lawyer in a large law firm specializing in construction industry matters. The partners of the two contracting firms were on a friendly competitive basis.

They negotiated a contract. Baker/Doyle said they wanted a lump sum proposal including a fair profit. Both firms were experienced in supermarket building and nobody knew any more about the business than they. The plans and specs were excellent, they had the permit in hand, so they expected no building department corrections. They knew the depth of fill, so the exact price could be determined. They agreed there were to be no extras. Both firms accepted this simple concept. Baker/Doyle would be ideal customers, since they would not be unreasonable like most of the other customers of both firms. Everything would go smoothly. There would be no justification for any extras.

They would prepare the contract on this basis. Lee-Noonan's skilled estimating staff carefully estimated all the quantities, called in their most trusted subcontractors, and computed their best overall price. Meanwhile, Baker/Doyle had their own estimating department figure the job. Upon detailed analysis of the Lee-Noonan price proposal, they felt that it was reasonable and they were happy to accept it. The price was right and they felt comfortable with Lee-Noonan.

Michael Noonan prepared the contract. He used Standard Form of Agreement Between Owner and Contractor (where the basis of payment is a Stipulated Sum), Twelfth Edition, AIA Document A101, 1987, General Conditions of the Contract for Construction, Fourteenth Edition, AIA Document A201, 1987, and incorporated Flynn & Hill's construction drawings and specifications. In A101, among other additional provisions, he added a clause that said, "It is contemplated between the parties hereto that there will be no extra costs arising from the work of this contract." This would guarantee a profit for Lee-Noonan and a stable price for Baker/Doyle. It was foolproof. How could anything go wrong?

* * *

At 5:00 PM on Friday, April 23, 1993, the four contractors met in Baker/Doyle's offices. They discussed the contract and their mutual pleasure on entering a contract with knowledgeable parties on both sides. All conditions were perfect. There should be no problem in meeting the time schedule as there would be no owner's changes to hold up the job and no extras. Charlie Doyle laughingly reminded Ken and Mike that Baker/Doyle would be watching them so there is no point in claiming changed conditions or constructive acceleration or impact or any others of the sometimes specious contractor excuses for extras. The four good humored contractors signed the contract and opened a bottle of specially chilled champagne. After lavish toasts, sincere congratulations, and pleasant banter, they went out to dinner, guests of Lee-Noonan Constructors. Ken and Mike were truly proud of being trusted so implicitly by their respected senior competitors, Baker/Doyle. Spirits were high and all were euphoric with optimism and the warmth of mutual respect and trust.

* * *

A little over a week later, on Monday, May 3, 1993, Lee-Noonan started mobilizing the job in the field. One of their best superintendents, Oran Peters, was available and was assigned to the job. He immediately got the site fenced, a job office, telephone, and chemical toilets installed, and temporary water and electrical facilities in place. He had previously ordered a large sign identifying the Frugalmart project with the names of Lee-Noonan, Baker/Doyle, and Flynn & Hill prominently displayed, and it was now sturdily installed.

Exercising caution, Lee-Noonan hired Quality Engineering, who had done the land survey, to locate and mark the property lines, lay out the building on the site, and establish a working bench mark for construction and grade levels. Ivan Jenson, low bidder on excavating, had moved onto the site with earth moving equipment and trucks. After completing the general site grading in less than a week, they started excavating for wall and column footings on Monday, May 10.

They dug down to the original undisturbed earth levels and continued on down a minimum of 18 inches more into the virgin soil. The fill depths proved to be exactly as predicted. After leveling off the bottoms of all excavations and cleaning out the loose earth, they started setting and tying the steel reinforcing in the footings. Dowels were set for overlapping with the masonry wall reinforcing to come above. Meanwhile, the steel column bolt templates arrived from the steel fabricator. By Friday they were ready for building department inspection of the footing excavation and reinforcing called for on the following Monday. After Monday's approval, they would start pouring the concrete footings on the following day, right on schedule.

* * *

On Monday, May 24, at 9:00 AM the Building Department Inspector, Ray Scott, showed up to look at the work in process. He had already briefly reviewed the drawings in his City Hall office the previous afternoon in preparation for this inspection. He introduced himself to Oran Peters and asked to look at the Inspection Record. They walked over to the footing excavation, paced solemnly around the entire building perimeter, then went inside to look at the two large isolated column pad excavations. The jobsite was neat, the work looked good, and Oran Peters was proud of it. He wondered what the inspector could find to criticize without nit-picking. Scott picked up an 8-foot length of #4 rebar and, using it, prodded the bottom of the excavation in various places. He commented, "It seems a little spongy. I'm not sure this soil would test up to 1500 psf. I may be wrong. Maybe it's okay. But I can't approve it. You'll have to get the soil tested to prove it's good for 1500 psf loading." Oran didn't know what to say and after some feeble defense of the soil, he gave up. Scott filled in a red tag and told Peters not to do any more work until the soil value was resolved between the structural engineer and the building department. He apologized to the deflated superintendent and left.

As soon as the inspector drove off, Peters telephoned the Lee-Noonan office and asked for Mike. Noonan's secretary came on the line to say he was busy in a conference. Oran ordered her to interrupt him. "This is a disaster!" Seconds later, Mike was on the line. Oran spilled out the whole story, finishing with, "Now we're closed down. What'll I do?" Mike told him to secure the site and come on in to the office.

Mike immediately dialed Baker/Doyle and got Charlie Doyle on the line. "The job's closed down by the Building Department. Something's wrong with the soil. We gotta get a soil report!"

Charlie snapped, "Who sez?"

"Our superintendent, Oran Peters, just called and said the Building Inspector closed down the job and demanded a soil report. I thought you guys had all the necessary engineering and approvals. What'll we do now?"

"I'm sure everything's all right. I'll look into it, Mike, and call you right back!" They hung up, both concerned and puzzled.

Charlie quickly briefed his partner Abe on Mike's phone call and Abe anxiously hovered while Charlie dialed Flynn & Hill's number. He asked for Gerry Hill and recapped the Noonan phone call. Gerry said, "Come over here this afternoon after lunch, say 2 o'clock. I'll have talked to Oran Peters and the Building Department before you get here."

* * *

Abe and Charlie arrived at Flynn & Hill's office promptly at 2 o'clock. Hill said that he'd talked to the job superintendent and one of the Building Department engineers about the soil problem. He explained that the Building Inspector, Ray Scott, had been prodding the natural bearing soil and that he felt that it might not test up to 1500 pounds per square foot. Hill explained, "This isn't a scientific method of testing soil, but now that he's raised the issue there's nothing left to do but order a soil investigation."

Charlie asked, "Can they do this to us? They approved the plans and issued the building permit! I thought we were home free!"

Abe, concurring, chimed in, "Yeah! Home free!"

Gerry replied, "There's no point in moaning about it. The Building Inspector is just being cautious and is within his rights. Let's call Tom Usher at GeoPhysical Technology and see how soon they can take some in situ soil samples. After we know some facts we can analyze the situation. Until then we're just irresponsibly speculating." The contractor partners nodded their assent.

So Gerry Hill telephoned GeoPhysical Technology and Tom Usher said he could take the samples in about 3 weeks. Hill earnestly explained the crucial urgency and got Tom to agree to take samples tomorrow, Tuesday, May 25, have tentative results late the next day, and a final report for the Building Department on Wednesday, June 2. Tom said this was ultrafast service for an old friend. GeoPhysical's fee arrangements were acceptable to Abe and Charlie. Gerry thanked Tom and told him go ahead.

Charlie said, "Maybe we should have gotten soil tests before the design was started?"

Gerry Hill reminded him, "Perhaps you've forgotten that was what I recommended at the time?"

"Yeah, I remember. You should have talked us into it!"

* * *

The preliminary soil report was hand delivered to Flynn & Hill's at 4:30 PM on Wednesday as Tom Usher had promised. It indicated that the soil is good for only 1350 psf but would be increased to 2000 by digging 2 feet deeper. Gerry Hill called another meeting with Baker and Doyle for Thursday, 10:00 AM, May 27.

* * *

Gerry started by showing Abe and Charlie the preliminary report and explaining the practical consequences. "It appears that the natural soil near the surface isn't very good. Although it would support the building if we merely widened the footings 10 or 15 percent, we'd have to alter the reinforcing to take care of the extra footing overhang. Alternatively we could deepen the excavations two feet to get into the higher value soil. In either case the reinforcing will have to be removed to redo the earthwork.

"In my opinion, it would be a better building to use the second option and found the footings in better soil. This would also entail adding 2 feet to the height of the exterior walls and a concrete pedestal at each of the interior columns."

After considerable discussion the partners decided that they should take Gerry's advice and build the better way. After all, they reasoned, they would own the building for a long time.

They authorized Flynn & Hill to proceed with preparation of the additional calculations and revised drawings. They'd submit the revisions to the Building Department when the final soils report arrived next Wednesday.

* * *

The final report was hand delivered to Flynn & Hill at 10:00 AM, Wednesday. It substantiated the earlier preliminary results. At 2:00 PM, Hill brought the whole redesign package to the Building Department and went over it with a senior engineering plan checker. He said to leave it so he could study it more carefully and review it with his superior. If it was okay, he would issue a revised permit tomorrow morning. At 10 AM, June 3, Hill picked up the approval and revised permit upon payment of the supplementary fee. He then delivered it to Baker/Doyle's offices.

* * *

Charlie Doyle had the revised drawings and Building Department approval delivered to Lee-Noonan's offices before noon.

Lee-Noonan reorganized the job and was back in full production the next morning, Friday, June 4. They removed most of the reinforcing that day. On Monday, June 7, they started deepening all the excavations, and by Thursday, June 10, they were ready to reinstall the reinforcing. However, this time they had made arrangements to have one of Tom Usher's engineering geologists on hand to examine the soil at the bottom of the excavations and approve it before replacing the reinforcing steel. The soil was okay and they commenced reinstalling the rebar.

By Monday, June 14, they were ready for Inspector Scott to re-examine their work before pouring the concrete. Early on Tuesday, June 15, Scott approved the excavations and reinforcing so they were able to pour concrete on Wednesday, June 16. Lee-Noonan had previously alerted XYZ Readymix that they wanted prompt service and they got all the footings poured that day.

Fortunately, the job continued smoothly from there on. No more glitches. They'd lost 22 days.

* * *

Life was back to normal at Baker/Doyle. All their jobs were running smoothly and they'd heard no more bad news from Lee-Noonan's Frugalmart job. Then, on Monday, June 28, a Lee-Noonan invoice arrived in the mail. It was for the extra costs of removing and replacing the reinforcing steel, deepening the excavations, disposing of the excess earth, the additional masonry wall height, the concrete column pedestals, standby costs of closing down the job for 22 days, overhead and profit, a total of $39,855.

Charlie, his face white, took the bill into Abe's office, threw it on the desk and flung himself into a chair. Abe read the bill and sat there with his mouth agape. The two partners were incredulous and outraged. The contract, clearly, was to have no extras. Everyone had agreed to this concept. Lee-Noonan is trying to screw us. They've got no ethics. Charlie then telephoned Mike and let him have it. The old camaraderie seemed to have vanished. Mike said it would be better to meet and talk it over, calmly and rationally. Why not meet at Flynn & Hill's office? So Mike volunteered to set it up.

* * *

The next afternoon at 4 o'clock the four partners of the two contracting firms were formally seated on opposite sides of the large table in Flynn & Hill's conference room. Architect Flynn was sitting at one end with Engineer Hill facing at the opposite end.

The contractors were poised, alert, grim faced, and quiet, so Gene Flynn tentatively assumed control of the meeting. Gerry Hill and Mike Noonan were at the ready with note pads and pens.

Gene started out by saying that Flynn & Hill has not been retained for contract administration nor dispute resolution services. He and Gerry now are just trying to accommodate old friends.

At this point, Charlie interrupted to say, "This'll be very simple. We have a contract here, signed by Lee-Noonan and Baker/Doyle, that says no extras! Now we've received a billing from Lee-Noonan for an extra 40 thousand dollars. Signed contracts don't seem to mean anything to some people." Abe concurred, "Yeah, some people!"

Gene, feeling his way, suggested that both sides should state their viewpoint. "Mike, please explain Lee-Noonan's position on this."

Mike Noonan cleared his throat and began, "We admit that the contract contemplated no extras. But what occurred was clearly not anticipated by either party. The soil turned out to have less bearing capacity than Baker/Doyle thought it had. It's their soil, not ours. The costs we billed were honestly incurred. We should be reimbursed in full for all our costs and expenses plus a reasonable profit."

Gene nodded to Charlie, who calmly replied, "We don't question the honesty of the billing. That's not the issue. We had a no extra agreement. No extras means no extras. We talked it over and agreed to it. That's all there is to it."

Abe sagely nodded agreement and added, "We owe nothing."

Gene Flynn, always optimistic, suggested that maybe this could be settled by compromising the bill on a 50/50 basis. All four contractors reacted vehemently, objecting.

Mike Noonan replied, "No. We spent the money. It's a fair bill and we'll collect it in full."

Charlie said, "No extras. We owe nothing!"

Gerry Hill suggested that they take a short recess for a cup of coffee. He arranged for coffee to be served and he and Gene went into an adjoining office, closing the door. They discussed the possibility of recommending mediation but didn't think it would be likely to succeed, since both parties wanted all or nothing. They felt that the spirit of compromise required in a mediation was completely lacking in this case. They decided to recommend arbitration as a more practical procedure under the circumstances.

The six reassembled in the conference room and closed the door. Resuming the chair, Gene invited suggestions. Abe immediately suggested where Lee-Noonan could put their bill. No one rose to the bait.

Mike Noonan reminded them that the Frugalmart contract included an arbitration provision in Article 4.5 of the AIA general conditions, AIA Document A201. He also pointed out that all references to the architect in the dispute resolution process in Articles 4.3 and 4.4 had been stricken as Flynn & Hill had not been retained for those services. Therefore, the architect's decision was not required as a precedent to arbitration.

Gene urged that the dispute be submitted to arbitration and let an experienced construction arbitrator decide the matter. After considerable discussion, the two firms of contractors agreed that this would be the only satisfactory course. They would wait until the end of the construction and submit this single issue for a final and binding decision.

Kenneth Lee said, "In the meanwhile, we might as well shake hands, be friends, and build the building."

Abe laconically replied, "Right."

Points of Law 1

The Foolproof Construction Contract
Who Pays for the Extra Work

1.1 What is Fair

The project owner, Baker/Doyle, rejected the architect's recommendation that the soil at the project site be tested to determine the allowable soil pressure for footings. Although Charlie Doyle knew that there was fill on the property, he decided to save the money that soil testing would cost by just assuming that the maximum allowable pressure would be 1500 pounds per square foot. Although this was a fairly conservative estimate, it turned out that the city inspector required soil testing and that the value of the existing soil was only 1350 psf. That made it necessary to remove rebar that had already been installed, dig deeper, and redesign the footings at a total cost, including overhead and profit, of $39,855.

This money was not included in the contractor's bid, so if the contractor is not paid for the extra work, the owner will be getting something for nothing. The owner would therefore avoid the consequences of the risk that it deliberately assumed in order to save the cost of soil testing. Following this line of reasoning, the arbitrator is not going to want to deny the contractor compensation for the extra work since it would offend an elementary sense of justice. The owner who got the monetary benefit of saving money on soil testing should bear the risk rather than the contractor who simply bid the job according to plans and specifications and, without complaining, overcame the soils problems and finished the job.

1.2 What is Legal

The other side of the argument would run like this: A construction contract is a document that is intended to, and by its very terms does, allocate risk. When the contractor signed the contract, it knew, or should have known, that no soil testing had been done and as an experienced contractor it should have known that there was a risk that extra work would be required if the soil bearing value was less than 1500 psf. If the contractor was willing to accept that risk, it must be assumed that the contractor included the value of that risk in the contract price and therefore the contract should be enforced.

This brings us to the language of the contract:

> "It is contemplated between the parties hereto that there will be no extra costs arising from the work of this contract."

According to the owner this language is sufficient to cast the risk of unanticipated underground conditions on the contractor. But that's not exactly what the language says.

The language falls short of a guarantee. It does not say that the contractor agrees that the contract price includes the cost of unknown and unanticipated underground conditions that might make the performance of the contract more expensive.

What the contractor has promised to do is build the project according to plans and specifications for the contract price, while acknowledging that the parties contemplate there will be no extra costs. According to the dictionary, the word "contemplate" has three meanings: 1) to look at something attentively and thoughtfully, 2) to consider carefully and at length, to meditate on, or ponder, 3) to have in mind as an intention or possibility.

In short, the parties do not expect that there will be any extra costs. But it's also true that the parties did not expect that the bearing value of the soil would be less than 1500 psf.

When we consider it carefully, then, this language does not specifically deal with what happens if the unexpected occurs! This language does not answer the question "who pays for the extra work?"

1.3 Quantum Meruit

This brings us to the rule of law embodied in the Latin term quantum meruit. This expression means "as much as he deserves." The doctrine of quantum meruit embodies the concept that a person who benefits by the labor and materials of another should not be unjustly enriched and that under those circumstances the law implies a promise to pay a reasonable amount for the labor and materials furnished even without a specific contract.

Therefore the arbitrator will decide that the contractor should be paid the value of the extra work, since although the parties did not expect extra work, it occurred! It would be unfair for the owner to get the benefit of the extra work without paying for it.

1.4 Following the Law

The dispute is going to be decided by an arbitrator. The courts in almost every state encourage arbitration as a speedy and economical means of dispute resolution. Arbitration is especially popular in the construction industry.

Arbitration awards are considered more final than judgments handed down by the courts since all judgments are subject to appeal, but the grounds for appeal of an arbitration award are very limited. One reason that arbitration awards are so final is that the courts will enforce them even if the arbitrator made an error in law. Arbitrators may make their awards based on notions of fairness and equity. They need not be learned in the law, and their awards cannot be overturned because of a failure to follow the law.

In our case if the arbitrator follows the law, the award will be in favor of the contractor for $39,855.

2

ABC Warehouse I
A Construction Defect

Construction of a tilt-up concrete warehouse for ABC Warehouse Company was going smoothly and progress was on schedule. There were very few changes and no unforeseen conditions, so the contract price and completion date are still intact.

Midway through the project, Allen Brady, president of ABC Warehouse Company, visited the jobsite with his warehouse manager, Carl Daly. They noticed slight cracking in the finish of the concrete floor slab and mentioned this to Ezra Field, job superintendent for Hyde Construction Company. Until this visit they had not gotten a really good look at the floor because of construction activity. Ezra defensively replied that the slab was as good as any he had ever seen, and besides, there was no such thing as a crack-free slab. Ezra thought he had pacified Allen and Carl, but they continued to discuss the matter while driving back to their office. Carl said, "You're paying good money for this job, Mr. Brady. We should get a perfect floor. There's no reason you should settle for less."

Allen replied, "I think you're right, Carl. I'm going to call Ivor. He's supposed to be a top architect. How could he let this happen? Why do we have to discover the cracks? What are we paying him for?"

Upon arrival at the office, Allen immediately telephoned Ivor Judge, AIA, senior partner of Judge & King, an architectural firm noted for the high quality of its construction documents and the recipient of numerous design awards.

Ivor had just returned from lunch with a new client and was very pleased with himself. He had an exciting new project ahead of him, and the ABC warehouse was progressing without problems. He was glad that Hyde Construction was the low bidder since they always did excellent work, and he considered himself lucky that Ezra Field was the job superintendent. Just then, Ivor's secretary interrupted his daydreaming and announced that Mr. Brady was on the phone. She volunteered, "He sounded very curt, not his usual friendly self. I hope nothing's wrong."

Ivor apprehensively answered his phone, saying tentatively, "Good afternoon, Allen. What's new?"

"Look, Ivor," began Allen, "I've just come back from the construction site and I'm concerned about the floor. It's all cracked and the finish looks lousy. It's a real mess. What are you going to do about it?" Allen was very demanding.

Ivor tried to calm him, saying, "I was just there yesterday, Allen, and the floor didn't look too bad to me. Sure, there are a few cracks and some minor crazing and maybe a little irregularity in the finish, but you've got to expect some imperfections. Concrete slab floors aren't always perfect. By the time the sealer is applied, you won't even notice the irregularities. It'll be even less noticeable when your merchandise is in the building."

Allen interrupted testily, "Yes, but this is a brand new building. It was going to be a model for our industry. It'll be embarrassing to us when people from other companies look at our new building and see a second-rate floor slab. I think Hyde should remove the slab and replace it with a proper floor with no cracks and a perfect finish."

Ivor, injecting a note of reason, said, "I'm sure it won't be necessary to replace the slab, Allen. It isn't that bad. And if we did replace it, it would set the construction back six weeks, maybe longer. You wouldn't be able to move in as planned."

Allen was becoming irate. He said firmly, "It sounds to me, Ivor, that you're just making excuses. I want you to look into this immediately and let me know what you and Hyde expect to do." And he slammed down the phone before Ivor could reply.

Ivor, shocked and dismayed, pondered the situation and then sought out his partner, Leo King. "Leo, we have a serious problem developing with Allen. He's unhappy with the new warehouse floor slab and wants it removed and replaced. He won't listen to reason."

Leo, concerned, said, "Wow, that really is a problem. Have you seen the floor?"

"Yes, I saw it yesterday. It looked okay to me. It did have a little more cracking than usual, and there is a certain amount of crazing. The finish is irregular in several places, too," said Ivor, trying to be objective.

Leo asked, "Do you think it's within industry standards?"

"I think it is, but it's borderline. I've seen better slabs, but this one should be acceptable. I wish Allen wasn't so upset. He's so worked up he won't listen to reason."

A look of anxiety crossed Leo's face and he asked, "Have you checked our drawings and specifications to make sure we haven't made a mistake that could have caused this problem?"

"No, Leo, I haven't had time yet. I'll do it right away and then have the testing lab make some cores. And I'll arrange for a site meeting with George Hyde and Ezra."

Leo asked if he intended to invite Allen to the meeting. Ivor replied, "I think I should. After all, it's his building."

Leo replied, "You're right," and suggested that Myles Nolan, their structural engineering consultant be present at the site meeting as well. Ivor agreed and spent the rest of the day and evening setting up the meeting, reviewing contract documents, and researching standards for concrete slabs. He got very little sleep, rose early, and skipped breakfast so he could be fully prepared for the meeting.

***

Ivor arrived early for the meeting and found George Hyde and Ezra Field already on the site. The group immediately began examining the floors, wandering about the warehouse with bowed heads and occasionally kneeling to examine the slab more closely. Over half the floors were nearly perfect, but the rest had more than the usual cracking crazing. Ezra tried to steer them into the better areas and verbally minimised the defects in a transparent effort to convince his boss that the floors met the usual high standards of Hyde Construction. George Hyde loudly agreed with Ezra, hoping to convince Ivor that these defects were minor and that the work in fact was better than average.

In the middle of their tour of the warehouse, Myles caught up with them. He had been the civil and structural engineering consultant on most of Judge & King's projects, including this one. After greeting Ivor, George, and Ezra, Myles wandered about the warehouse, noting the condition of the slab and occasionally drawing sketches.

A few minutes later, Allen and Carl joined the group. Ivor, having called the meeting, spoke first, and explained to George that Allen was concerned about the floors. Allen Brady interrupted, "The floor is one of the most important parts of a warehouse. This is the poorest looking floor I've ever seen. Something has to be done about it. I simply won't accept it." He was firm.

George Hyde's face and neck were getting red and he stated flatly, "Allen, there's nothing wrong with this floor. On average, it's as good as you'll find in this county. There may be some minor imperfections, but on the whole this is an acceptable warehouse floor." He, too, was firm.

Ivor turned to Myles and asked for his opinion. Myles, referring to his notes, said, "The cracking and surface defects are limited to less than twenty percent of the total area. The cracking appears to be superficial and should have no structural significance, but I'd like to review the drawings and specifications, cylinder tests, and slump reports before finalizing my opinion. Right now, I'd say that the floors could be easily repaired, and after the sealer is applied you'll hardly notice the repairs."

Allen said, "Well, I'm glad to hear there's no structural problem, but I'm still upset we're getting a substandard floor that already needs to be repaired in a brand new building that costs over $2 million. We'll always be able to see the unsightly repairs."

Carl Daly whispered to Allen, "I still think they should tear out and replace all the defective floors. We should have a perfect floor like we paid for."

George, hoping to hurry the meeting along, said, "We can start the repairs right away and everything'll be alright."

Ivor replied, "George, don't start any repair work until we decide exactly what should be done. We'll work up a repair specification right away."

Allen, feeling he was being railroaded, said, "Wait a damn minute. Let's not get carried away on repair specifications. I want that shoddy floor removed and replaced with the quality I'm paying for. Ivor, tell 'em to do the job right and forget the damn repair specification."

Hyde, getting redder, barked, "Ivor, there's no way we're going to remove this damn slab. It's there to stay. Work up the repair spec and we'll get on top of it. But it'll be a cold day in hell when we remove that slab. That's final. You can bet on it."

The meeting had broken down. Ivor concluded by saying, "Okay, gentlemen, we'll adjourn for now. But I'll give this my immediate attention and try to get this resolved. I'll keep in touch with everyone. Meanwhile, George, according to the contract, you must continue with the construction while this dispute is under consideration. And also, Allen, you must keep making any payments which have been certified. We don't want to cause any delay in the construction since there are liquidated damages of $1500 per day." George and Allen looked at each other incredulously.

* * *

Leaving the meeting, Carl congratulated Allen, saying, "You really told them, Mr. Brady. Now they know who's boss."

Allen replied, "I hope I said the right thing. If the floor has to be removed and replaced, completion'll be set back six weeks, according to Ivor. We have to be out of the old warehouse before June 1 or it'll cost us plenty. They've already leased it to a new tenant."

"But, Mr. Brady, the contractor would have to pay you liquidated damages of $1500 a day, and that'd amount to—that's over $10,000 a week. So six weeks would be $60,000. That should easily cover our costs," said Carl, trying to sound authoritative.

"I'm not sure it would. I still have an uneasy feeling about all this. George Hyde is a hard headed businessman, and he isn't going to pay for replacing all those slabs and pay us $60,000 without one helluva fight. As soon as we get back to the office I'll call our lawyer."

Allen called Philip Quinn, ABC Warehouse Company's attorney and explained the situation. Philip reviewed the contract over the phone with him and confirmed that Allen had to keep making payments pending final resolution of the dispute.

"Oh. It just seemed to me that it would be unbusinesslike to continue paying out money when our dispute must involve a fairly large sum of money. The payments have been around $350,000 per month."

"Well, it's not all one-sided, Allen. After all, the contractor must keep working and committing more funds to the job. If you quit making payments that are certified by the architect, you'll have breached your contract, thereby justifying the contractor's quitting the job. Naturally, Hyde would have to observe the formality of a written notice first, but he could then claim all the money he has coming as well as any proven losses plus damages. Allen, I strongly suggest that we conform to the contract and wait a few days to see if Ivor can resolve this dispute as quickly and painlessly as possible. Incidentally, Allen, don't forget that you're holding over $120,000 retainage from Hyde Construction so far. And Ivor is authorized by the contact to withhold the value of uncorrected defective work from future payment certificates."

Allen began to question the wisdom of his position and asked, "Phil, do you think I'm unreasonable in demanding that the defective floor slabs be replaced, rather than repaired?"

"I don't know, Allen. I'm just your attorney. I know very little about construction. Why don't you take your architect's advice?" Allen said he'd think about it.

"Oh, Allen, before you go, let me know if you receive any correspondence from George or Ivor on this subject and, for God's sake, don't you write any letters without checking with me first. Okay?"

"Okay, Phil."

"One more thing, Allen. Dig out your performance bond and send me a complete copy of it right away. I'd like to review it again."

"Okay, Phil."

* * *

At the warehouse, George and Ezra discussed the best way to handle the situation. They concluded that the most economical way to repair the slab would be to inject an epoxy filler into the larger cracks, ignore the hairline cracks, and grind off some of the worst irregularities with an angle grinder.. The originally specified slab sealer would tend to disguise these repairs. If they purchased the materials, rented the equipment, and assigned two laborers, the repairs would cost under $4000. They agreed that this was the best solution, but George cautioned Ezra to wait until they had talked to Ivor before they began the repair work.

* * *

Upon leaving the meeting, Ivor and Myles conferred privately about the floor slab. Myles reconfirmed that the problems were cosmetic and not structural, but cautioned Ivor to wait until he could review the test results to confirm this conclusion. Ivor asked Myles to expedite things: "Myles, please stay on top of this and report to me the minute you have a firm opinion. As you probably know, I'm obligated to make an interpretation and a decision on this dispute. It's required of me by the Owner-Architect Agreement, AIA Document B141, subparagraph 2.6.15 through 2.6.19. The owner and contractor have agreed in their contract to abide by my decisions. This is in the AIA General Conditions, AIA Document A201, paragraphs 4.3 and 4.4." Both of these documents were 1987 editions.

* * *

Friday morning's mail brought what Ivor was waiting for. The test reports from Reliable Testing Laboratories. The cores tested proved that the slab was in excess of the specified compressive strength. The 28-day compressive strength specified was 2500 pounds per square inch. The cores tested, approximately 70 days old, were 2817, 2745, 2685, and 2640 psi. A fat envelope from Hyde Construction Company contained the laboratory tests of the 28 day old cylinders, observations of slump tests, the ready-mix concrete delivery tickets, and the job superintendent's daily reports from the days when concrete slabs were being poured and finished. Ivor had previously telephoned George Hyde's secretary and requested all these items.

Later on that morning, Myles Nolan arrived to review and discuss all the assembled data with Ivor. Myles spent an hour alone in the conference room with the contract documents and all of the engineering data and jobsite information. Then Ivor joined him and they reviewed the situation anew. Myles said that all of the laboratory tests and the concrete delivery tickets revealed that the concrete was in compliance with the specifications. The spacing and configuration of the control joints were proper in the drawings and in the construction. The steel reinforcing members had been noted as proper in the architect's and the engineer's field inspection reports. The contractor had used the proper sprayed-on curing compound.

The only anomalies discovered in the whole array of data were to be found in Ezra's daily reports, which revealed that some of the concrete had been poured on hot, dry days. This could easily explain the problems with rapid surface drying, thereby interfering with finishing operations and the normal curing process. If the curing compound was not applied in time, this could also contribute to the problems.

Myles, referring to a growing pile of notes, said, "Ivor, I'm convinced there's nothing structurally deficient with those slab floors. They should give many years of good service. The cracks and irregularities are slightly beyond what would normally be expected, but they can be effectively repaired, although the repairs probably will show. However, the sealer will lessen the contrast, so the floors should look alright for a utilitarian building."

They concluded the best way to repair the floors would be to enlarge the larger cracks by sawcutting and sealing them with an epoxy filler, leaving the hairline cracks untouched. The high rough spots could be smoothed with a terrazzo grinder, and then the hardener-sealer would be applied, completing the repairs.

After Myles left, Ivor began to write his preliminary decision and to reach certain conclusions under the AIA General Conditions. First of all, he recognized that he had the authority to reject work that does not conform to the contract. This power is conferred by subparagraph 4.2.6. This also authorizes the architect to require additional testing. In this case, the slab cores tested were in conformance with the contract, so the cost will be borne by the owner. The slab repairing, however, will have to be paid for by the contractor.

The question asked by the owner was whether or not the slab was to be removed and replaced. Ivor noted that Allen had raised this question in a timely manner, that is, within 21 days of claimant's first recognition of the condition, as required by subparagraph 4.3.3.

Ivor noticed that subparagraph 4.4.4 provides that his decision will be final and binding on the parties, but is subject to arbitration if contested by either party. However, subparagraph 4.2.13 provides, "The Architect's decisions on matters relating to aesthetic effect will be final if consistent with the intent expressed in the Contract Documents." This put Ivor is in a quandry. The question of removing the slabs was probably arbitrable, but the question of how to conduct repairs was a matter of aesthetic effect.

Ivor was required to make a preliminary response to Allen's claim within 10 days. Ivor was given five options: "(1) request additional supporting data from the claimant; (2) submit a schedule to the parties indicating when the Architect expects to take action; (3) reject the Claim in whole or in part, stating reasons for rejection; (4) recommend approval of the Claim by the other party; or (5) suggest a compromise. The Architect may also, but is not obligated to, notify the surety, if any, of the nature and amount of the Claim."

Ivor reasoned that what he has in mind is a compromise, option (5). He drafted a joint letter to ABC Warehouse and Hyde Construction suggesting that they compromise by allowing the anomalous slabs to remain and repair them in the manner he and Myles devised. He also stated that the final decision would be made within seven days.

* * *

When George received Ivor's preliminary decision, he immediately asked his estimating department to figure out what this method of repair would cost him. Out of curiosity, he also asked what it would cost to remove and replace the irregular slabs. The estimating department soon responded. Repair of the slabs using Ivor's technique would cost Hyde Construction $12,230; removal and replacement of the slabs would cost $105,680.

When George contrasted a $12,230 repair bill with his own $4000 scheme, he immediately called Ivor and said, "Hell no, we won't go for this gold-plated repair job. It costs too damn much and is totally unnecessary. We can do it much easier and cheaper and it will be plenty good enough."

Ivor tried to explain that he had to be fair to both the contractor and the owner and ended the conversation by reminding George that the final decision would be out in a few days.

* * *

Allen's response to Ivor's preliminary decision was an irate phone call, demanding removal of the slabs. Allen concluded by stating, "I haven't changed my mind. I'll have my lawyer write you a letter confirming this decision."

Allen immediately called Phil, fuming, "Ivor stabbed me in the back. His preliminary ruling is against me. He wants me to accept a hokey repair scheme. I paid for a perfect concrete slab floor, and that's what I should get. I want you to write him a letter rejecting this ridiculous compromise and directing him to proceed with my original demand for a new, perfect floor. You might also explain to him who his client is and who's paying his fees."

Phil agreed to do so, pointing out that the letter would only reject the compromise and reinstate Allen's original claim. Phil added, "It'd be highly improper and inadvisable to pressure him as you suggest." Phil asked Allen to send him a copy of the preliminary decision. Allen agreed to do so, wondering how much Phil was going to charge for all these extra services.

* * *

The next day, Ivor realized that the situation was not improving. George Hyde and Allen Brady were both highly displeased with his preliminary decision. Apparently no decision would be acceptable to both of them. Thus Ivor decided to disregard George or Allen and instead to design a solution that he personally believed was both fair and technically correct.

His final decision was the same as the preliminary decision except that it stated this decision would be final and binding but subject to arbitration as provided in subparagraph 4.4.4. He also pointed out that if the decision were appealed by either side, an arbitration demand would have to be filed with the American Arbitration Association within thirty days.

* * *

Upon receiving Ivor's final decision, Allen became furious and immediately called Phil, saying, "Phil, the final decision just came. It's the same as before. Ivor is still only asking for repairs of our floors. Hyde's getting off too easy. What can we do now?"

"Now, Allen, calm down. Ivor didn't have much choice in this matter. In law there is a general principle called substantial performance. Even if Ivor had required complete replacement of the affected concrete slabs an arbitrator would probably reverse the decision. The floors, after repairs, will perform their intended purpose. Ivor must feel that the contractor has substantially performed the contract."

Allen asked about appealing the decision, and Phil replied, I don't think it would be wise to appeal. It would take a lot of time, there would be costs of expensive expert witnesses, and in my opinion, you'd probably lose. It seems to me that you're getting a pretty good building at a reasonable price and the floors will be serviceable despite some slight imperfections."

"Well, maybe you're right, Phil. Maybe I was paying too much attention to Carl, my warehouse manager. He had his mind set on a perfect new warehouse. I can see now it's somewhat impractical in the real world. I appreciate your good advice."

* * *

When George received the final decision, his first reaction was disappointment. He walked into his chief estimator's office and said, "I just reviewed Ivor's final decision. We're going to have to spend $12,230 on flooring repairs. Thank God he didn't side with Allen or we'd be looking at $105,680. Maybe if we went for an appeal we could get them to cut the repairs down to $4000 doing it our way. What do you think?"

The chief estimator replied, "Well, George, if we go to arbitration, we could easily spend the $8230 difference in lawyer's fees, loss of your time, Ezra's time, and my time. We might even need an expert witness, an architect or engineer. And we might not win. There's also the risk that the arbitrators might see some merit to Allen's idea of replacing the slabs. No, George, I vote for accepting the decision and hope that Allen doesn't go for an arbitration appeal."

George, philosophically and with resignation, commented, "I guess I should be grateful I have an analytical chief estimator and a fair minded architect on the job."

* * *

This is the end of the ABC warehouse floor controversy. The floors were repaired, and they look much better than Allen and Carl expected. Allen also learned from a realtor friend that the blemishes would have no adverse effect on either the rental value or the resale value of the property.

The new building was completed before the contracted completion date, thereby enabling ABC Warehouse to vacate their old premises on time. They were well settled in their new building in time for their grand opening party on June 1. Allen and Carl were proud of the new warehouse when their customers and warehouse industry leaders showed up at the celebration.

Aside from the flooring problem, Hyde Construction suffered very few setbacks, completed the building a few days early, and nearly made their scheduled profit. George and Ezra toasted themselves with champagne at the opening party and happily accepted well-earned congratulations.

Ivor and Leo were pleased with the final result, and can now count both Allen and George among their loyal friends. At the opening celebration they met some excellent prospects among the leaders of the warehouse industry. The completed project will be a worthy entry for the next AIA chapter award program. And they all lived happily ever after.

* * *

Moral: Firm, fair, and prompt contract administration in the hands of a competent architect will go a long way in settling construction disputes before they get hopelessly out of hand.

* * *

Points of Law 2

ABC WAREHOUSE I
Allen Gets a Cracked Slab

2.1 Breach of Contract

The topic of concern in this episode is breach of contract. A construction contract consists of a set of undertakings in which the promise of the contractor to build the project in accordance with the drawings and specifications is supported by the promise of the owner to pay the contract price. By custom, the contract price is paid in monthly increments against the progress of construction. The owner holds back from the progress payments a retention, usually ten percent, as security for the contractor's proper performance of the work. Let's assume, for example, that construction has progressed to a point where thirty percent of the value of the work has been completed and accepted by the architect. The architect would issue a certificate that the contractor is entitled to be paid thirty percent of the contract price, less 10 percent retention. As an example, if the contract price were $2,000,000 then 30 percent completion would entitle the contractor to be paid ninety percent of $600,000, or $540,000.

For a concrete example, suppose that the slab has been poured and Hyde has achieved thirty percent completion and is therefore entitled to progress payments of $540,000, but Allen is dissatisfied with the slab. It would cost $105,600 to remove and replace the slab. Since the retention is only $54,000, the retention alone does not supply adequate security for the correction of such a major problem. On the other hand, if Ivor's "fix" is agreed to, the cost of repair will be only $12,230, which is less than the retention.

Allen is dissatisfied with the slab. He wants a perfect floor, and considers that under the contract documents he is entitled to it no matter the cost. Granting the good faith of this position, let's examine Allen's options.

2.2 Remedies for Breach of Contract

The harshest option, and the most dangerous from a legal point of view, would be for Allen to order Hyde Construction to remove and replace the slab. If Hyde refuses, Allen could eject Hyde Construction from the jobsite, and employ another contractor to remove and replace the slab and complete the project. Upon completion, Allen could sue Hyde Construction for damages for breach of contract. The amount of damages sought would be the difference between the contract price and the actual cost to Allen of completing the project, including the cost of removing and replacing the slab. Allen could also recover damages for delay in the completion of the project. Those damages would be measured by the reasonable rental value of the project.

Now let's examine Hyde's options. One option would be to obey the order and proceed, under formal protest, to remove and replace the slab and finish the project. Hyde Construction could then file suit against Allen and his company for breach of contract. In such a lawsuit, Hyde Construction would attempt to convince the court that it had not committed a breach of contract and that the slab, as built, complied with the requirements of the contract documents. If successful, Hyde Construction would then recover from Allen and his company the cost of removing and replacing the slab.

Rather than obey Allen's order to remove and replace the slab, Hyde Construction might decide to ignore the order and proceed with the project. This, in turn, would give Allen two options: 1) declare Hyde Construction in default and eject it from the jobsite, or 2) allow Hyde Construction to proceed with the construction of the project but withhold from progress payments the estimated cost of removing and replacing the slab.

If Allen exercised the second option, it would again give Hyde Construction two options: one would be to proceed and finish the project and sue for the amount withheld, and the other would be to declare Allen in default for failing to make progress payments, stop work, and file suit to recover the cost of its partial performance plus the profit that it would have earned if Allen had allowed Hyde Construction to proceed with the project.

2.3 Interpretation of Contracts

Each party is contending that the other has breached the contract. Which is right? This is a question of the interpretation of contracts.

A well prepared construction contract includes detailed drawings and specifications describing the size, quality, and placement of reinforcing steel, the degree of compaction of the supporting earth, the components of concrete mix, and the strength of the concrete (determined by measuring the pounds per square inch of pressure that test samples can sustain before failure). Concrete specifications also describe the process by which the concrete is finished.

In our story, Hyde Construction appears to have installed the right mix and to have achieved proper placement, quantity, and amount of rebar but nevertheless the slab isn't "perfect". The mere presence of cracks and crazing does not necessarily mean that Hyde Construction breached the contract. If Hyde Construction followed to the letter and with utmost precision all of the requirements of the contract documents and yet the result is imperfect and unsatisfactory to the owner, Hyde Construction has not committed a breach of contract because when an owner supplies drawings and specifications to a contractor the owner impliedly warrants that if the contractor faithfully follows the drawings and specifications the results will be satisfactory.

2.4 Trade Practice

Another element comes into play: good trade practice. Ivor Judge seems to think that cracking and crazing occurred because portions of the slab were poured on hot windy days. Even though the contract documents may not specifically prohibit a pour on such days, contract documents are held by the courts to incorporate good trade practice. If cracking and crazing occurred because of the violation of good trade practice, Hyde Construction is in breach of contract.

Trade practice is established by the testimony of expert witnesses: architects, engineers, prime contractors, and concrete subcontractors whose business it is to mix concrete, pour it, and finish it. They are familiar with established trade practices in the area where the project is located.

2.5 Rescission

Let's assume that Hyde Construction Co. refuses to remove and replace the slab but continues with the work. This would present Allen with three potential legal strategies: 1) declare a default and rescind the contract, 2) declare a default and eject Hyde Construction from the job without rescinding the contract, or 3) declare a default and withhold from Hyde Construction's next series of progress payments a sum sufficient to remove and replace the slab. (Allen might also continue making progress full payments and sue Hyde Construction for breach of contract.)

Allen would not be likely to rescind the contract. The concept of rescission is similar to the concept of annulment of a marriage: the law indulges in the fiction that a contract never existed. If two parties found themselves in the position of Allen and Hyde Construction with no contract, the course of the law would be to restore the parties to the same economic position they would have enjoyed if no contract had ever been formed. This would be done by determining whether a party has been unjustly enriched, and requiring that party to disgorge. The court would decide the value of the work and materials supplied by Hyde Construction, subtract the amount paid under the contract, and award the difference to Hyde Construction or to Allen. By rescinding the contract, Allen would lose the benefit of the provisions that are written into the contract for the express purpose of protecting the owner, such as the contract price, the schedule, and the requirement that the contractor will strictly comply with the provisions of the contract documents.

Allen may also hesitate to keep the contract alive and eject Hyde Construction from the job. To do so would be to stake a great deal on the ability to convince a court that Hyde Construction committed a material breach of contract. If the court should decide that Hyde's performance lived up to the standards of the trade, it would be a material breach of contract for Allen to throw Hyde Construction off the job.

Allen is most likely to select option 3 by withholding from Hyde Construction's progress payments a sufficient sum to remove and replace the slab. This would be something in excess of $105,600. The down side of that strategy is that while it puts pressure on Hyde Construction, Allen had better be right, because if the court decides that Hyde Construction did not breach the contract then Allen's failure to make full progress payments would itself be a material breach of contract that would justify Hyde Construction in rescinding the contract and pulling off the job!

2.6 Substantial Performance

In deciding what course of action to take, Allen should consider the doctrine of substantial performance. A party that has fulfilled all the obligations of a contract is said to have performed the contract. Failure to perform is classified as a breach of contract unless performance, for some legally recognized reason, is excused. (Performance can be excused, for example, by impossibility.)

There are occasions when the law will accept less than the strict fulfillment of a contractual obligation as performance. Let us assume, for example, that the contract requires the installation of a built-up roofing system that consists of layers of felt mopped in with 1A020 tar. Unknown to the prime contractor or the subcontractor, the tar supplier substitutes, without authority, 1B020. The 1B020 product is slightly less resistant to sunlight and therefore the useful life of the roof might be expected to shorten 10 percent. The cost of removing and replacing the roofing system would be $300,000. An appraiser testifies that the market value of the building is unaffected by the substitution. An economist testifies that the economic value of a 20-year roof exceeds the economic value of an 18-year roof by $20,000. The project owner refuses to make any further payments to the prime contractor unless the roof is removed and replaced using 1A020 tar.

Normally a breach of contract on one side excuses further performance on the other side, so the substitution of 1B020 for 1A020, if a material breach of contract, will excuse the owner from making further payments to the prime contractor.

The doctrine of substantial performance may apply when a contractor has performed in good faith, and to insist on full and exact performance would result in economic waste, and the other party to the contract can be compensated for the discrepancy by an award of monetary damages. The doctrine does not apply when deviations are serious.

2.7 Economic Breach of Contract

Some courts also require that deviations be unintentional. This element reflects a philosophical evaluation of the very function of contract law. The function of criminal law is to protect society by punishing wrongdoing. The function of tort law is to deter misconduct by making wrongdoers pay compensation for the harm they cause. The function of contract law is economic in nature: economic prosperity depends on enforcement of contracts.

Some courts reason that if contract law exists to promote economic objectives then a breach of contract is not blameworthy if it promotes economic efficiency.

Let us say that a contract calls for a mix that will produce 2500 psi concrete. Research develops a mix that will produce the same strength and quality of concrete with greater energy efficiency and at lesser price. A contractor willfully substitutes the new mix for the old, thereby failing precisely to comply with the formula specified by the contract. Economic efficiency is promoted without any tangible damage to the owner. Should the willful nature of the breach in such a case prevent application of the doctrine of substantial performance? Some legal thinkers condone "efficient" breaches of contract and others do not.

2.8 No Breach

Now let's go back to Allen and Hyde Construction. The requirements for the application of the doctrine of substantial performance are fulfilled because the performance is the virtual economic equivalent of the performance required by the contract documents. Any deviation can be readily compensated by a monetary award. To remove and replace the slab would cause economic waste. The deviation was unintentional. Therefore, even if we assume that there was a violation of trade practice when the slab was poured on a hot windy day, the doctrine of substantial performance should apply. The only question should be how much, if anything, should be deducted from the contract price in order to compensate Allen for the discrepancy.

Since the doctrine of substantial performance applies, Allen would not be justified in rescinding the contract or in terminating Hyde Construction's performance of the contract or in withholding progress payments. By doing any one of those three things Allen would, himself, commit a material breach of contract.

2.9 Measure of Damages

On such issues, a "battle of expert witnesses" could be anticipated. Appraisers would differ as to whether the cracking and crazing would have any effect on the market value of the building. Warehousing experts would differ as to whether the cracking and crazing could affect the efficiency of warehouse operations.

2.10 Contract Provisions

Until now we have been discussing breach of contract and default issues as if there were no governing contract provisions. Well-articulated construction contracts include provisions that deal with rescission, default, and dispute resolution.

2.11 Notices and Procedures

Many construction contracts require a party to give written notice to the other party before declaring a default, stopping the work, or rescinding the contract. One typical provision requires the contractor to give seven days written notice to the owner before pulling off the job. The notice must describe the nature of the owner's alleged breach.

A similar typical provision requires the owner to give seven days written notice of breach before ejecting a contractor from the job.

2.12 Prevention of Performance

The purpose of the written notice provision is to establish a "cure period": a period of time when the party in default can cure the default, or at least start to cure the default.

Courts insist that the default provision be observed with the utmost punctilio. One reason for strict observance is that an owner, by removing a contractor from the job, prevents the contractor from curing defaults. If an owner ejects a contractor from a job without permitting a cure, then the contractor's breach may be excused by application of a doctrine known as prevention of performance. This doctrine is nothing more than the common sense notion that a party can't prevent another party from performing a contract and then sue for it.

Some breaches of contract cannot be cured in seven days, or in the even more stringent forty-eight hour or twenty-four-hour notice period used in some contracts. In such cases the contractor may cure the default by promptly starting corrective measures.

2.13 Alternate Dispute Resolution (ADR)

The construction industry is an enthusiastic user of alternate dispute resolution.

The main reason parties resort to ADR is to avoid getting bogged down in the court system. Court proceedings are expensive and they take a long time. AIA document A201 provides three levels of ADR. First, the architect decides. Second, the contractor and the owner mediate. Third, the dispute is arbitrated under the rules of the American Arbitration Association.

The decision of the architect as to matters of aesthetic effect is final. The decision of the architect resolving other disputes between the owner and the contractor is subject first, to mediation and, after mediation, to arbitration. If the parties accept the architect's decision, or do not appeal it within 30 days (under the AIA general conditions), then the architect's decision is final.

An agreement to arbitrate is binding and enforceable. An arbitrator behaves much like a judge, giving the parties a fair hearing, listening to their testimony, and allowing them to cross examine witnesses and argue their cases. The arbitrator then issues an award that is enforceable just as if it were a court judgment. In fact, an arbitrator's award is somewhat more powerful than a court judgment because it is not subject to appeal!

Mediation is an entirely different process. A mediator cannot impose a decision on the parties, but does try to help the parties negotiate a resolution to their dispute that is acceptable to both sides. The outcome of a successful mediation is a signed settlement agreement.

2.14 The Architect's Decision

In our story, Ivor Judge's decision is not completely acceptable to either side. Hyde Construction thinks that it should not have to spend $12,230 to fix the crack when a satisfactory repair would cost a lot less. Allen is not satisfied because in his view he will still be getting a second class slab and not the perfection that the contract documents led him to expect. Nevertheless, a careful evaluation of the potential costs and benefits of submitting the case to arbitration convinces both parties that it is better not to go to war over the slab. By accepting Ivor Judge's decision the job can remain on schedule and both parties avoid attorneys fees and the hazard of an unfavorable decision.

3

ABC Warehouse II
Winding Up the Job

It was June 4th, the Monday morning after ABC Warehouse Company's grand opening celebration. ABC's president, Allen Brady, was comfortably planted in his imposing new office chatting elatedly with Carl Daly, his gloomy warehouse manager. Allen crowed, "Well, the party was a resounding success. Did you see the envious looks of our competitors? And our regular customers were really impressed. I'm sure business will be booming!"

Carl, dampening Allen's enthusiasm, complained, "Yeah, the party was great, but look at all the unfinished construction work. There are a million loose ends and Hyde's people are all over the place getting in our way. We can't occupy all of the warehouse yet and their job shack, chemical toilets, and trucks are scattered all over our truck turning area."

Allen suggested, "Why don't you phone Hyde and tell them to get off the dime? After all, with the final payment and retention, we're holding over $500,000 of their money. That should get their attention!"

* * *

At Hyde Construction Company's offices, George Hyde was at his desk and had his ABC job superintendent, Ezra Field, on the phone. He was laying down the law, "Ezra, I want all those punch list items finished this week. We're submitting our final payment request tomorrow and I don't want any flack from Allen or Ivor about unfinished work."

Ezra replied, "OK, Boss, you got it! I'm on top of it! Most of the subs are here now and I'm sure we'll be out of here by Wednesday, no later."

Hyde's last shot was, "Well, you better be!" and hung up.

Then he strode to the nearby office of Ulysses Vance, his chief financial employee. "Vance, how're you coming with the ABC final billing? I want it out early tomorrow morning for sure!"

Ulysses replied, "I'm working on it, George. Say, what should I do with all these bills for extra concrete and forming where we over-excavated? It's over $6,000!"

Hyde, frowning, said, "Let's bill it as an extra. It might slip through. After all, we did the work and spent the money. Also, add in something for our labor and supervision, and don't forget overhead and profit."

"Okay, yes, sir, will do."

* * *

Ivor Judge was having coffee with his partner Leo King at the offices of Judge and King, AIA, Architects. They were gratified with the design and construction of the ABC Warehouse and had thoroughly enjoyed themselves at the grand opening party. They'd made a few good contacts and were now hoping for some exciting new warehouse projects. Ivor seemed concerned however and said, "Leo, last Tuesday's punch list was a mile long and I'll be checking progress at the site tomorrow. They only just made it into the building for the party and now the work will go slower with the warehouse full of stuff as well as having to work around ABC's people. I hope Hyde doesn't lose momentum."

Leo suggested, "Keep the pressure on Hyde and they'll get it finished. I hope we don't have to drag out the close-out procedures too long. We've spent entirely too much time on this job's construction administration already."

"Well, Leo, the Certificate of Substantial Completion is out of the way as we issued it on May 28th prior to ABC's moving in. I understand that the building department made their final inspection on the same day and that the power company installed their permanent power and light meters on May 31st."

"It looks like you're making good progress, Ivor. How about the Notice of Completion?"

"It's been filled in except for the completion date, Allen's signature, and notarization. Even though the building is in use there's still too much work going on to consider the project complete. I'll be checking tomorrow to see if it can be filed. If so, I'll hand the notice form to Allen and ask him to sign it right away."

* * *

At 9:00 AM Tuesday, Ivor was at the jobsite conferring with Hyde's superintendent. Corrective work was being accomplished by an electrician, two painters, and a finish carpenter. After reviewing the punch list with Ezra, very few items remained. Ezra said, "We'll be completed with all punch list work by quitting time tomorrow. You can count on it. We're moving the job shack and chemical toilets out now. There was some minor damage done during the owner's moving in but it's all repaired now. We won't be billing for it as the cost was negligible."

"I'm sure the owner will appreciate that as you'd certainly have been justified in charging for it." Ivor and Ezra both doubted that Allen would be all that appreciative.

"At this point, Ivor, all I want to do is get this job wrapped up and get outta here. This time next week I'll be relaxing on the Colorado River and hauling in fish."

Ivor then left Ezra and went in to the new offices to see his client, Allen Brady. After mutual pleasantries, Ivor put the Notice of Completion on Allen's broad oak desk, explaining, "Allen, Ezra says that the remaining punch list work will be completed tomorrow for sure. If it is, then tomorrow's date, June 6th, should be entered as the completion date. The notice must then be filed within 10 days thereafter, on or before June 16th, in the office of the County Recorder of this county. This is very important and is for your protection as it will have the twofold effect of establishing a definite completion date and of limiting the time periods in which mechanics' lien claims may be filed against your property. Subcontractors and suppliers will then have 30 days after the notice is recorded and Hyde construction will have 60 days in which to file their claims. You must sign the notice before a notary public. The sooner it's recorded the better it is for you."

Allen nodded in assent, "I understand. I'll take care of it."

* * *

When Ivor returned to his office just before noon, he found that Hyde's final payment request had just been hand delivered. Ivor immediately reviewed it line by line and noted that all line items in the Schedule of Values were now advanced to 100 percent. That's okay, he thought, since all that're left are minor pick-up items. He spot-checked the arithmetic with his desk calculator and found no errors.

But--what's this? A new Change Order #15, for extra concrete as a result of over-excavating, $11,353.38. That was Hyde's fault for not checking the plans and watching their levels.

And another, Change Order #16, for wood paneling in Allen's office, $14,661.35. Allen never actually agreed to this charge, but I know he asked for the work, so there'll be some heated discussion on this, I'm sure.

All the other change orders appear to be in order. Wait—there's no final accounting for the allowances. I'll have to talk to George Hyde about this. So he dialed Hyde Construction's number and got George on the line.

"George, Ivor here. I just finished reviewing your final bill. It seems to be pretty well in order except for a couple of small matters. First, we'll need your final accounting on the allowances."

"Okay, Ivor. No problem. I'll have Vance get right on it."

"Then, we must discuss Change Order #16 since it was not actually agreed upon. The field work got ahead of the change order paper work. We'll need to see some back-up to support the charges. Material bills, time cards, wage rates, and so forth. You know what's needed. But I'm sure Allen will pay it since he asked you to do the work."

"Well, okay. More work for Vance."

"And finally, as to Change Order #15, $11,353.38. That was for inadvertent over-excavation and cannot be billed to the owner. You will have to ..." He wasn't allowed to finish the sentence.

George's pressure had been building up and now exploded, "Ivor, what the blazing hell are you talking about? We spent that money and it's in the job! Whadda ya mean we won't get paid?"

Ivor was holding the phone about 12 inches from his ear and didn't actually hear the last of it before George angrily slammed the receiver. So, Ivor called back and asked Hyde's secretary to have him at a meeting in Allen Brady's office tomorrow at 11:00 AM. "And tell him to bring the final accounting for the allowances."

Ivor then called ABC Warehouse and confirmed the appointment with Allen to discuss the final billing with George Hyde. Ivor thought, this will be a memorable meeting.

* * *

Ivor arrived at the ABC Warehouse an hour early on Wednesday morning so he could check on the progress of punch list work before meeting with Allen and George. Greeting Ezra at the warehouse entrance, he quickly reviewed the list and they discussed the few remaining items. Ezra was right, they could easily finish the work by mid-afternoon today. He and Ezra went around verifying and checking off what had been completed. He was satisfied with the quality and with Ezra's efficiency in organizing the work of the various trades. He finished with Ezra just in time to walk into Allen's office at 11.

George and Allen were both seated and were chatting with an air of uneasy formality. They both had their copies of the final payment request in front of them but had not started to discuss it. As soon as Ivor was seated and had his copy on the desk, George Hyde blurted out, "Let's cut through all the crap and get right down to the important matters! Allen, Ivor wants to swindle me out of my rightful money! I hope you're not going to go along with that, too! Let's talk about those items first."

Ivor, trying to restore order, "Now, George, calm down. We'll discuss everything, but let's do it in a logical order." He was keeping notes.

George conceded, "Okay, Ivor. But I'm not going to sit here and get swindled."

Ivor spread out Hyde's final bill which was on AIA's standard form, Application and Certificate for Payment, Document G702 and the Continuation Sheet, Document G703. The total amount now requested was $353,511.24. This still left $221,220.94 in retainage to be requested later. Ivor was thinking, George might be an outspoken, irascible, old curmudgeon, but his paper work was always in order. He rearranged his papers and note pad on the desk and addressed Allen and George, "The Payment Request shows that all work items are completed to 100 percent and I'm in agreement with that. All the physical work appears to be satisfactorily completed except for the few remaining minor punch list items that are expected to be completed today. This leaves two items to be discussed, Change Orders #15 and 16."

George interrupted, "You're damn right we'll discuss them!"

Ivor, ignoring the interruption, continued, "Change Order #15 is for additional concrete and forming to correct a problem caused by the contractor's inadvertent over-excavation at the southeast corner of the loading dock. The excavation subcontractor misread the plans and cut too deeply. To maintain structural integrity of the footings and foundation walls, additional concrete had to be poured. This involved additional forming, reinforcing steel, and compacted fill. The total cost for materials, labor, overhead, and profit is $11,353.38."

George erupted, "And we spent every red cent of it! We should get paid!"

Allen replied, "This is the first I've ever heard of this. I didn't sign any change order!"

Ivor went on, "I can't approve this change order. It's clearly the contractor's responsibility. George, you're going to have to resolve this one with your excavator. It's not the owner's responsibility."

George knew Ivor was right and had suspected it would turn out this way. He didn't pursue it any further. Better to save the effort for Change Order #16. I'm not going to give in so easily on this one, he resolved. He sat grim-faced, alert, and poised for action.

Ivor started explaining, "Allen, Change Order #16 is for the enhancements to your private office, in which we're now sitting. The original plans and specifications call for 5/8-inch gypsumboard walls with a texture coat, two coats of paint, and a vinyl carpet base. During construction, you and I decided to upgrade it to book-matched oak paneling with solid oak raised mouldings, an oak crown molding at the ceiling, and a solid oak two-piece base. The two room doors had to be changed to oak to match."

Allen interposed, "I know all that. And when we asked George to do the work, he said okay. He didn't say anything about charging for it. I thought he would just throw it in as a friendly gesture. After all, this project is costing me over two million dollars! Besides, I didn't sign anything and the contract says that all change orders must be in writing!"

George Hyde's knuckles were white as he grasped the chair arms and his face was reddening, "How could you think I wouldn't charge for all this? If I knew you would welsh on this I wouldn't have done the work! You have the fanciest office in the county and now you don't want to pay for it!"

Ivor, trying to regain control of the meeting, "Allen, I'm going to have to decide in George's favor on this one as you asked for the changes and you knew the work was being done. You also approved the revision drawings. I'm sorry if you thought it would be free but, on reflection, surely you must now realize that wasn't a reasonable presumption."

Allen, thoughtfully reconsidering his position, "Well, I guess that was an unrealistic expectation on my part and I'm more than pleased with the outcome. But the price looks high to me. Do I have to pay whatever he asks?"

Ivor replied, "No, Allen, the price must be reasonable and must be supported by invoices for materials, time cards, payroll costs, and invoices showing credits for the omitted texturing, painting, vinyl base, and the originally specified doors. I've compared these costs with similar work and they seem in line. I've already asked George to submit all the back-up documentation."

George confirmed, "We're working on it and I'll have it delivered this afternoon."

Ivor, summarizing, "So, I'll approve Change Order #16, assuming that the back-up supports the figures. I'll therefore ask George to revise this payment request, eliminating Change Order #15, and I'll approve it."

Ivor continued, "Allen, after you pay this installment, all that remains is the retainage. It now amounts to $220,085.61 and can be applied for, George, as soon as you can get all your documentation in order. Then, Allen, according to the contract it'll be due and payable 35 days after the Notice of Completion is recorded. I presume you'll take care of it and have it recorded today as we discussed?"

"Yes. It's already dated, signed, and notarized. I'll have it delivered to the County Recorder first thing after lunch this afternoon."

Ivor added, "One other thing, Allen, if you wish to have your accountants audit the payment requests, retainage, payments on account, and the balance due, this would be the ideal time for that to be done. They should also review Hyde's reconciliation of allowances. They'll have over a month before the retention payment will be due."

"Yes, that would be a good idea."

Ivor resumed, "George, you'd better get started accumulating the documents and other things which must accompany your request for the retention payment. Some of these are required by the General Conditions, AIA Document A201, Fourteenth Edition, 1987. Others are itemized in the construction agreement. Documents required by the General Conditions, Subparagraph 9.10.2 are as follows:

(1) An affidavit that payrolls, bills for materials and equipment, and other indebtedness connected with the Work for which the Owner or the Owner's property might be responsible or encumbered (less amounts withheld by Owner) have been paid or otherwise satisfied.

(2) A certificate evidencing that insurance required by the Contract Documents to remain in force after final payment is currently in effect and will not be cancelled or allowed to expire until at least 30 days' prior written notice has been given to the Owner.

(3) A written statement that the Contractor knows of no substantial reason that the insurance will not be renewable to cover the period required by the Contract Documents.

(4) Consent of surety, if any to final payment.

(5) If required by the Owner, other data establishing payment or satisfaction of obligations, such as receipts, releases and waivers of liens, claims, security interests or encumbrances arising out of the Contract, to the extent and in such form as may be designated by the Owner. If a Subcontractor refuses to furnish a release or waiver required by the Owner, the Contractor may furnish a bond satisfactory to the Owner to indemnify the Owner against such lien. If such lien remains unsatisfied after payments are made the Contractor shall refund to the Owner all money that the Owner may be compelled to pay in discharging such lien, including all costs and reasonable attorneys' fees.

"The following are required by the construction agreement, supplementary conditions, and specifications:

(1) All of the building keys and master keys and the keying schedules.

(2) Operating instructions for all the mechanical and electrical equipment. You will also be required to instruct Allen's maintenance personnel in the operation of all equipment.

(3) All of the specified spare parts for mechanical and electrical equipment.

(4) All specified extra tiles and other materials needed for future maintenance.

(5) All specified equipment lists and color and material schedules needed for future maintenance.

(6) An updated list of all subcontractors and suppliers.

(7) A complete set of the record drawings and specifications marked to show all as-constructed conditions where they deviate from the contract documents.

(8) Return all excess plans and specifications to the architect's office, retaining one complete set for the contractor's records.

(9) All specified warranties and Hyde Construction Company's written warranty.

(10) A Mechanics' Lien Guarantee from a title insurance company showing that there have been no liens recorded within the 30 days after the Notice of Completion was recorded.

(11) Hyde Construction Company's unconditional waiver and release of mechanic's lien.

"George, this'll keep your office busy. I suggest you get right on it so the retention payment won't be held up."

Allen was overwhelmed by the large amount of administrative technicality required in getting the project wound up. "George, I'll have your check ready in a couple of days and in no event later than 10 days from today as required by our agreement."

George, warmly, "Thank you, Allen, I'll appreciate that."

Allen added, "By the way, George, I noticed a burned out light bulb in the hallway outside of my office. Will you take care of it?"

"Sure, Allen, I'll have it replaced this afternoon. But from now on you gotta take care of the building yourselves. We'll be finished by the end of today. Don't call us back unless it's to correct defective construction. From now on normal care and maintenance is your problem."
Ivor confirmed this, "Allen, he's right."

Allen acquiesced, "Oh."

They then all stood up, shook hands, and the architect and contractor left.

* * *

The three reassembled a month and 3 days later, on July 9th, in Allen's office. George had just emptied the contents of a large cardboard box onto Allen's clean desk. "Well it's all here, Ivor, everything required by the contract documents. However, we do have one small problem."

Allen and Ivor exchanged glances and looked up apprehensively.

George continued, "Yeah, well when the Mechanics' Lien Guarantee arrived yesterday, we noticed a lien claim had been recorded by W & X Air Conditioning Company, in the sum of $900. We had backcharged them for some carpentry they were responsible for and now they don't want to pay it. We deducted it from their last progress payment and that stubborn old Charlie Woods filed a mechanic's lien claim."

"Why don't you just pay it?" asked Ivor.

"I don't want to give in to Charlie! He can go fly a kite. We've worked on dozens of jobs together but he isn't getting any easier to deal with."

"Why do you keep doing business with him?"

"Well, he's the most reliable air conditioning contractor in the business and besides he's one of my oldest friends and poker-playing buddies."

"Well, you and your buddy are going to have to straighten out this mess before I can approve payment of the retention."

George asked, "What'll I have to do?"

"That'll be up to Allen's lawyer, Philip Quinn. Allen, will you please call Phil and ask him how this can be resolved?" Allen, becoming irritated, said, "Ivor, why should I incur more legal expense? Why can't you make this decision? Why should I spend money because George wants to play stupid games with his poker crony?"

Ivor replied, "Allen, this has become a legal matter and, according to our agreement, you're obligated to furnish any legal services as may be necessary at any time for the project. This is in subparagraph 4.8 of AIA's Owner-Architect Agreement, Document B141, Fourteenth Edition, 1987."

George expansively offered, "Don't worry, Allen, I'll pay Phil's fee on this one."

So, Allen dialed Phil's number, put the call on the speaker phone so all could hear, and asked Ivor to explain the situation. Phil's immediate reply was, "Essentially, George has two choices, either pay W & X and get a release of this claim, or post a bond with the title company until the lien claim is resolved."

Allen asked, "Is that all there is to it?"

"Yes, but everything has to be done properly so the lien will not attach to your property. This could foul up your application for refinancing. And George should get his own attorney to handle all of the technical details. I'll review it to protect your interests."

Allen concluded the call and turned to George, "The ball's now in your court."

"Okay. I'll take care of it. But I'm not going to pay that pirate, Charlie. He'll have to sweat a little more. I'll bond around it."

Ivor, summarizing, said, "Well, Allen, as soon as George has this legal matter resolved, I'll approve your paying the retention of $220,085.61 It'll be due July 11th and must be paid within 30 days thereafter."

Ivor continued, "The general contractor's one year warranty period started on May 28th, the date of Substantial Completion. It'll be important to report promptly any defects noted during this period and Hyde Construction will take care of them. Early next May we should make a comprehensive examination of the building and grounds so that Hyde's people can rectify any defective work that becomes apparent during the warranty period. The business of this meeting is now concluded. I'll send a copy of the minutes to each of you as usual."

The three shook hands and parted. Hyde was elated with the prospects of imminently receiving his retention payment and keeping the needle in his old poker-playing associate.

Ivor departed resolving to compile Judge & King's final bill as soon as he reached his office so he could get on to other projects.

Allen was again confirmed in the wisdom of his choice of architect and comfortable in the feeling that the bidding process had produced a fine, reliable, although extremely stubborn, contractor. And, most of all, he was uncommonly proud of his new building.

* * *

Moral: Construction contracts are not as simple as they look. They are far from self-executing and require reasonable and appropriate attitudes and a fair amount of skilled administration.

* * *

…

Points of Law 3

ABC Warehouse II
Job Close-Out

3.1 Extra Work

Hyde is submitting two claims for extra work: one for additional excavation, rebar, and concrete necessitated by inadvertent over-excavation ($11,353.38) and another for upgraded wood paneling in Allen's office installed at his request ($14,661.35).

Hyde has no chance of getting paid for the extra rebar and concrete. As a matter of law, the prime contractor is responsible for the mistakes of the subcontractor and therefore Hyde must bear the additional expense caused by the negligent over-excavation.

Allen receives no economic benefit from the additional money expended for rebar and concrete because he ended up getting nothing more than the building he bargained for. The fact that some extra money had to be spent on the foundation work will not add anything to the market value of the building.

The upgraded wood paneling is another matter. It was not called for by the original contract and it does increase the value of the building. It cost Hyde $14,661.35 to perform the upgraded work. As things stand Allen got the benefit and Hyde got the expense. An inequitable situation!

The contract provides that the contractor is not entitled to extra compensation absent a written change order signed by the owner. Although Allen asked Hyde to perform the upgrade work, Hyde never got around to processing the change order. Will this technicality allow Allen to escape, legally, from paying for the extra work? Allen argues that although he asked for the extra work, he never promised to pay for it. Hyde argues, logically, that Allen could not reasonably have expected that a gift was intended!

Most courts, nationwide, will decide this one in Hyde's favor. Simple justice requires that the work be paid for by the party who got the benefit of it.

3.2 Claims

The AIA general conditions call for the architect to rule on claims for extra work. Ivor Judge has made his ruling. The ruling, however, is not final. It is subject to arbitration under the rules of the American Arbitration Association. The decision of the arbitrator is final, binding, and enforceable. In fact, decisions of arbitrators are more final and binding than those of trial court judges, since they are not subject to appeal.

3.3 The Retention

After Allen pays for the extra paneling, he'll still be holding a retention of $220,085.61. Most construction contracts permit the owner to withhold a retention from progress payments in order to give the contractor adequate motivation to finish the job and also to secure the owner against defective construction and mechanics liens.

3.4 The Mechanics Lien

Hyde is in a snit fight with his poker playing buddy and air conditioning subcontractor over a $900 backcharge.

As long as the mechanics lien is on the record, the final financial close-out of the job will be delayed. Everybody knows the lien will probably expire because it's not worth hiring a lawyer to file a foreclosure suit for $900. Nevertheless, in order to get his final payment, Hyde is going to have to get rid of the lien. He can either pay it off or bond it.

In some states the effect of a lien can be removed by persuading a title company to write a title policy that does not disclose the lien. If Hyde has a good relationship with his title company, he may be able to persuade it to do this by signing an agreement to indemnify the title company against loss. Otherwise, he can remove the effect of the lien by filing a corporate surety bond in the county recorder's office. The premium charged by the bonding company will probably be a couple hundred dollars, and in addition to that Hyde will have to put up collateral (say, a certificate of deposit) for a couple thousand dollars. Once the bond is recorded, the title company will issue the mechanics lien guarantee and the way will be clear for the disbursement of the final payment.

4

ABC Warehouse III
A Maintenance Failure

The ABC Warehouse Company had been in their new offices and warehouse for over a year and they still found it exhilarating in such a comfortable, efficient, and attractive facility. They appreciated the fruits of competent architectural and engineering design and skillful construction. They were justly proud of their brand-new premises and were gradually getting used to taking care of it. Hyde Construction had been diligent in correcting the few minor defects that had surfaced during the first year and now the contractual warranty period had expired.

Allen Brady, ABC's president, had assigned Carl Daly, his thorough and hardworking warehouse manager to be in charge of property maintenance. Carl had hired a landscape gardener whose capable crew attended weekly and kept the planted areas looking beautiful and well groomed. Carl also had a list of reliable service companies specializing in industrial plant maintenance that he'd call upon the few times he would need an electrical, plumbing, or other technical repair. He had signed an annual air conditioning maintenance contract that provided for regular monthly inspection and filter service as well as emergency calls. In addition, a couple of his warehouse employees were handy for doing miscellaneous repairs and upkeep such as replacing light bulbs and fluorescent tubes and for performing minor painting jobs. He purchased a sturdy metal toolbox with an assortment of small tools like screwdrivers, wrenches, pliers, and a good hammer. He also bought a solid 18-foot aluminum ladder to facilitate light bulb replacements and other high repairs. He felt secure that he had all bases covered and that his ever-demanding boss, Mr Brady, could not criticize him for not taking proper care of the new building and grounds.

Every 5 or 6 weeks Carl would have one of his men go up on the roof to gather and remove any debris that would have been blown up there by errant wind currents, carried by industrious birds, or thrown by irresponsible passers-by. Al Barker, one of Carl's warehouse workers, was always surprised at the diversity of stuff that he would find on the roof. He regularly found pieces of newspapers, cigarette and candy wrappers, bits of cardboard, leaves and twigs of shrubs and trees, string, plastic grocery bags, and fast food wrappers. Al once found a tennis ball and, another time, a Frisbee. Usually the debris would accumulate next to the parapet walls, on the flow lines, and at the drains where it would have been carried by rain and wind. After collecting all the debris in a plastic bag, he would individually check all of the roof drains to make sure they were clean and clear. Al then brought the bag of debris with him, descended through the roof hatch, relocked it, and disposed of the trash. Afterwards, he always reported back to Carl for his next assignment.

On Friday of a dry sunny week in early summer, Carl instructed Al Barker to go up on the roof to remove the debris and check the drains. After collecting the usual windblown and thrown objects,

Al started in on the roof drains. In order to clean out the accumulation of silt from one of the 4-inch diameter roof drains, he had to remove the cast iron high dome strainer. This involved removing four wing nuts which he carefully placed on the roof next to the open drain. He was kneeling on the roofing gravel and when he shifted to find a more comfortable position, his knee accidentally brushed against the wing nuts. He instinctively grabbed at them but was not fast enough. He watched all four of them quickly disappear down the gaping open pipe. He uttered a suitable exclamation, gathered up the trash bag and dome strainer, and went back inside the building.

He immediately reported the problem to Carl and handed him the strainer. Carl said, "Nothing to worry about, Al. I'll get some more wing nuts this afternoon. They're only about a nickel each. You can reattach the strainer first thing tomorrow. But right now I want you to go help Nigel move those television cartons."

So Carl placed the cast iron strainer on top of a filing cabinet in his office and promptly forgot about the wing nuts. He had more pressing matters. He thought about getting the wing nuts 2 or 3 times during the next month but never did quite get around to it.

* * *

ABC's business was steadily increasing and Allen was sure the new facility was attracting customers they had never seen before. He had just that morning told Carl about a large shipment of Danish furniture that would be arriving around noon. They would have to offload it at their receiving dock, store it for about two weeks, and then ship it out gradually over 10 days to a local hotel that was undergoing a major refurbishment. The total value of the furniture was approximately $600,000. Carl would have to clear out and consolidate about 5000 square feet of warehouse space to keep this shipment all together. It took most of the morning to clear the needed space and then they stopped for lunch.

After the warehouse crew had finished its lunch break, the Danish furniture shipment started arriving. It took them nearly 4 hours to get it all unloaded and neatly stacked in Sections 18 through 28 of Bay 12 of the warehouse. Shortly thereafter, at 4:30, they closed down the warehouse and secured it for the weekend.

Carl went around and made a careful final check to be sure all doors were securely locked, set the burglar alarm, and hurried home so he wouldn't be late for his weekly bowling session.

* * *

It turned out to be an unexpectedly windy night and upon arriving home from bowling he was concerned about his trees. With the wind gusting and whistling, he went to bed. After a busy day at the warehouse and a vigorous bowling session he quickly fell asleep. During the night the wind slackened and was replaced by a heavy and persistent rainfall.

Saturday morning it rained continuously. It rained steadily all afternoon and evening. On the 6 o'clock news, the weather forecaster said that over 2 inches of rain had already fallen and there would be considerably more before the storm moved on.

ABC Warehouse III

* * *

It was a busy graveyard shift at the master control console of Continental Security. Donna Evans, one of a two-person team, was kept busy monitoring the central control system to which most of the local industrial burglar alarms were connected. The winds and heavy rains had set off several false alarms but who can tell the difference from the real thing? All default signals must be taken seriously.

When the automatically dialed call came in from System Number 9327 she called up the identity information and standing instructions on the VDU screen. It was ABC Warehouse. The instructions were to first call the warehouse office on the chance that someone there had set off the alarm accidentally or there was an armed robbery in progress. That's not too likely, Donna thought, it's three in the morning. She dialed, waited, and got no answer. The second call was to the local police station. The police operator said all their available cars were out on traffic accidents and false burglar alarm calls. "We'll get there as soon as possible. I can't say when." So then Donna made her third call, to Allen Brady.

* * *

Allen couldn't imagine who would be calling at this ridiculous hour. He turned on the bedside lamp, noted the time as 3:15 AM, and answered the phone. He could hear the noisy downpour of rain pelting down on his adjacent patio. Donna said, "Sorry to call so late, Mr. Brady, this is Continental Security. We've received a signal from your intrusion alarm system. I've already called the police. It may be a false alarm, as we've received a rash of them in the last few hours on account of the wind and rain."

"Okay, thanks for calling." Allen was apprehensive but somewhat relieved since it probably was only a false alarm. He then dialed Carl Daly's number. Finally, Carl answered. Allen commanded, "Carl, will you go to the warehouse? I just got a call from Continental Security that our alarm is signaling. They've called the police. I'll meet you there."

Carl, now alert, said, "Yes, sir, I'll get there as fast as I can."

* * *

Allen lived closer than Carl so was first to arrive even though he had had to drive slowly on account of the heavy rain and flooding. As he pulled into the ABC parking area he was puzzled, as everything looked normal except for the alarm bell ringing and the flashing blue strobe lights near the top of the warehouse parapet. He carefully parked in the stall marked Mister Brady, got out, and locked the car. He dashed through the rain to the front entrance door of the office wing and was standing on the covered porch shaking off rain and wondering if he should enter and risk meeting an armed intruder.

Just then Carl arrived, parked at an untidy angle in the aisle near the entrance, and ran to the porch. They went in together, turned on lights, and went directly to the main electrical switchgear room where the burglar alarm control panel was located. They both knew the proper code number but Carl got there first and keyed in the number to turn off the bells, sirens, and flashing exterior lights.

It then seemed unnaturally quiet. A quick check of the offices revealed nothing out of place. They stayed together and started for the warehouse door. It was then that they noticed a sparkling trickle of water seeping in at the doorsill. As Carl opened the warehouse door, backed-up water gushed into the office corridor. Carl then hit the lighting circuit contractor for the first two warehouse bays. It immediately flashed and half of the lights went out again. There was still enough light so they could see what had happened. The warehouse looked like a great shimmering lake. There was water all over the place. It went up the aisles in every direction. It was dammed up against and running out through the small crack under the loading dock doors. They could see that a large piece of the roofing and roof-framing members were dangling down into the warehouse. A broken 2-inch diameter main water pipe was disgorging spouting and spraying water. Lighting fixtures and conduits were drooping and crazily swaying as the water kept them moving.

"The roof collapse must have triggered the burglar alarm," reasoned Carl. After a second of shocked realization, his brain shifted back into high gear and he shouted, "We've gotta get the water turned off." He ran for a wrench to turn off the valve at the meter. Where the hell's the damn water meter, he was thinking...Oh, I remember, it's next to the loading ramp entrance. He opened one of the freight dock roll-up doors and dammed up water flooded out over the dock into the truck well. He made a dash for the meter and found the valve. He fumbled, skinned his knuckles, but finally got it turned off and stumbled back through the rain to the warehouse. He was wet to the skin. He found Allen standing in front of the Danish furniture shipment. Most of the cartons were soaking wet, soggy, and dripping. Allen looked overwhelmingly desperate, shocked, and devastated.

Carl, now thinking clearly, said, "We'd better call some of our people and get them down here so we can start mopping up this mess and save our customers' storage." So they made their calls. Gradually the crew arrived and by 4:15 a force of six had assembled and they spent the next several hours squeegeeing, mopping, sweeping, drying, and cleaning floors, moving cartons and crates, and trying to restore order.

Rain was still falling through the gaping hole in the roof. Added to this was rainwater running down the roof's flow lines seeking the missing drain. So Carl's crew erected plastic sheeting like a tent and channeled the intercepted water to a 55-gallon barrel that had to be bailed out with a bucket every 20 or 30 minutes. By 5 o'clock in the morning the rain had stopped and the storm seemed to be moving on. Carl and his men had not stopped to rest and still had a lot of mopping and drying left to do. They had no idea yet how much damage had been done to their customers' goods.

Shortly after the clean-up operation had been organized, Allen dialed George Hyde of Hyde Construction Company and roused him out of a sound sleep. George was not at all pleased to hear from anyone at 5 o'clock on a Sunday morning.

Allen quickly explained the situation and George promised to hurry over to see what could be done. So George called Ezra Field his valued job superintendent and asked him to be there too. George was naturally concerned and curious about what could possibly have caused this catastrophe so thought it best to call Ivor Judge, ABC's architect. We may need some ideas on how to repair the building and how to protect our interests. If we're not careful, there'll be a lot of finger pointing and unjust apportionment of blame.

George Hyde and Ivor Judge arrived at about the same time, Ezra shortly thereafter. Ivor couldn't help noticing that most of the cars in the lot were parked helter-skelter with no regard for the carefully planned and laid out parking arrangement. He shrugged and parked similarly.

It was around 5:30 and the fresh new day was well under way. When George, Ezra, and Ivor arrived inside they received a complete run-down and a tour of the disaster zone by Allen and Carl. George and Ivor were extremely apprehensive as to the cause of this very costly looking mishap. They were both furiously churning their minds for possible lapses in their design or construction operations, fearful that their own contributions might have been tested and found wanting.

With considerable trepidation and ill-foreboding, the contractor and the architect scaled the stationary steel roof ladder, Ivor leading, through the hatch that had to be unlocked, and out onto the roof. Ivor was carrying his flash camera and a note pad. Two side bays adjacent to a roof drain had collapsed. They could see the roof drain that had ridden down with the subsided section of roof framing. That section of the roof structure was hinged down about 6 feet. They could clearly see that the drain's dome strainer was missing and there appeared to be something plugging the drain.

To get a closer look they had to return to the warehouse floor and, using Carl's 18-foot ladder, Ivor climbed up through the hole formed by the sagging portion of the roof. He examined the drain. In it was a round waxed paper milk shake carton wedged tightly into the drain as neatly as a cork. It must have been blown up onto the roof, he thought, and been carried by the first flow of rain into the drain. This stoppage would only cause rainwater ponding to a depth of 2 or 3 inches, not enough weight to collapse the roof. Then the overflow scupper would come into play. Why didn't it take the overflow? He climbed up another rung to where he was better able to see the remnants of debris that had so effectively closed the overflow outlet--newspaper pages, leaves, a rubber ball, a wrecked kite, some small sheets of waterlogged padded packing material, some strips of carpet padding, all wedged into and over the 4-inch x 3-inch opening where it was carried by the rainwater flow. He carefully photographed these conditions and climbed back down to the warehouse floor. Then, while George was up the ladder, Ivor made extensive notes and sketches of all that he had seen, painstakingly inventorying the debris.

He conjectured that the weight of the accumulated water must have collapsed a glu-lam purlin and when it suddenly gave way and dropped, the water pipe was broken. The fractured roofing membrane simultaneously liberated several thousand gallons of ponded water which emptied into the warehouse. Electrical fixtures, conduits, ventilation ducting, and insulation had all come down with the roof.

Meanwhile, Ezra busied himself listing the types and quantities of lumber, other materials, and equipment that would have to be brought to the job Monday morning to start the rebuilding. He was also listing the various subcontractors that would have to be put on call.

A few minutes later, Allen, Carl, George, and Ivor were drinking hot coffee in Allen's office. Carl was utterly exhausted from all the physical labor he had performed, but he did not sit down, as he was so wet and grimy. He seemed to be extremely uncomfortable and somber. Allen was concerned and absorbed and appeared to be in a state of shock.

George, in his usual direct mode of expression, said, "Well, this is one that can't be hung on old George Hyde. We build 'em right. Solid as Gibraltar! Never known to fail!"

Allen stirred from his preoccupation with customer losses and possible insurance coverage and replied, incensed and indignant, "What do you mean they don't fail? It did fail! No doubt about that! Hyde really screwed up this time! If it's not your fault, then Ivor must have botched the design. All I know is, decent buildings don't fall apart like this!" His voice was rising and unsteady.

George kept his cool and sat there smugly waiting for Ivor to speak up. Carl, distraught, stood uneasily, eyes downcast, as he now fully realized what must have happened. He wished he could disappear.

Ivor started to explain the situation to Allen. He described the entire physical condition as he found it and all of his presumptions. He explained, "If the strainer was in place none of this would have happened." Then he added, "There's only one thing I don't understand. Why was the cast iron dome strainer missing?"

Ivor continued, "I specifically remember seeing the strainers on all of the roof drains on my last inspection of the building just before the warranty period expired. What happened to that high dome strainer?"

Carl Daly looked up. "I can explain. I know what happened. It was my fault." He then recounted how the strainer happened to be missing. He was embarrassed and miserable.

After a few moments of silence while the import of Carl's admission permeated everyone's mind, Ivor gently suggested, "Right now, let's concentrate our efforts on cleaning up, getting this repaired, and keeping the rain out. Hyde construction can remove all of the collapsed and damaged building elements and reconstruct the whole thing in accordance with the original plans and specifications. As far as I know the original design is adequate. But, just to make sure, I'll have our structural engineering consultant, Myles Nolan, check it anyway. Also, we'll need a new building permit, so I'll get right on that, too. Reconstruction drawings and specifications will have to be prepared."

Allen started to consider his position. If this fiasco was caused by my own employees then ABC might get stuck with the cost of rebuilding the building. And what about the warehouse contents? We have over 3 million dollars worth of customers' goods including all that Danish furniture! I'll be ruined! All caused by that damn nitwit, Carl.

Well, nothing can change that now. It's done. We'll have to accept it and learn to live with it. Looks like he's really taking it hard. He probably thinks I'll fire his stupid ass. That's what I oughtta do. Damn him! Well, he's too loyal, conscientious, and diligent to let him go. He'll work even harder after this. I'll let him suffer first, though.

Allen was discouraged and beat. He turned to Ivor, "What do we do now? What'll it cost? Do you think my insurance will cover this?" What'll our customers say? Our competitors'll have a field day."

Ivor, thoughtfully, "You've raised some serious questions here, Allen, that I can't answer. I'd suggest that you call your lawyer, Phil Quinn. Right now. See if he can come right over here. I'm sure he isn't very busy at 6 o'clock on a Sunday morning." So Allen called Phil, briefly described the situation, and Phil readily agreed to come right over.

While they waited for Phil, they went out into the warehouse to take another look at the ruins and the industrious mopping up operation that was still in full swing. Carl quietly rejoined the warehouse crew and pitched in again on the physical labor. He still looked distressed and dispirited.

* * *

Phil Quinn, with the firm of Quinn & Quinn, had been handling Allen's legal problems since inception of ABC Warehouse nearly 20 years ago. He made good time on the deserted streets and smartly pulled in to ABC's now chaotic parking lot. By now, almost 7 o'clock, the sky was clear and developing into a beautiful sunny day.

Casually dressed, Phil strode briskly through the office and into the warehouse where he was stunned by the discouraging scene of building destruction and flood damage. Allen told him he should have seen it a couple of hours earlier when it looked even worse.

Allen led the way back to his office followed by Phil, Ivor, and George. Carl saw them leave but didn't attempt to join them. He went back to squeegeeing. Ezra was still organizing the reconstruction.

Back in his office, Allen invited Phil to sit in his tall-backed black leather swivel-chair behind his broad oak desk, a generous gesture he considered to be the ultimate honor. Allen then asked Ivor to recount to Phil the physical circumstances, his previously stated conjectures, and Carl's incredible admission.

After Ivor's concise elucidation, Allen turned to Phil, "What'll we do now? How'll our insurance carrier react to all this? Will we lose our coverage? What'll we do about our customers' damage? What if some of them are unreasonable? What'll this cost us?" His mind was overflowing with these and similar questions. He was deeply concerned.

Phil and Ivor were both taking copious notes. George Hyde now considered himself fortunate indeed to be merely an innocent bystander, as he clearly shared none of the blame for this disaster. Furthermore, he could look forward to a profitable cost plus contract starting first thing tomorrow morning. He could use a nice little fill-in job right about now. He just sat and made himself comfortable and available to answer questions.

Phil, finishing his notes, and after some thought, replied, "Well, Allen, first, continue your clean-up activities and do your best to limit any further damage to your customers' stored property.

"Secondly, authorize your architects, Judge & King, and Hyde Construction to proceed immediately with reconstruction of the building and the restoration of weatherproofing and security.
"Thirdly, we must notify your insurance brokers and all of your storage customers of the situation.

"Right now, I want you to get me all of your insurance policies and customer storage contracts. I'll take them with me today. After I've reviewed the policies and contracts and evaluated them we'll decide how to proceed from there. I'll call you early tomorrow afternoon."

Allen was relieved that Phil was now firmly in charge and felt confident that all would be promptly resolved in the best possible way.

George and Ivor rose and walked out together. They each had an indefinable lightness in their strides and a profound sense of freedom and relief.

* * *

Moral: Construction defects are not always the fault of contractors and architects. Sometimes owners' imperfect maintenance programs cause them.

* * *

Points of Law 4

ABC Warehouse III
Plugged Roof Drains

4.1 Cause of the Loss

As a general rule water intrusion losses can be divided into four types: those caused by faulty construction, faulty design, faulty materials, and faulty maintenance. The liability for losses caused by water intrusion, and the prospective insurance coverage for them, depends upon this initial classification.

In this story, the acknowledged cause of the loss is faulty maintenance: ultimately, the negligence of the warehouse employees who were in charge of maintaining roof drainage. It was through their negligence that the four wing nuts entered the drain, the dome strainer was removed and not replaced, and the overflow scupper was clogged.

4.2 Property Insurance

Among several types of insurance policies on the market the two here implicated are property insurance and liability insurance.

An easily recognized type of property insurance policy is fire insurance, which insures a building owner (and mortgage lender) against loss or damage to a building caused by fire. Modern property insurance policies cover more than fire; in fact, they cover "all risks of physical loss" and are known as all risk policies because they cover all risks of loss that are not specifically excluded from coverage. (Common exclusions are for flood, earthquake, and tidal wave.)

Many all risk policies cover not only loss from damage to the building itself, but also are extended to cover loss or damage to the contents of the building. ("Contents" would include goods stored in the warehouse.)

4.3 Liability Insurance

Property insurance is known as two party insurance because it deals with the legal relationships between only two parties: the insurer and the insured. Liability insurance is called three party insurance because it deals with the legal relationships between three parties: the insurer, the insured, and the claimant. Liability insurance protects the insured from loss and expense caused by claims asserted by third parties.

It is obvious, then, that Allen will report the water intrusion both to his property insurance carrier and to his liability insurance carrier: to the first because his property (the warehouse building) has been damaged and to the second because he is or may be liable to the owners of property damaged while stored in his warehouse.

4.4 Homeowners Insurance

Not so obvious, perhaps, is the possible coverage afforded by warehouse manager Carl Daly's *homeowners policy*.

The typical homeowners policy is a package deal that is intended to protect the homeowner both against damage to the home (property insurance) and liability to third parties (liability insurance).

There is great variety in amounts and types of liability coverage provided by different homeowners policies. But under the facts of this case the manager, who forgot to replace the dome strainer, has potential liability, and therefore Allen's lawyer should take a look at Carl's homeowners policy too!

4.5 Water Intrusion

Water intrusion is the construction defect that comes to the attention of lawyers, arbitrators, and judges more than any other. There are reasons for this. For one thing, water is just hard to keep out! For another, installations that look as if they should be waterproof oftentimes are not! For a third, the roof is a remote area, difficult of access to inspectors and superintendents. Substandard work often goes undetected. And, as our story illustrates, substandard maintenance is hard to prevent.

Roofing consultants offer inspection and maintenance services. Employment of such services could have avoided such problems as that related in this story.

5

The Tulare Job
A Costly Labor Relations Problem
A Disastrous Subcontract

Three exhausted men were sitting dejectedly in an untidy smoke-filled room. They were talked out. Piled papers and discarded coffee cups littered every surface. No, they were not picking candidates for political office. They were endeavoring to piece together the remnants of a construction subcontract that had gone seriously astray. They were striving to understand how such a promising contract had gone so sour. They couldn't believe the dismal figures they were reviewing. They had met early on a Saturday morning, talked all day, discussing the same old points over and over, and were now disheveled, disheartened, and exhausted.

The Power brothers, Michael and Tracy, were the owners of Power Electrical Corporation, a fairly successful electrical contracting firm with offices, shop, and yard in Santa Monica, California. They had built the business from scratch during 15 years of diligence and hard work. Seated with them was Ward Montgomery, their superintendent on the Tulare job, and a trusted key employee. Monty had been with the brothers for over 10 years and knew the ins and outs of Power Electrical as well as Mike and Tracy.

The project they were so earnestly dissecting was just completed and their figures showed an out of pocket loss, so far, of over $45,000. There was no profit, no recovery of overhead, and they had actually spent over $45,000 more than they could ever collect. They were in mortal shock. Although the work in the field was completed, except for minor pick-up and a one year warranty, they still had to face up to an arbitration, lawyer's fees, and a lot of lost time for Mike, Tracy, and Monty. And they could easily lose the arbitration. Their position wasn't that good. All of this for a year's work on a contract of over half a million dollars and an expected profit of nearly $75,000.

How could this have happened? The explanation starts with events set in train over a year ago. There are a lot of details, but I'll try to be brief.

* * *

Michael Power made all the decisions about which jobs to bid. It was early in March, 1986. He was sifting through the Commerce Business Daily looking for interesting work to bid and spotted a possible live one. It was a State of California job with an architect's estimate of $12,275,000. One of the listed general contract bidders was their old friend, Mammoth Construction Company in Northridge. Through the years they had done a lot of work with Mammoth and were on good friendly terms with Graham Elliott, Mammoth's president. A phone call to Graham got Mike the promise of the drawings and specifications for two days, enough to take off the job so they could prepare a bid.

When he met his brother in the coffee room an hour later, Mike said, "Tracy, we're going to bid the Juvenile Rehabilitation Center in Tulare County for the State of California. Mammoth is one of the generals. We'll have the plans and specs next Wednesday and Thursday. We can have our bid prepared by the following week."

Tracy looked questioningly at Mike over the rim of his coffee cup, "Isn't Tulare a little out of the way for us?"

"Well, it is a little far but we need the work and I think we can cope with it. We'll have to do a little research on the labor situation up there, but there should be enough money in it to make it worth our while."

"It won't hurt us to look at the plans and specs anyway. Do whatever you think is right, Mike." Tracy trusted Mike to make the right decision. So far he had led the company into a most profitable electrical contracting business.

* * *

Mike Power had the plans and specs picked up at Mammoth Construction and had agreed to return them before nine o'clock Friday morning. That gave them two full days to make their take off.

Upon delving into the documents, Mike found that the project was a Juvenile Rehabilitation Center in Tulare County, about 25 miles from Sequoia National Park and about 160 miles from home base in Santa Monica. The project consisted of several buildings, all one story. There were 24 barracks style dwelling buildings of 3000 square feet each, a recreational-educational building of 6000 square feet, a workshops building of 9000 square feet, and an administrative unit consisting of kitchen, dining hall, infirmary, and offices of 12,000 square feet, a total of 99,000 square feet.

All of the buildings were to be of UBC Type V, frame and stucco construction, with gypsum wallboard interiors. The electrical portion was simple and, in addition to the buildings, included some grounds lighting and on-site underground electrical distribution inter-connecting the 27 buildings.

It didn't take long to make the take off as the residential portion consisted of 24 identical buildings. All that was left was to price it out, estimate the labor, and add in the overhead and profit. The plans and specs were returned to Mammoth on time as promised.

Mike discussed the labor problem with Tracy and Monty. The electrical union surcharges for out of town travel and subsistence made it impractical to use their regular Los Angeles electricians. So, they'd send Monty up to Tulare to run the job. He'd stay in a motel in Weed Creek, a woebegone settlement about eight miles from the jobsite. He'd obtain as many electricians as he'd need from the union hiring hall in Fresno, about 70 miles from the site. Mike had already phoned the secretary of the local electrical union, George Black, and obtained all the details. Their labor costs would be little different from local jobs.

Monty said, "That'll be great fun spending my evenings in Weed Creek. I'll bet the night life there is hilarious."

Mike said, "You won't have to worry about the night life there, Monty. You'll need your evenings for record keeping, phoning back here to let us know how the job is coming, ordering materials, planning the next day's work, and getting to bed early. Don't worry, you'll get a good cut of the profit to make it worth your while."

Tracy's parting shot as he dashed off to a jobsite meeting, "Well, Monty, don't worry too much about the night life. We haven't got the job yet!"

* * *

Mike and his estimating assistant, Martin Bradley, worked over the figures until they got them just right. High enough to make a profit and low enough to get the job. That's the magic formula. Their Bid was $561,401. They double checked everything and sealed the envelope. Martin delivered the bid to Mammoth Construction's office in Northridge on Wednesday morning, March 26. The general contractors' bids were due at the State Building at 1:00 PM the following day.

* * *

Mammoth's bid was low in a field of seven bidders at $12,120,000. When Graham Elliott, Mammoth's president, called Mike Power, he was elated, "We've got the job! And you're the low electrical bidder so you've got the job! You'll be hearing from us soon. We've got to get started on time if we're to avoid paying any liquidated damages!"

Mike and Martin immediately re-checked all their bidding papers and couldn't find anything wrong so they then broke the good news to Tracy and Monty.

* * *

Mammoth immediately prepared all of their subcontracts and held them until after they received their signed contract from the State. It was signed on April 17, 1986. Then they started mobilizing the job in earnest. Jerry Shore was just finishing a school district job in Yorba Linda and he'd be the right man for job superintendent. He was a bachelor and wouldn't mind spending months away from home in the Weed Creek Motel. He could return to Los Angeles on weekends.

They got the job started on May 5, 1986 and, allowing 330 calendar days for construction, they would have to achieve substantial completion by March 30, 1987. Not too hard to do according to the detailed CPM construction schedule prepared by Mammoth's chief estimator, Rivers Owen. They had gone over it with their main subcontractors and suppliers and everyone agreed it was practical, reasonable, and attainable.

Monty initiated the electrical work by setting up a portable job office and secure storage container. He made arrangements with George Black at the Fresno Electrical Union for four electricians and they started work immediately on the underground electrical distribution system. Then, with a couple more electricians added, they began installing the underfloor conduits in each building. They were meshed in with Mammoth's concrete crews as they moved from building to building forming and pouring the foundations and floor slabs. Jerry Shore was starting one building every week. He was proceeding well on schedule. This was construction efficiency and cooperation at its best.

The schedule called for phased construction of the buildings, starting with Administration Building A, Shops Building B, Recreation Building C, and the Residential Buildings D1 through D24. As soon as the first concrete foundation was completed, carpentry crews commenced, followed by plumbers and the other trades in scheduled order. As soon as Building A was ready for above slab electrical work, Monty had started increasing his requisition for electricians from the Fresno union hall. He added even more electricians as the job picked up momentum.

Monty was pleased to report to the head office in Santa Monica each evening that everything was progressing in accordance with the schedule. They would easily finish within the time schedule and safely under budget. Monty and Mike were enthusiastic. Mike told Tracy, "Monty's on top of the job. We'll make a few bucks on this one!" Tracy shared his brother's enthusiasm.

***

The crew of electricians was up to 22 now and the men appeared to work harmoniously. Many of them knew each other as they had worked together on previous jobs in the Fresno area. One of them, Amos Benton, seemed to be a natural leader and the others paid attention to what he said since he sounded so authoritative. He gave some of them tips on how to read the plans and how to do their work more efficiently. He knew everything there was to know about the electrical trade.

When they got to the conduits for feeding the ceiling lighting fixtures in Building A, Monty started a crew of two in installing 2-inch by 6-inch wood blocks between the ceiling joists. According to Section 16.57 of the Electrical Specification, the fluorescent lighting fixtures were to be installed after completion of the ceiling by screwing through the gypsumboard into the wood blocking. There were to be two blocks per fixture. He set up a portable work place for measuring the blocks, cutting them with an electric skillsaw, and carrying them into the building to be nailed into place at previously measured and marked locations. The wood blocking operation was efficiently under way for about three hours when the electricians stopped as usual for lunch.

The electricians usually gathered to eat their lunches in the shade of the Power Corporation office and storage container. As usual, most of them were raptly listening to Amos Benton expounding authoritatively on some arcane subject in which he had extraordinary in-depth knowledge. After the subject began to pall and he sensed the defection of his audience, he zeroed in on the wood blocking crew. With elaborate disdain and his usual air of authority, he smirked, "Boy, you guys are some great macho electricians. Doing carpentry work. Carpenters should be doing that kind of work. Not electricians. You'll never catch me doing carpentry work. I'm too valuable as a highly trained and skilled electrician! You chumps are pathetic!"

Crestfallen, the wood blocking crew went back to work, measuring, cutting, carrying, and nailing. Around mid-afternoon, Monty checked with them and found them dissatisfied with their assignment. They recounted Amos' lunchtime comments and Jack Russell, the disgruntled blocking crew leader, said, "Monty, isn't this carpentry work? Why are we doing it? We're electricians! Highly trained and skilled!"

Monty patiently explained, "Look, you guys, this blocking is in the Electrical Specification, Section 16.57. Power Electric's bid includes everything in the Electrical Specification. So, we have to do it. Y'understand?" They went back to work.

* * *

The next day at the electricians' lunch, Amos Benton was back on the subject of electricians doing carpentry work. "It isn't right. Electricians should do only electrical work."

Jack Russell countered, "Yesterday, Monty told us that the blocking is specified under electrical work and we've got to do it because Power's contract includes everything in the electrical section. There's nothing we can do about it."

The charismatic Amos persisted, "Oh, yes, there is! We can refuse to do the blocking and refuse to work in any building where electricians have installed any blocking. That blocking's tainted. It should be removed and be reinstalled by carpenters."

They all looked up in disbelief as Amos' amazing theory unfolded. Nevertheless, it started them thinking about it.

The next day at lunch, the authoritative Amos declared that electricians doing carpentry work was against union rules. "We should walk off the job right now. Then Monty will come to his senses and get some carpenters. It won't take long!"

Jack Russell protested, "Look, Amos, I need this job. I've got bills. I don't mind doing a little blocking work."

Most of the others sided with Jack but Amos persisted, "If we let them break this rule none of our rules will mean anything! I say we should teach them a lesson. Then they'll respect us. Let's start now!"

A few agreed with Amos and they started working on the others. Finally they bullied the rest into supporting a united attack on Monty.

When lunch time was over, Monty noticed that the electricians were still sitting there with no move back to the job. He approached the group, "Time to go back to work, men. No time to waste. We're on a tight schedule and can't afford to fall behind!"

Amos said, "We're staying right here until you call off the carpentry work. It's against union rules. We're electricians you know!"

Monty replied, "We've got to do the blocking. It's in our contract!"

"Well, we're not going to do it. Go and get some carpenters!" Amos retorted. No one moved.

* * *

Monty then left and returned to his office. He telephoned Mike Power to inform him of this unexpected glitch and to get his advice. Mike suggested calling George Black.

Monty dialed the Fresno Electricians Union and got George on the line, "George, your guys won't work. They're just sitting here. They object to nailing in blocking to support lighting fixtures. They say it's against union rules." George said, "I'll be right out. I'll be there in less than two hours!"

Then Monty went to Mammoth's site office to talk to Jerry Shore. After bringing Jerry up to speed on the whole situation, Monty finished with, "Jerry, can't you get your carpenters to do this blocking? We'll pay you for it."

"No, Monty, it's in your contract. I've no carpenters to spare. You've got to do it. And you better do it now. We don't have any extra time for screwing around arguing about nonessentials. You go back and tell those asinine electricians of yours to get off their dead butts and go back to work!" Jerry was clearly unsympathetic. Monty left without replying. He was getting worried. He'd expected some help from Jerry. He went back to his office to wait for George.

When the union secretary arrived, it was a little after three o'clock. Monty met him at the site gate and quickly told him what had transpired. George said, "Let me talk to them by myself." He walked across the site to where his members were still sitting around arguing among themselves. They looked up expectantly when he showed up. He exploded, "What the hell do you guys think you're doing?"

Amos, the self-appointed spokesman, calmly and deliberately replied, "We're union electricians. We don't have to do carpentry work. It's against the rules. They should get carpenters!"

George said, "You're right about the rules, but this is no way to go about it. You guys go back to work and I'll take care of this the proper way! We've got to negotiate."

Amos, the weight of leadership going to his head, said, "No way! We'll go back to work when the wood blocking is settled. Then they'll respect us!" The oracle had spoken. They all sat where they were. Nobody moved. George stalked off to find Monty. They discussed it together for a while and then went to Mammoth's office.

Jerry looked up, exasperated, "What is it now, Monty?"

Monty said, "This is George Black with the electricians' union in Fresno. George, this is Jerry Shore, Mammoth's Job Superintendent."

George, trying to sound authoritative but reasonable, explained, "Jerry, our rules don't allow electricians to do carpentry work. We have an understanding with the carpentry union. We don't do their work and they don't do ours."

Jerry said, "The State's specifications put the blocking in the electrical section and that's the way we let the subcontracts. I don't hear any carpenters complaining about your doing their work. This is a two bit phony dispute. It's an illegal work stoppage. Why don't you get your guys back to work so we can finish this job?"

George, starting to see the impossibility of changing Jerry's mind, hardened his position, "It's a matter of principle. We gotta uphold the rules. Our electricians won't do any more blocking. In fact, we won't work in any building where electrician-installed blocking has been installed. It must be removed and reinstalled by carpenters." He started out of the office, followed by a bewildered Monty. Monty looked back over his shoulder to catch Jerry's cynical eye.

George and Monty went back to talk to the electricians. George explained their current position to his members. Amos crowed smugly, "See, I told you guys!"

Monty said, "If you're not going to work you might as well go home. We're not paying you to sit here."

* * *

So, no electrical work at all was done for about a week and a half. Meanwhile, higher union officials from both involved trades were discussing the matter and decided to submit the problem to the Joint Conference Board in Washington, D.C. This is a body consisting of representation from all of the national and international unions. They meet periodically to settle issues of jurisdiction between the various unions. Their protocol was that there should not be strike action. The work should proceed on the job by the trade assigned by the general contractor pending the Board's final ruling.

The managers of Mammoth Construction decided that the work should be completed by Power Corporation as it was already in their contract and the contract had been let in accordance with the State specifications.

So Monty's wood blocking crew resumed work and the electricians went ahead with installation of conduit and boxes. However, Amos kept the pot boiling by telling his associates that they were right and would win the argument in the long run. Amos and a few of his supporters constantly ridiculed anyone who appeared to be working with any speed or relish. Some of the electricians refused to work in areas where wood blocking was being done and they were steadily falling behind schedule. The initial momentum of early weeks on the job was completely lost. Production was perilously falling off and the electrical work was seriously out of synchronization with Mammoth's general construction schedule.

The State construction inspectors and administrators were continuing to apply pressure on Jerry Shore and he, in turn, had repeatedly warned Monty about keeping up to schedule. Finally, Mammoth's home office served written notice on Power Corporation that they would be held financially responsible for all costs of the electrical slow-down as well as liquidated damages.
Jerry Shore was seriously concerned about his overall schedule and shuddered to contemplate what would happen to him if Mammoth had to pay $2000 a day liquidated damages. He called Monty to his site office repeatedly to discuss the problem. It was time to start drywall work in several buildings where the electrical rough-in was not yet completed. In fact, in some buildings it wasn't even started. After several days of holding off the drywall contractor, Jerry notified Power Corporation, and Monty personally, that they would proceed in scheduled order on each building whether the electrical was in or not.

Monty did his best to get his electricians motivated and keep the job moving. But Amos always undid everything Monty accomplished with a few pointed smart aleck remarks when Monty left the scene. Monty tried to get Amos and some of the other die hard militants off the job but George Black insisted that union hall hiring is by seniority, not by discrimination or favoritism. He said he would talk to Amos and his followers but it did no good. George acted like he was afraid of Amos. After all, his job depended on maintaining the good will of the members.

Jerry kept his promise and went ahead with installation of the gypsum wallboard walls and ceilings, starting with Building A. The drywall crews were specialized, so the men who hung the board continued on to the next building while joint tapers would then take over. The texturers and finishers would follow on. The wallboard hangers were already up to Building C, while the tapers were in B and the texturers and finishers were starting in A. With several skilled crews they were moving right along.

Meanwhile, the electricians were concentrating their efforts on the last buildings, D15 to D24, as they were not yet tainted by electrician-installed blocking.

As the electricians ran out of work to do in the last of the D buildings, they started working closer to the wallboard hangers and finally caught up. From then on, wherever they encountered gypsum wallboard already in place, they would have to break holes in the wallboard to install their conduits and boxes. This slowed them down even more and the drywall contractor was constantly complaining about the damage. Jerry Shore told them to keep track of all repair costs as they would be backcharged to Power Corporation.

The job became a nightmare for Jerry Shore and the Mammoth Construction people and it was no better for Monty and the Power brothers.

The original general contract completion date of March 30, 1987 had been extended to April 21, 1987 by approved change orders. Power Corporation finally completed their wiring and the drywall patching was completed on May 15. The other follow-on subcontractors still had at least 45 more days of work.

The date of substantial completion was July 10, 1987. Eighty days late!

* * *

Eventually, the decision was handed down by the Joint Conference Board. By then the disputed work had been long completed. It was finally decided that the blocking work should have been done by carpenters. The Board confirmed, however, that the general contractor was entirely justified in making the work assignment in accordance with the State specification's allocation of the work. The Board's decision would apply to all future cases.

* * *

A few days after Monty had closed down the Tulare operation and was back in the Santa Monica office, Power Corporation received an AAA Arbitration Demand from Mammoth Corporation.

They were asking for:

Liquidated Damages, 80 days @ $2000 per day	$160,000
Backcharges from the Drywall and Painting contractor for patching and repairing gypsum wallboard walls and ceilings	$30,500
Extended supervision and overhead, 80 days @ $1145 per day	$91,600
Total	**$282,100**

Plus attorney's fees, interest, and costs.

* * *

Mike Power had discussed Mammoth's arbitration claim with Glenn Landis, Power Corporation's attorney, who then immediately filed a counter-claim with the Los Angeles Office of the American Arbitration Association. Power Electrical Corporation's counter-claim asked for:

Additional electrical labor made necessary by Mammoth's ordering the gypsum wallboard walls and ceilings installed out of normal and expected scheduled sequence	$181,900
Extended supervision and overhead, 80 days @ $550 per day	$44,000
Total	**$225,900**

And denial of all of Mammoth's claims. Plus attorney's fees, interest, and costs.

* * *

Which brings us up to the all day Saturday marathon discussion at Power Electrical Corporation.

Mike finally rose from his chair and said to Monty and Tracy, "This isn't the end of the world, men. We'll live through it somehow. This is the contracting business. Let's go home and rest up over what's left of the weekend and come back here Monday to sort this out. We'll leave it in the capable hands of Glenn Landis and concentrate all our efforts on finding more work and keeping this ship afloat."

* * *

Points of Law 5

The Tulare Job
Problems with Labor Relations

5.1 Who's to Blame?

Here, the unanticipated labor relations problems of a subcontractor loused up the scheduling, efficiency, and productivity of a job causing losses to the prime contractor and also to subcontractors. Nobody seems to be morally to blame except, possibly, the trouble-making senior electrician who persuaded his colleagues to refuse to perform work that would normally be done by carpenters. The resulting job action has some characteristics of a jurisdictional strike or slow-down based on disputes between unions as to the classifications of work to be performed by their members. This is not a typical jurisdictional dispute, however, because it is not the carpenters (whose work is being performed by electricians) who are complaining – it is the electricians who spurn carpenter work!

5.2 Scheduling

The electricians really have a grievance with their employer. They object to the fact that the employer is requiring them to do work that would normally be performed by carpenters.

It is crucial to the efficient performance of a construction project that the work of subcontractors be scheduled and coordinated so they don't interfere with each other. The very epitome of poor coordination is illustrated by the drywall crews getting ahead of the electricians so that the electricians have to break through finished gyp board in order to wire the job.

The upshot is cross-claims between prime contractor and subcontractor, each seeking from the other damages of more than $200,000 for delay, inefficiency, lost productivity, and the cost of repairing damage done by electricians to the work of the drywall crews.

We will apply two legal doctrines in assessing the strength of these claims. The first is *impossibility* and the second is *implied covenants of cooperation*.

5.3 Impossibility

Legal impossibility excuses the performance of a contract. The prime contractor accuses the subcontractor of failing to install the electrical system on schedule. On schedule performance would be excused if it were impossible. An example of impossibility would be destruction of a building by fire. It would be literally impossible for an electrical subcontractor to install wiring in a building that did not exist because it burned down.

Power Electrical Corporations's difficulties do not rise to the level of legal impossibility. Prompt installation of the wiring system was not made *impossible* by the lack of cooperation of the journeyman electricians even though it was made more difficult and expensive.

5.4 Implied Conditions of Cooperation

Subcontractor signs a contract under which it promises to install an electrical system on schedule. Job action by electricians prevents the subcontractor from fulfilling that obligation. If the subcontractor expected to be excused from that obligation by such a contingency, it should have provided in the contract that timely performance would be excused by such job action. Absent such language, the law will hold the subcontractor to its obligation.

The prime contractor is backcharging the subcontractor for repairing gyp board that was damaged when the drywall crews got ahead of the electricians. The subcontractor could contend that the prime contractor should have ordered the drywall crews to wait on the electricians. But there is no specific provision in the subcontract that would require the prime contractor to reschedule the drywall crews for the benefit of the electrical subcontractor. This is where *implied covenants of cooperation* come in.

5.5 Express Covenants and Implied Covenants

Every construction contract contains both express covenants and implied covenants. The express covenants are to be found within the four corners of the instrument. In addition, the law recognizes implied covenants of cooperation, to the effect that neither party will do anything that would deprive the other party of the benefits of the contract. For example, even though there is no specific provision to that effect, the prime contractor has an obligation to make the jobsite available to the subcontractor when it is needed for the subcontractor's work.

5.6 The Schedule

Can such an implied covenant be stretched to cover an obligation of the prime contractor to order the drywall crews to wait on the electricians? Many subcontract documents authorize the prime contractor to establish a schedule and require subcontractors to follow it. Such a provision would give the prime contractor the legal power to require the electrical subcontractor to install its work ahead of the drywall crews.

5.7 Trade Practice

Absent such a directly controlling provision, the courts would resort to trade practice to determine the obligations of the parties. In this dispute, trade practice favors the prime contractor because it is customary for the prime contractor to schedule electrical work ahead of drywall work and for the electrical subcontractor to take care to stay ahead of the drywall crews. Therefore, although it's true that the prime contractor is bound by an implied covenant to cooperate with the electrical subcontractor in scheduling the work. that obligation does not go beyond trade practice, which would be for the electrical subcontractor to advance the job on schedule.

6

ABC Warehouse IV
The Mysterious Retaining Wall Collapse

When Allen Brady first saw the property in late 1988, he fell in love with it. It was in a modern landscaped master planned industrial park in northwest San Fernando Valley. The site was ideal, but he was concerned that the eastern 75 feet was practically unusable. It sloped up sharply to the east property line where a chain link fence marked the beginning of the neighboring DEF Gasket Corporation's property.

The site was almost 10 acres, 802 feet wide and 510 feet deep, generally level, except for the east end, and some noticeably low spots. It was a perfect site for the 100,000 square foot warehouse Allen wanted to build and there would be plenty of space for offices and warehouse expansion. For the first time in the life of his company he would have enough space for truck parking and maneuvering and a decent loading dock.

In spite of his initial reservations about the 75 feet of unusable hilly space, he went ahead with his offer to purchase. After a lengthy exchange of offers and counter offers, his offer was ultimately accepted. He felt that at the finally negotiated price, he hadn't really paid very much for the sloping section.

* * *

He'd heard about Judge & King, AIA, an architectural firm noted for outstanding modern efficient industrial buildings, always within the owner's budget. When he first met Ivor Judge and Leo King, he liked them and knew that he would get along with them, so he signed them up.

They discussed the sloping section of the property. Ivor felt initially that they had a large enough site to ignore the difficult 75 feet. But, as time progressed, ABC's program grew. So, they had to consider how to use at least part of the sloping section. They finally decided to erect a 12 foot high retaining wall, 300 feet long so they could extend the level site an additional 24 feet. They weighed the wall cost versus the site expansion advantages and decided to go ahead with it. They concluded it would be more economical to build the wall with the initial warehouse construction rather than as part of a future expansion program. Part of the retaining wall cost would be offset by using the excavated earth to fill and reclaim the low spots northerly of the first phase warehouse.

The selected development scheme placed the 5000 square foot office wing about 100 feet from the east property line. The 12 foot high retaining wall was about 50 feet from the office entrance. The intervening space was to be developed as an attractive entrance courtyard, with customer parking hidden by landscaping beyond.

Above the retaining wall, the land angled upward on a 2:1 slope, rising 25 feet to the east property line. DEF Gasket's 8 foot high chain link fence was on the line with a 10 foot planting area adjacent and then their parking area. The sloping bank would be improved with suitable groundcover, trees, shrubbery, and an automatic sprinkler system.

* * *

In one of their early design conferences, Ivor Judge discussed the retaining wall with his structural engineering consultant, Myles Nolan. They agreed that the wall would be constructed of concrete block masonry.

A few days later, Myles disclosed the details to Ivor. They would use smooth, lightweight concrete blocks in natural grey granite color. All masonry units would be 16 inches long by 8 inches high, nominal sizes. The vertical wall would be constructed of 2 rows of 8 inch thick concrete blocks for the first three courses, thereafter 12 inch thick blocks for the next 6 courses, and 8 inch thick blocks for the remaining 9 courses. The 18 courses would produce a total height of 12 feet above the footing.

The concrete footing would be 8 feet 4 inches wide and 16 inches thick. It would have a 16 inch by 16 inch concrete key on the bottom to help prevent the wall from sliding. The wall would be located 12 inches from the back of the footing with 6 feet of the footing in front of the wall. The concrete key would be directly under the stem wall.

The wall would be reinforced with deformed steel reinforcing bars sized in accordance with the engineering calculations. All cells would be grouted solid. The earth side of the wall would be dampproofed to minimize unsightly efflorescence on the face of the wall. There was no point in incurring the considerably higher cost of a full waterproofing treatment as it was not protecting a building interior.

Weep holes at 32 inches on center would be provided by omission of the head joint mortar at every other block in the first course. Continuous behind the wall at the line of the weep holes would be 2 cubic feet of gravel per lineal foot to allow any water in the soil behind the wall to flow horizontally and be released at the weep holes.

Ivor decided that the mortar joints would be concave tooled as this would yield a strong compacted joint and allow the blocks to be seen individually. The wall was carefully designed and the specifications would be quite thorough in respect to all the materials and workmanship including the concrete, steel, masonry units, mortar, and grout.

* * *

ABC Warehouse IV

The entire project was completed on time and within the budget on June 1, 1990. Allen Brady and all at ABC Warehouse Company were proud and elated.

During the next two years the landscaping flourished and the premises looked better every day. Fortunately their warehousing business was on a steady increase and they were making a neat profit, far better than Allen had ever dreamed.

Every time Allen looked at the space behind the customer parking area he felt that it would be a good idea to develop it as an executive parking area for his and the office employees' cars. He also dreamed of a beautiful landscaped park for his employees to eat their lunches on pleasant sunny days. He visualized a heavy timber pergola with beautiful bougainvillea vines and maybe a few shade trees. He talked it over with his secretary, Teri Unger. She was enthusiastic about it and encouraged Allen to go ahead with the project. She asked, "Do you want me to call Ivor Judge to get the design started, Mister Brady?"

"No, Teri, there's no point in making a big deal out of this. We won't need any architects or engineers or contractors. They'd just make a simple project complicated. Besides, we'll save their fees. We can figure this out for ourselves."

"Are you sure, Mister Brady?"

"Sure. It's simple. Nothing to it. What could go wrong?"

He had Teri dig out the site plan from the Judge & King warehouse construction drawings and make a few copies on the office copier. It would be simple to lay out the parking area and the lunch park. He started sketching his visions on one of the copies and then asked, "Teri, what scale is this anyway?"

"I think it's one inch equals 40 feet, Mister Brady. That's what it says under the drawing. Just get a ruler and you can work it out. A half inch would be 20 feet and so on."

"Oh. That's a good idea. Do you have a ruler in your desk, Teri?"

After working with the site plan and the ruler for about a half hour, he called Teri back into his office. "I think this drawing is too small. It's hard to see the detail. How can we enlarge it?"

Teri suggested, "If you just redraw it, doubling all the distances with the ruler, it ought to be okay, I think."

"Teri, I've got to leave for an appointment right now. Do you think you could do the enlargement?"

"Sure, Mister Brady, it'll be fun." She doubled the drawing in a couple of hours and made some copies so Allen could continue his layouts.

The next morning Teri showed him the enlarged drawing and he was duly impressed. So, between them, they designed the parking area for 8 cars and a lunch park. It had a patio and pergola, planting areas, grass, shade trees, drinking fountain, and 2 concrete ping pong tables. When it was completed they would furnish it with a couple of picnic tables and some redwood benches. The drawing was a masterpiece, with different colors for each of the elements. They were proud of their work and posted it on the coffee room wall. It was admired by all in the office.

Allen happened to see a landscape contractor working in the vicinity and stopped to talk with him. He was installing a concrete walk and preparing planting beds at a nearby plastic molding factory. Allen asked him, "Are you interested in submitting a bid for some landscape work?"
"Sure!"

"Can you figure it all out from a drawing?"

"No problem. When can I see it?"

Allen gave him directions for finding ABC and left. So, Leonard Martinez, Landscape Contractor, came by to see the drawings. He pointed out a few omissions on the drawings, such as irrigation sprinklers and lighting, and submitted an overall bid to build the whole thing including all hardscape and landscaping. At first, the price seemed very high to Allen and Teri, but after much discussion and a few concessions offered by Leonard, they agreed on a final price of $31,200.

* * *

After several weeks' use of the new parking and alfresco lunching facilities, Allen sensed a discordant visual note. He invited Teri out in the park to have another look. He asked, "Teri, everything's beautiful except for that ugly concrete block retaining wall. What do you think?"

"I think you're right, Mister Brady! It's horrid. It detracts from the beauty of the park. It should be painted. A restful soft mauve would be lovely."

Allen agreed, "A simple paint job shouldn't cost too much. I'll contact a painting contractor right away."

When he got back to his desk Allen remembered a young painting contractor that had been in recently soliciting maintenance painting. He rummaged around in his desk and found the business card. He asked Teri to call him and arrange for him to come in and submit a bid on painting the wall. "And ask him to bring in some paint sample charts so we can pick the color."

"Yes, Mister Brady!" Teri got right on it. First thing the next morning an eager Nels Olson was waiting in the parking lot for someone to show up. A little later when Allen arrived he couldn't help noticing the blue 1979 Chevy pickup truck splattered with various colors of paint. Its roof rack was piled high with ladders. The pickup was loaded with paint cans, brushes, rollers, and drop cloths. All the trappings of a veteran painter ready for action.

Allen showed him the concrete block wall and invited him in to the office. They went to the coffee room to discuss the bid over a cup of coffee. Nels was eager for the job as he had his eye on the future maintenance painting at ABC. He left some paint color charts and said he would go out and measure the wall and prepare a bid. In less than a half hour he was back with a handwritten bid which he left with Teri. She told him they would call him to let him know if he had the job. Teri studied the proposal: "Paint 300 foot long concrete block wall, 12 feet high. Surface preparation, one undercoat, and one finish color coat. $1750. Payable $600 in advance and the balance on completion."

She went into Allen's office and showed him the bid. Allen thought it was a little high and suggested that Teri get a couple more proposals. She obtained two more bids from painters she found in the yellow pages. They were both over $2000. So she and Allen concluded that Nels really wanted the job. Allen instructed Teri to call Nels and get him started on the wall. Nels said, "I'll be there first thing tomorrow morning!"

***

Allen and Teri went out together to see how Nels was getting set up. He'd parked his truck in the executive parking area so it'd be conveniently close to the work. The truck doors and tail gate had the same message as Nels' business card:

> **NELS OLSON**
> Painting Contractor
> Licensed, Bonded, Insured
> C-33, License # 564728

This gave Allen a warm feeling of confidence in his painting contractor. "Look, Teri, licensed, bonded, and insured. He must be okay!"

Nels explained to them that the job would take three days. The first day was the most important and would be devoted to surface preparation. The second day would be for application of the undercoat while the final color coat would be applied on the third day.

Nels and his two painters were industriously preparing the surface for the undercoat. They were knocking off the high spots caused by mortar droppings and filling in the chipped blocks and other surface defects with a portland cement and sand filler. Nels explained that this would produce a more perfect job. "For the undercoat we'll use a heavy sand-enriched coat to fill in minor imperfections. Then we'll finish with a heavy-bodied color coat. Allen and Teri were impressed. Teri enthused, "He seems to know what he's doing."

Allen said. "Yeah, I agree. Teri, give Nels our color selection."

She handed Nels the paint color chart with their selection boldly circled: 6673 Heavenly Lilac. It was a subtle pastel rosebud mauve. Teri had picked the color. She was thrilled when she visualized the result. She visited the job at least six times a day to confer with Nels and to make suggestions for improvement and refinement.

When the painting was completed, ABC employees agreed that Allen and Teri had made a fantastic color selection and that Nels had done an outstanding job. Allen promptly approved his check and promised Nels that he'd be first in line for ABC's maintenance painting.

* * *

About four months later, in February 1993, Southern California had several days of heavy rains. Early on a Monday morning, Carl Daly was the first to arrive to get the warehouse opened and ready for the day's activity.

As he drove onto the property, heading for the new executive parking area, he was suddenly shocked by the utter disarray. He expected to see the elegant lush landscaping on his right, backed up by the tasteful mauve retaining wall and landscaped bank, surmounted by DEF's tree-lined parking area. The new office building would be on his left. He stopped the car to allow his senses to adjust to the astonishing sight of destruction laid out before him. His bruised senses were unable to grasp the enormity of the disaster. He had difficulty in comprehending what could have happened. The mauve retaining wall was missing. The landscaped bank was gone and all of its greenery was mixed up in the all-pervasive brown mud. ABC's customer parking and the new lunch park were covered in several feet of mud and rocks. The east side of ABC's offices had stopped the mud flow and the water saturated soil was two to three feet up the wall sealing the front doors closed. Mud was oozing down the driveway into the street. He couldn't see back to the executive parking area. He was aghast and overwhelmed with the immensity of this unexplained catastrophe.

He gathered his senses, backed his car out into the street, drove down to the truck entrance, and reentered the property. He stood unsteadily by his car and waited until a few more warehouse employees arrived. He stationed one at the office parking entrance to divert cars to the warehouse parking area. He directed two more to erect a barricade to keep cars out of the disaster area. He then drove around to the north side of the warehouse to the employees' parking area. He parked and opened the rear entrance to the warehouse. He walked through the warehouse to the door leading to the office building. When he reached his office he sat down to figure out what to do next.

Others started to arrive for work and found Carl in a catatonic state. He didn't know what to do. Teri suggested calling George Hyde of Hyde Construction Company. "He's a practical man. He'll know what to do." Carl said okay, so she dialed his number. He was on the scene within half an hour. Meanwhile, the office employees, gradually arriving, were mostly drinking coffee, talking in funereal tones, and looking out the windows at this incredible sight. No one could think of working.

When George Hyde arrived, he suggested notifying their easterly neighbor, DEF Gasket, at the top of the bank to forestall any further damage or injury. The mudslide had slid out from under part of their parking lot paving. Their chain link fence was suspended in mid-air. Luckily no cars had been parked up there. Teri called her counterpart at DEF and relayed the warning that their parking area was in danger of collapse. The DEF president, David Flynn, directed his plant manager to set some barricades and immediately called his attorney and insurance broker.

* * *

Allen and his sales manager, Vernon Williams, were in Denver trying to sew up a new piece of business at Star Manufacturing. He was upset with Teri when he received a call at Star, as he had told her not to call under any circumstances unless it was extremely urgent. He picked up the phone apprehensively, "Hello. Allen here."

Teri said, "I've got bad news, Mister Brady." He steeled himself as she unloaded all she knew at the moment. Allen dropped everything and rushed home leaving Vernon to finish selling the deal.

Allen and Vern had left their cars parked in the ABC executive parking lot when they flew to Denver, so Allen had to take a taxi from the Hollywood-Burbank Airport. When he arrived at ABC, he was astounded at the chaotic appearance of the property that was so perfect only a few hours ago when he and Vern had left for Denver. It was far worse than Teri had described in her phone call. Everything in sight was the drab color of mud. He made his way through the warehouse and into the offices where he found Carl in a state of shock and inaction. Teri had taken responsible charge. She greeted him and quickly brought him up to speed, "We just discovered some more bad news, Mister Brady! Your and Vern's cars are under the retaining wall. George Hyde says he saw portions of them sticking out of the mud. We don't know if any other cars are under there or not! Mister Hyde is still outside probing around!"

Allen went outside to talk with George Hyde. He found George climbing around in the mud and walked over to join him. Allen was ruining his suit and shoes but ignored them. When he caught up with the builder, he could no longer suppress his anger. "Look at this fantastic mess, George! Your wall collapsed. You've ruined us! What did your guys leave out? I expect you to do the honorable thing and clean up this mess and rebuild the wall. You also owe us two expensive cars and a landscaped park. How could you do this to us? While you're at it, you might also think about restoring DEF's property to the way it was."

"Like hell, we will! You must be outta your mind! It wasn't our fault. We followed the plans and specs to the letter. There must've been something wrong with the design." George, irate, stalked off with no further comment, stamping the mud off his feet before getting into his car.

* * *

George Hyde, back at his office, immediately started marshalling his advisors. His insurance broker, his attorney, his cost estimator, and Ezra Field, his superintendent on the ABC Warehouse job.

George didn't know why the wall collapsed. The only thing he was sure of was that Hyde Construction had followed the plans. They always did. He also called his masonry subcontractor who had actually built the wall, Harvey Irwin. Harvey, Inc had built dozens of masonry structures for Hyde and they always did an honest skillful job. Maybe Harvey could figure out what went wrong.

* * *

Allen went back into his office to confer with Carl, who was useless, and Teri. Teri'd been keeping notes. They decided to call ABC's insurance broker, Brian Cullen, who told them he'd immediately notify their insurance carriers. "I'm fairly certain we have coverage on this but I won't know for sure until we know exactly what happened and what caused it."

Allen, concerned, asked, "Brian, are you saying we might not have coverage?"

"No, Allen. I'm saying we don't know yet. Keep me informed."

Allen, visibly shaken, directed, "Teri, get our lawyer Phil Quinn on the line. He'll know what to do."

Allen explained, "Phil, we have an absolute catastrophe here. Our property's ruined." He described as best he could the present sorry state of the situation and concluded with, "Brian Cullen isn't sure we have insurance coverage. Can you get over here right away?"

"Okay, Allen. I'll drop everything and run right over!"

They also called their architect, Ivor Judge, of Judge and King, and he in turn phoned their structural engineering consultant, Myles Nolan. Ivor and Myles were on the site shortly and were trying to piece together what could possibly have caused this terrible catastrophe. They waved to Phil Quinn when he drove onto the property.

Phi went directly into the office and greeted Allen with suitable expressions of shock, dismay, and empathy. He offered to put all involved parties on notice. He recommended that they get Hyde Construction to start cleaning up and have everyone with any suspected liability invited to observe the clean-up operation. He advised that it would be best to clean it up, limiting any further damage, determine the costs of reconstruction, find out who is at fault, research insurance coverage, and then make appropriate claims. He suggested having Frank Grimm, an independent forensic architect, on hand to make sure that the true cause of the collapse is positively determined, and a photographic record made.

* * *

Allen asked Phil and Ivor to sit in on a meeting with George Hyde to negotiate a contract for cleaning up the premises. A glowering George Hyde immediately made his position abundantly clear. He'd assume no responsibility whatsoever for the wall failure and the resultant damage. He was willing to organize the cleanup work but would not quote an estimate or guarantee a maximum price. He'd proceed only on the basis of cost plus a percentage fee with payment in full, without retention, on completion and rendering of the final bill. This was his non-negotiable position.

Allen invited George to leave the room for a few minutes so he could confer privately with Phil and Ivor. When George left, Ivor said, "No knowledgeable contractor would give you any better deal."

Phil, agreeing, said, "George has always treated you fairly in the past, Allen. Under these conditions we need someone we can trust. I think you should go ahead with him."

Allen said, "But what if his people caused the problem in the first place?"

Phil replied, "That's why I want Frank Grimm present all during the uncovering and cleanup operation."

Allen called George back into the office and told him to go ahead with the cleanup. A contract reflecting his required terms would be signed.

* * *

Hyde Construction was hard at work early the following morning. They had trucks and loaders to scoop up the mass of ooze and muck mixed with landscaping, lighting conduits, and sprinkler piping. They had deactivated all water and electrical services in the troubled area. Ezra Field, after conferring with Ivor and Myles, decided to get rid of the mud and not reuse it in the final backfilling, as it would take too long to dry out. He found a place to dump it that was only 8 miles away. George negotiated a favourable price for dumping it.

Hyde's earthwork subcontractor estimated that about 3200 cubic yards would have to be hauled away in addition to the masonry wall structure. They started removing the saturated earth and uncovering the ruined wall. The wall was so well constructed and reinforced that it did not break off at the footing as had been originally supposed. They found that the wall had slipped horizontally in some places and, where sliding did not occur, the stem wall was deflected and deformed. Nothing was salvageable.

It seemed at times that there were more observers of the cleanup operation than there were workers. Ezra Field was in constant attendance and in charge of the work for Hyde Construction. George was on the job parts of every day but could accomplish more in the office. Allen came out of his office several times a day to talk to whoever was in attendance but mostly he just stood around with a great lump in his throat.

Ivor and Myles, as architect and engineer of record, were there every morning. They had a keen interest in knowing how the failure occurred and whether there had been any shortcomings in their own professional services. They'd already been over the engineering calculations, drawings, and specifications a dozen times. They'd also pored over their construction observation reports and could not figure out what had gone wrong. They suspected the heavy rain and mud but knew that the design had anticipated such conditions.

Harvey Irwin, the masonry contractor, and his field superintendent, came to look a few times as they were under a certain amount of suspicion as the builders of the wall for Hyde Construction. They looked nervous and never stayed long.

Ivor had asked Larry Martin, Engineering Geologist with Lawson-Martin, Geotechnical Consultants, to drop by to get the flavor of the situation as he would be asked to help design the reconstruction of the failed earth bank.

Phil had asked Frank Grimm, the forensic architect, to observe the cleanup operation, take photographs from time to time, and to determine if possible the cause of the failure so that the culpable parties could be identified. Frank spent three to four hours a day observing the operation, making notes, and taking photographs.

At one time the design architect and engineer, the soil engineer, and the forensic architect were there simultaneously. After much discussion, analysis, and speculation, they agreed that the wall should not have failed, that the soil was not draining for some inexplicable reason. Allen Brady didn't understand most of their arcane technical discussion.

When the muddy soil was removed and more of the wall could be seen, remnants of the weep hole gravel could be seen and it seemed sufficient in quantity and properly located. It wasn't until the wall was being broken up with pneumatic jack hammers that Frank noticed that there were no weep holes. This was enough to explain the collapse, he said.

Myles, puzzled, said, "I know there were weep holes. They were on the design drawings and I saw them in place. I've some construction progress photographs in the file back at the office. I'm sure they'd show!"

Just then George Hyde and Harvey Irwin arrived and joined the architects and engineers examining the section of wall with no weep holes.

Harvey said, "This is weird. I know we built the weep holes. Now they're gone. I laid them out myself. We specifically cleaned them out before the gravel backfill was placed."

Hyde said, "Sure, I saw the weep holes myself."

Allen heard all of this discussion and quietly left. He went into his office to think about what he'd heard.

Hyde Construction finished the wall demolition and cleanup. They had to break up the wall and footings with jackhammers and cut the steel with welding torches. The two expensive new cars were total write-offs. They were loaded and hauled away by an auto salvage company. None of the landscaping remained. The picnic tables and benches, ping pong tables, and pergola were all smashed and ruined and had to be junked. The trees, ground cover, and shrubbery were not reusable. The parking area paving was ruined by the heavy duty wrecking and trucking operations. Altogether it took two weeks to get everything trucked away and get down to a clean site restarting point.

* * *

Allen hired Ivor and Myles to redesign the retaining wall, parking area, lunch park, and landscaping and obtain the proper building permits. They also had to seek recommendations from Larry Martin for the reconstruction of ABC's damaged bank. Martin had confirmed that the muddy soil couldn't have been reused for backfill or reconstruction of the bank. Hyde Construction rebuilt the concrete block retaining wall, park and lunch area, the executive parking area, and restored DEF's chain link fence and parking area. George made arrangements with Leonard Martinez to replace all the landscaping features and the irrigation system. It took another 6 weeks to restore the entire premises to its former glorious state of tranquil beauty and perfection.

A few days after completion of reconstruction, Hyde came in to the ABC office and presented an itemized bill to Allen Brady in the amount of $217,900. After George left and Allen had recovered from the shock, he phoned Philip Quinn to come in to talk over the situation. Phil suggested that Allen also invite Ivor, Myles, and George. Phil said he would have Frank Grimm present.

At the meeting in Allen's office, Allen led off with, "Someone will have to pay for this." He had Hyde Construction's bill on the desk before him. He was clearly outraged. Ivor and George exchanged glances.

He continued, "The wall must have been improperly constructed or it wouldn't have collapsed. Any fool can see that!"

George, not one to be lectured to, again vociferously clarified his position, "We followed the plans and specifications exactly. We installed every piece of steel specified. The building inspector saw to that. Also, Judge and Nolan carefully observed our work and must've been satisfied that we followed the plans and specifications. Otherwise they'd have said something at the time. We saved all of Judge's jobsite observation reports. They said nothing negative about our construction of that wall."

Allen, seeing the logic of George's remarks, turned to Ivor and Myles and said: "Then there must have been something wrong with the structural design of the wall! Nobody else's wall collapsed! Just mine! Properly designed walls don't just suddenly collapse."

Myles started explaining, "It's standard engineering practice to design retaining walls to support drained earth. In the design we provided for weep holes every 32 inches. The weep holes were in the form of mortarless head joints 5/8 inch wide by 8 1/2 inches high, the full depth of the wall. Behind the weep holes was two cubic feet of gravel per lineal foot so the water could readily migrate horizontally and drain out the weep holes, thereby preventing the earth from becoming saturated. It's a standard way of designing retaining walls in this region. Based on the assumption of drained earth on a surcharged bank, we treat the soil pressure as if it were a hypothetical liquid with an assumed weight of 43 pounds per cubic foot. The actual weight of saturated earth would be over 100 pounds per cubic foot. That kind of increased soil pressure load could easily collapse the wall. And that's what we believe actually happened."

Allen said, "Well, why didn't you design it for the load imposed by saturated earth? Then we wouldn't have this mess today?"

Myles chimed in, "The cost would have been considerably higher. We avoided that substantial extra cost by designing in weep holes that are, for all practical purposes, free."

Phil looked questioningly at Frank who confirmed that Myles' explanation was correct. Frank said, "I have my preliminary report here. Complete with photographs. Someone has filled in all the weep holes with cement mortar and that is why the wall failed."

Allen quietly concluded, "Our painter must have filled in the weep holes."

* * *

The next day Allen and Phil met with Nels Olson. Phil asked him if he had filled in the weep holes. Nels proudly admitted, "Yes, that is what I was paid to do. I always do an honest proper job."

Phil and Allen exchanged serious glances. Phil requested, "Allen, will you please ask your secretary to come in here to take some notes?" Allen called in Teri who sat next to Nels.

Phil then re-asked the question. "Nels, you just told us that you filled in the weep holes in the concrete block wall. Is this true?"

"Yes, with portland cement and sand mortar. There's nothing better. That's the right way to do it."

Teri backed him up, interjecting, "Well, naturally. Those unsightly holes that Hyde Construction left all along the base of the wall had to be filled. I told Nels he ought to do it. He didn't charge any extra. He said he'd throw it in gratis as he expected to get all our future maintenance painting." Nels was smiling and nodding, relieved that the secretary was corroborating his account. Teri was glad to help.

Phil and Allen exchanged horrified glances.

* * *

Moral: Contractors should always be carefully selected and even seemingly simple contracting scenarios require some skill in administration. As it turned out in this apocryphal tale, the contractor that was licensed, bonded, and insured had the statutory $7,500 state license bond and his insurance was a $5,000/$10,000 public liability policy on the blue 1979 Chevy pickup truck. He had little, if any, other assets.

Points of Law 6

ABC Warehouse IV
Retaining Wall Collapse

6.1 Liability and Insurance

The retaining wall was properly designed and built. It collapsed because weep holes were plugged by a painting contractor. It looks like the painting contractor is probably uninsured, so Phil will take a careful look at Allen's property insurance policy. He will also want to examine Allen's liability policy, since damage has been done to the property of third parties: DEF Gasket Corporation and the automobiles belonging to Vernon and Allen himself.

Under an ancient common law doctrine, a landowner is strictly liable for damage done by dirt, mud, or water that escapes from its property. This doctrine applies to damage to the automobiles. If Allen's automobile belongs to ABC Warehouse Company, the potential for coverage will be affected by whether the negligence of ABC caused the loss. If the automobile belongs to Allen, on the other hand, the negligence of ABC would be irrelevant.

6.2 Property Coverage

Allen's property insurance probably covers not only loss to the warehouse building itself, but also to other structures including the pergola, retaining wall, landscaping, irrigation and electrical installations. Whether coverage is afforded by the policy will depend on the specified exclusions from coverage.

Common exclusions that could apply include *landslide, earth movement, faulty design, faulty workmanship*, and *latent defect*. The *latent defect* exclusion would not apply because the defect is obvious to reasonable inspection (at least a person with a rudimentary understanding of the laws of nature). It doesn't really take an expert to understand the purpose of weepholes at the bottom of a retaining wall. It is to prevent the wall from becoming a dam subjected to the pressure of rainwater.

The *faulty workmanship* exclusion could apply since the painting contractor performed the work of plugging the weep holes. Courts, however, interpret insurance policies in favor of coverage and would probably accept the argument that the workmanship itself was good since the painter did an excellent job of plugging those weep holes: it was not faulty workmanship that caused the collapse, but the fact that the work was undertaken at all.

6.3 Landslide and Earth Movement

The *landslide* and *earth movement* exclusion, at first blush, appears to remove coverage since the earth certainly moved. Here, however, the loss must be examined to determine its efficient, moving, proximate cause. A court might well determine that the efficient moving cause was the negligence of the painting contractor who filled the weep holes. Since contractor negligence is not excluded the policy would afford coverage.

6.4 Collapse Coverage

Many property insurance policies offer *collapse coverage* for which an additional premium is paid. Since the retaining wall did collapse, collapse coverage may apply to the loss.

6.5 Liability Coverage

ABC might have a good liability claim against Nels, the painting contractor – but there's a fly in the ointment. The decision to fill the weep holes was partly made by Teri, who is an employee of ABC. Allen can't recover compensation from Nels for a loss that was actually caused by the negligent conduct of his own employee!

6.6 Comparative Fault

Here, we must deal with the concept of comparative fault. It often occurs that the fault of two or more persons together causes a loss. Nels and Teri are jointly responsible for the decision to fill the weep holes. Since Teri is an employee of ABC, Nels' liability will be reduced by a factor that measures the degree of his fault against hers. The law will certainly place a higher degree of responsibility on a licensed contractor than an inexperienced property owner: nevertheless, Allen's recovery would probably be reduced by 10% or 15% to take into account the negligence of his own employee.

6.7 Employee vs. Independent Contractor

Why isn't Allen also responsible for the misconduct of his painting contractor? The reason is precisely that Nels is a *contractor*, and not an *employee*. Companies are responsible for the misconduct of their employees but not of independent contractors.

The main distinction between an employee and an independent contractor is that an employer exercises supervision and control over the activities of employees, while an independent contractor is responsible only to achieve results, but is not subject to close supervision.

6.8 Damage to DEF Gasket Corporation Parking Lot

The final problem is the claim DEF Gasket may have for damage to its parking lot. Nels is uninsured and apparently has few resources other than his Chevy pickup. ABC Warehouse Company, though, carries liability insurance. Nels and Teri are both responsible for the loss, but between the two of them Nels' degree of fault outweighs Teri by about 85 to 15. Does this prevent DEF Gasket from obtaining 100% compensation from Teri's employer. In most states, the answer is "no." DEF Gasket can collect 100% of its damages from ABC Warehouse Company, and ABC Warehouse will then have the right to recover 85% from Nels under principles labeled "*contribution and indemnity.*"

7

ABC Warehouse V
Earthquake Damage or a Latent Defect?

Just before dawn, at 4:31, on a crisp, bright Monday morning, an unforgettable cataclysmic disaster struck in the San Fernando Valley, to be forever after referred to as the Northridge Earthquake.

The new premises of ABC Warehouse Company were located worryingly close to the epicenter of the 6.7 Richter Scale magnitude temblor. Thus, when Allen Brady arrived at work that morning, he was extremely apprehensive about the fate of his precious new warehouse and offices.

Driving onto the property, he was relieved to see everything apparently intact, even the new retaining wall looking as solid as ever. All seemed normal as he entered the building. The office staff was animatedly exchanging tales of their own experiences that morning. Everyone was there and the scene was safe and comfortable. He proceeded through the offices and into the warehouse. There he found Carl Daly supervising his dock crew busily restacking a couple of dozen crates and cartons that had toppled. Carl reported, "Everything's okay, Boss. We're putting these cartons back into neat stacks where they were. I haven't seen any other damage."

Allen was thankful that ABC had come through the quake that easily. Back in his office, he turned on the TV to catch the latest in news coverage. There was an engrossing interview with a prominent structural engineer who was urging close inspection of the structural components of buildings. This would be important in all types of buildings, he said, but especially pertinent to multi-story structures or those with long roof spans. "That would include our warehouse," reflected Allen.

Allen immediately punched in Carl's intercom number and said, "Carl, will you please come to my office right away?"

Minutes later Carl showed up, wondering what he had done wrong this time. Allen motioned him to a seat and issued instructions, "Carl, I want you to look over all the structural components of our warehouse building and let me know if everything is all right. See if there is any quake damage."

"Okay. I'll get right on it. Uh, what, exactly, is a structural component?"

Allen thought it over. "Well, I'm not sure. I'd think it would be the walls, columns, roof beams, rafters, and that sort of thing. See if there are any cracks or signs of movement. Use your own judgment. Let me know if everything's all right."

"Okay, Boss." Carl went back to the warehouse, relieved that he wasn't being criticized for something.

Carl, back at his warehouse office, instructed Al Barker, one of his dock crew. "Al, when you get some free time I want you to go around and look at all the structural components of the warehouse. Use that 18-foot aluminum ladder so you can get a close look at all of the roof-framing members. But, first, we gotta finish loading this truck."

"Okay, Carl, I'll do it first thing after lunch."

* * *

All that afternoon, Al Barker diligently scrambled up and down the ladder examining rafters, beams, piping, lighting fixtures, conduits, and anything else in sight. Everything looked all right to him. After his mid-afternoon coffee break, he resumed his climbing, inspecting, and ladder moving operation. His leg muscles were getting increasingly sore and he was becoming rapidly fed up with this fruitless, seemingly useless, boring assignment. Late in the afternoon, he was about ready to throw in the towel when he noticed something strange about the side of one of the large laminated wood beams. There were horizontal splits in the side of the beam at the end next to the concrete wall column. It appeared that the individual planks of wood that comprised the beam were coming apart along the glue lines. Al did not see the significance of this condition but felt that it was probably important. So, he immediately reported the situation to Carl Daly, "Say, Carl, there may be a problem in the middle of the back wall in Section 12-G, behind Star Industries' merchandise. Do you want to see it?"

Carl looked up from his paperwork. "Yeah. I'll look at it first thing tomorrow morning. Right now I gotta get these damn shipping manifests checked. You can start getting the warehouse locked up so we can leave on time for a change."

So Al left the ladder where it was and began locking up the warehouse. Tomorrow's another day.

* * *

On Tuesday morning, after Carl got the crew started on their assignments, he checked to see what Al Barker was going on about. He scaled the ladder and peered at the side of the beam. He didn't think it was too important as it didn't look dangerous and nothing appeared to have shifted. But he felt it would be a good idea to report it to the boss. He told Al to continue examining the beams to see if any of the others had the same splits. He then returned to the loading dock to keep the shipping operation moving.

After lunch the beam again crossed his mind so he went into the office to see Mr. Brady. He described the situation to Allen who accompanied Carl out to the warehouse to have a look himself. Al moved the ladder back into position and held it while Allen clambered clumsily to the top. Allen scrutinized the beam splits and wondered what it meant. He returned to his office to think about it. Who should he call? If he called his architects or engineers they would probably tell him nothing is wrong, these are normal occurrences, and send him a fat bill. So, he decided to call Hyde Construction Company. George Hyde would know what it is and was not likely to send a bill if there was no work to do. He rang Hyde Construction and asked for George Hyde. After Allen described the situation, George immediately barked, "I'll be there, Allen, in less than 20 minutes!" Allen asked, "Does it sound important?"

George replied, "I'll tell you when I see it!" He hung up and made a dash for his car.

Allen concluded that George must be hungry for work.

* * *

George arrived promptly and continued on to the warehouse where Carl was waiting with the ladder in place. George hoisted his ample frame up the ladder, took a good look, and came down. "Carl, this should be shored up right away. No point in taking unnecessary chances. Then we can figure out what to do next."

Carl had been accused recently by Allen of throwing money around recklessly, so he asked, "How much'll this cost, Mr. Hyde?"

"I think it could be done by two carpenters in less than a day. The labor and materials would be about $1200 more or less. We should start this afternoon."

Carl said, "Now, wait a minute. I'll have to get authorization for that kind of money. I'll phone you tomorrow." He felt George was too eager to get his foot in the door and start running up an outlandish bill.

As George was leaving, he said, "Well, don't wait too long. This could be dangerous!"

Carl went into Allen's office. Allen agreed with Carl that $1200 was a lot of money for less than a day's work. He suggested that Carl could buy some wood and nails and have a couple of the dockhands build the shoring in their spare time. No trouble at all. "We have to pay their wages anyway and the wood'll be cheap."

As Carl was leaving to go back to the loading dock, he assured Allen, "I'll get the materials organized tomorrow morning and we can start building." Allen thought, that Carl Daly isn't so stupid after all.

* * *

The following morning Carl sent Al Barker to the local lumberyard with a purchase order to buy materials necessary for shoring up the beam. Al went into the sales office and took a number. Most of the standees, also waiting, were macho construction types. When his number was eventually called, he hitched up his trousers, strode up to the counter, lowered his voice a half octave, and requested, rather authoritatively, "I need some shoring materials."

The lumber clerk, pencil and order pad at the ready, replied, "Okay, Ace, tell me what you want."
"I need some shoring lumber."

"What size and length?"

"The usual size."

"There is no usual size."

"Oh? What do most people buy?"

"Well, what do you want to do with it?

"Shore up a beam."

"What kind of beam? How high is the beam off the floor? How much load must be carried? And how do you plan to spread the load on the floor? What do you want to use for bracing? Do you need any nails?" This overwhelmed Al.

After considerable discussion, Al and the lumber salesman agreed on a lumber list and a quantity of nails. By late morning he was back to the warehouse with a load of lumber and nails. He handed the bill to Carl. $659.80. Carl asked, "Why'd you buy so much wood? And all those nails! Well, unload the truck. You and Mike can start building the shoring tomorrow morning."

* * *

Unfortunately, the warehouse was extremely busy all day Thursday and Carl wasn't able to spare Al and a helper to work on the shoring.

On Friday morning Carl came in a little early to get all the work assignments and his paperwork straightened out. But first, he went over to have another look at the beam where the shoring would have to be built. When he got there and looked up there was an unexpected bright gash of blue sky showing through at the edge of the roof next to the wall. The laminated beam had broken into splinters about two feet from the end and had dropped about a foot. It would have dropped more but the mangled beam end was wedged into the top corner of the concrete column. The bolts had pulled out of the beam. It looked precarious and he was shocked at the chaotic appearance.

Carl dashed into the office just as Allen was arriving. He quickly filled Allen in on the details and, now breathless, said, "We gotta call Hyde and get him out here right away. The whole damn roof could collapse!" He was furiously punching in numbers on the phone.

* * *

When George arrived he had his superintendent, Ezra Fields, in tow. Ezra looked at the damaged beam and the pile of shoring materials. "We'll have to send this stuff all back to the lumber yard and get some proper materials. I'll phone them and have them deliver what we need right away. He directed Carl to move all the stored merchandise out of the area as it would be in the way of the shoring and scaffolding operation.

After a short while on the phone Ezra had the job organized and they were in production within an hour. Carpenters had to be commandeered from another job. They built a sturdy scaffold that could withstand the load of a housemover's jack to safely raise the beam up to its former position. They spread the load across the floor with some large timber beams. They independently shored the purlins that carried the rafters and roof, enabling later removal and replacement of the beam. They temporarily patched the damaged roofing membrane and spread polyethylene sheeting on the roof. This would protect the building and its contents over the weekend. They'd have to come up with permanent solutions next week.

Ezra and his carpenters were ready to call it quits for the day. George and Allen were standing on the warehouse floor studying the sturdy maze of temporary construction. George asked Ezra what could possibly have caused all this trouble. Ezra said, "I don't know for sure. Probably the earthquake. Who knows how these things happen?"

George told Allen, "We'll have to get our electrician and plumber out here to check for damage to their work and see what'll have to be done. I'll also call my roofing contractor to see what can be done about the roof membrane."

Allen was getting nervous. "This looks like a lot of money, George. How much have you spent so far?"

George said, "I don't know. I can have a bill for you on Monday."

Allen added, "How much will it cost to repair the beam and finish up the whole job?"

"I'll have an estimate for you Monday."

Back at his desk, Allen called his insurance broker and said, "Brian, we have some earthquake damage and I'll be wanting to make a claim."

Brian Cullen replied, "But, Allen, you don't have any quake coverage!"

"Whatta you mean?

"Don't you remember our discussing this? The premium was so expensive that you decided against it."

"Yeah. I remember now."

* * *

On Monday morning when Allen arrived, none of the Hyde people were working in the disaster area. About nine o'clock George Hyde showed up. Allen asked accusingly, "Where are your men? We've got to put the building back together!"

George, unperturbed, answered, "We don't have a deal yet." He threw a sheet of paper on Allen's desk. It was a detailed compilation of costs incurred on Friday for the shoring operation. Labor, materials, tool rental, supervision, overhead, and profit, $2732.89. This is our bill for work to date. It's now due and payable."

Allen almost exploded, "I thought you said it would be around $1200. Are you trying to capitalize on my bad luck, George?"

"Things are different now that the beam has failed. I had to use four carpenters and it took more shoring lumber and bracing. We had to rent a housemover's jack to raise the beam up into its proper position. We also had to make temporary roofing repairs. That is what it cost. Now, here are my estimated costs for finishing up the repair." He threw another itemized breakdown sheet onto Allen's desk. The bottom line showed $16,500.

Allen's face turned white. "Dammit, this is ridiculous. Where do you get these numbers, George?"

George patiently explained, "These are our best estimates for each item. We're assuming that the beam must be removed. Beams this large are not in stock so we must wait until it has been custom fabricated. We'll try to do all the work from under the roof so we won't have to remove and replace roofing and sheathing. We'll also have some plumbing and electrical work to do. Roofing, sheet metal, and painting are also involved. We'll need new steel connections and column tops. The fire sprinkler piping must be checked. It's difficult to estimate this kind of work. We try to anticipate everything that has to be done and realistically estimate what it will cost. This is our best estimate."

"Can you guarantee that the work won't exceed $16,500?"

"No, Allen. We'll only work on a cost plus fee basis with no guaranteed maximum."

"I'll think about it." Allen felt engulfed in the circumstances.

George left the bill and the estimate on Allen's desk and departed, saying, "Well, don't wait too long, Allen. We've got to get the beam ordered. The situation is precarious. There could be another earthquake or a serious aftershock."

Allen hurriedly left for a Rotary Club meeting. He contrived to sit next to Ivor Judge, ABC's architect, during the luncheon. He told Ivor about the beam problems and George's unconscionable bill and ridiculous estimate. He then told him about Ezra's comment about the beam problem being caused by the earthquake. Allen complained, "I'm going to have to foot the whole bill for this. We have no earthquake insurance."

Ivor told him about some of the terrible quake damage he'd seen the last few days. "Most people I've talked to don't have any insurance. Many of those who do have insurance won't collect anything because their loss won't exceed the ten percent deductible."

Allen said he was going to go ahead with Hyde Construction even though the estimate seemed high. Ivor agreed that Hyde always did a good job even if it might seem expensive at first. In the end, his bills were always fair.

* * *

When Ivor arrived back at the offices of Judge & King, AIA, Architects, he told his partner Leo King about Allen's earthquake problem with the beam. Leo thought about it, looked puzzled, and inquired, "How could an earthquake cause a beam to delaminate? Something else must have caused it. Maybe you ought to call Allen and run out to take a look before everything is torn out and rebuilt."

"You're right. I'll call Allen now."

He got Allen on the line and recapped his discussion with Leo. He told Allen he would be there in half an hour. Allen at first said there was no use in coming out to look at it, that George Hyde had the matter well under control. Allen was concerned about unnecessarily running up consultation fees. He finally acquiesced when Ivor seemed so insistent and concerned.

Allen greeted Ivor in the office reception area and they immediately went out to the warehouse. While walking, Allen handed Ivor a copy of Hyde's repair cost estimate. Carl Daly put the ladder in place so Ivor could climb up and examine the damaged beam.

Ivor was equipped with a heavy-duty flashlight, flash camera, and notepad. At the top of the ladder, he noticed that the beam was delaminated from the end and at several levels. The beam had broken roughly along the line of the end of the delaminating, about two feet from the end. The beam, when it failed, had pulled out from its restraining bolts and the steel column plate was grotesquely distorted.

Aided by his flashlight, Ivor could see considerable evidence of repeated water staining on the beam, the column top and sides, and on the concrete tilt up wall panel. He took a few pertinent photographs. He then descended to the warehouse floor and asked Carl to move the ladder to the roof access scuttle.

Ivor, Carl, and Allen went up the ladder single file and out the hatch onto the roof. Ivor led the group over to the parapet wall above the failed beam. He carefully scrutinized the galvanized iron parapet cap and the mineral surfaced cap sheet wall covering. He noticed that the sheet metal coping at that point was not sloping toward the roof as is customary and as shown on the construction drawings. The metal cap was not only close to level, but was cupped thereby creating a generous channel to direct rainwater, or even a heavy dew, directly into the joint between coping sections. A coping joint was directly over the beam location. It was obvious that rainwater had been leaking into this joint for a good period of time, probably for the whole four years of its existence. Penetrated water would flow behind the parapet wall roofing and drip down to empty onto the top and end of the laminated wood beam. He removed a portion of the temporary polyethylene sheeting and stripped back the wall covering. He saw numerous stains and discoloration indicating water intrusion at dozens of separate times. He took copious notes and more photographs.

Ivor led the small delegation back to the roof scuttle and down the ladder. Carl followed last to lock the hatch and remove the ladder.

Walking back to the office, Ivor said, "Well, Allen, it seems that the earthquake had little or nothing to do with this problem. It was a flashing leak."

Allen's brain was churning at this point. "If the earthquake's not at fault, then what or who is?"

Ivor, preparing to leave, replied, "That's a complex question, Allen. Let me do some research and think about it. I'll call you back tomorrow." He could hardly wait to get back to his office to check the drawings and specifications and the construction observation reports. On the way he stopped off at the one-hour photo shop to get his film developed.

Allen, now back in his office, immediately rang Phil Quinn, ABC's attorney. "Phil, I just found out that our beam failure was probably not caused by the earthquake."

"That's interesting."

"So, I'd like you to come over here tomorrow morning and we can discuss it. I'll have George and Ivor here so we can reach some realistic conclusions."

"Okay, Allen. I can be there at 11 o'clock."

* * *

The key players were assembled in Allen's office at 11. Ivor had brought Myles Nolan, his structural engineering consultant, with him. Also present were George and Phil. Allen, perched imperiously behind his great oaken desk was presiding. He offered coffee and all accepted. The lawyer, architect, and structural engineer were all poised with pens and notepads.

Allen started the ball rolling by directing his first salvo at Hyde Construction. "Nice try, George. Getting your superintendent to blame the beam failure on the earthquake."

George, taken unaware, blurted, "What do you mean by that?"

Allen, ignoring George's obvious umbrage, continued, "Ivor, tell us what you discovered yesterday."

Ivor unpacked his briefcase and started explaining, "Well, it's obvious that the earthquake wasn't the real reason that the beam failed. The earthquake just triggered the inspection and discovery of the problem." He passed out four sets of prints of the photographs he had taken yesterday.

While the others were examining the photos, he started describing what he had seen. He explained how rainwater over a long period must have leaked into the building and soaked into the glued laminated wood beam. "The glulam beam was large enough to soak up quite a large amount of water. Most of the balance would have soaked into the concrete column and wall. The small amount left over would flow down the wall and be absorbed without becoming obvious on the floor. The merchandise stacked close to the wall would have hidden the watermarks on the wall and floor.

"The water entering the wooden beam eventually attacked the glue joint planes and as delaminating progressed, the beam's capacity to absorb water increased. Capillarity would have drawn water into the interlaminar interstices. Wood that has become wet and dried repeatedly will attract and hold increasingly larger amounts of moisture. The moisture appears to have attacked the interface between the glue and the adjoining wood in addition to dissolving the glue itself." The photos showed the delamination and the water stains.

Allen interrupted to ask, "But, Ivor, how did the water get into the building in the first place?"

Ivor pointed out the section of sheet metal parapet cap that was not properly sloped. He continued his explanation, "Instead of draining impinging rainwater harmlessly onto the roof, the coping channelled and directed it right into the joint. The water apparently penetrated the joint and a steady flow of water would enter the building. The joint would not have allowed this if the adjacent parapet cap was properly sloping."

Allen then asked, "How did the delamination cause the beam to fail?"

Ivor explained that the gluing together of numerous small planks of wood creates a large beam of a strength far exceeding the sum of all of the small pieces acting individually. "So when the glue joints fail, it allows each of the laminations to act alone as a collection of small flat inefficient beams. The beam simply broke at the point where the soundly laminated beam ended and the partially delaminated beam started. This was about two feet out from the column."

Phil asked, "Why did you specify laminated beams if they are subject to delamination?"

Allen pointed out that the forty-foot span was excessive for a natural wood beam. Steel could have been used but laminated wood was more economical. The laminated beam was 12" by 36" and consisted of 26 laminations. The beam's function was to carry 4" by 12" timber purlins 8 feet apart. A laminated beam would be of higher quality than a natural beam because smaller pieces of higher graded wood can be used. Laminated wood beams are considered to be quite acceptable in interior dry environments. Even high humidity would not harm a glulam beam although it could not tolerate an environment of constant soaking and drying cycles. Myles Nolan nodded his approval of the explanation.

Allen then asked, "Why was the improperly built parapet coping not noticed during construction? I thought George and Ezra were supervising the work of the subcontractors and you, as the architect, were observing the construction. And how about the city building inspector?"

At this point, George chimed in, "You're damn right we supervised the work of our subs! Ess/Em Sheet Metal did a great job. As all our subs do! I don't know how that section of coping got damaged. It must have happened after we left the job. Your maintenance people must have done it!"

Ivor continued, "As architects, we don't examine every lineal foot and each square inch of every part of the building. If this piece of coping was built this way, we didn't see it. The building inspector only spot checks. And, as George says, it could have happened later."

George asked Allen, "I thought you had a man up on the roof several times a year to clean the roof. Why didn't he see this problem?"

Phil answered for Allen, "Come on, George, ABC's man wouldn't have the expertise to recognize this kind of technical problem. He just cleans the drains and removes wind-blown debris."
George said, "Well, he could have damaged the parapet cap. Or it might have been damaged by ABC's air conditioning repair people."

Allen asked, "How do we know this isn't a design defect?"

Ivor rolled out the construction drawings and referred to the parapet cap drawing, Detail 13 on Sheet Number A-13. "This is the way it should have been built. Most of the parapet cap was built this way. If it was built in conformance with these drawings, and then remained that way, the beam problem couldn't have occurred."

Allen directed his next question to Phil, "What can we do now? I don't want to have to pay close to $20,000 for someone else's problem."

Phil had been jotting notes during the interchange of ideas. He started slowly, gaining speed as he warmed to the subject, "Well, Allen, I'll be perfectly candid in my comments. Ivor and George may not like to hear some of my thoughts, but I think we can solve this problem better if everyone is acquainted with all the facts and theories and knows their responsibilities. We've all worked together long enough to do that.

"This problem was discovered three and a half years after completion of the building. I'll have to do some research to determine if it falls within Hyde's warranty."

George interrupted, "Whoa! Wait a damn minute. We only give a one-year warranty on our work! This is almost four years."

Phil said, "Don't get excited, George. I'll have to research this and see how it squares with latent and patent defects. If you're found responsible, we may have to ask you to look into possible insurance coverage. If Ess/Em Sheet Metal did the work, you may have recourse to them and their possible insurance coverage." George wasn't pleased with this explanation.

Turning to Ivor, Phil continued, "If there's anything wrong with your design or your construction period services, well, then, it's only fair that we talk to your professional liability insurers." Ivor had a pained look on his face.

Myles, who had been sitting quietly listening and doodling on his sketch pad, offered, "I've been thinking about the problem and I think Ivor and I could design a less radical beam repair that wouldn't involve removing the beam. The damaged beam end could be cut off and be replaced by a steel restraining saddle that could resolve all of the structural loads. I've just been reviewing Hyde's figures and I think that the repair cost could be reduced to about $5,000 or $6,000."

After George, Myles, and Ivor left, Phil remained, thumbing through his notes. Allen, with a troubled look on his face, was absently sipping his now cold coffee.

Finally, Allen suggested, "Phil, maybe the best thing to do will be to have Frank Grimm, the forensic architect, look at this and study all the evidence. Maybe he can figure out who's responsible for all this."

Phil replied, "Not a bad idea, Allen. But there are some risks. There doesn't seem to be any evidence, so far, that the architect or contractor are responsible. Frank may find that your employees or repair people caused the problem. Then you'll have to pay the whole thing, and with no help from insurance. There's also the problem of your not getting the shoring up in time even after being warned about it."

Allen interrupted, "I'd also have to pay Frank Grimm's fee and your legal fees. With Nolan's suggestion for cutting the repair cost, I might be better off to cancel the inquisition and just pay Hyde Construction to fix up the building."

Phil agreed. "You may be right, Allen."

* * *

Points of Law 7

ABC Warehouse V
Glulam Beam Failure

7.1 Warranty vs. Statute of Limitations

Three-and-a-half years after the completion of construction, a glu-lam beam fails. The prime contract included a clause by which the contractor warranted the building against defects for one year. Is the contractor liable for the loss?

George Hyde says that his company only warrants the building for one year. That fact, however, does not excuse the contractor for a breach of contract that is discovered more than one year after the completion of the building.

The statute of limitations for breach of contract causing property damage varies from one state to another. In most states the limitations period for such a claim is three or four years from discovery. The "one-year warranty" provision does not shorten this statute of limitations. Therefore Hyde is still potentially on the hook. The "one-year warranty" provision of the contract is no defense.

From Allen's point of view there are four potential sources of recovery for the cost of repairing the damage to beams. It is possible that the problem is the fault of the architect either for improperly designing the sheet metal capsheet or for negligence in inspecting the construction.

It also seems that the sheet metal subcontractor may not properly have installed the capsheet. If that were the case, both the subcontractor and the prime contractor would be liable for breach of contract and for negligence.

7.2 Insurance

One other potential source of recovery would be Allen's property insurance policy which, in all likelihood, covers the loss.

If Allen collects on his insurance, then the insurance company will be entitled to recoup from any party whose negligence may have caused the loss which would potentially include the architect, the prime contractor, and the subcontractor. If the architect has errors and omissions insurance, its insurance carrier would respond. If Hyde Construction and the sheet metal subcontractor carry liability insurance policies, as they probably do, those liability insurance carriers would respond.

If the cost of repair were to be $50,000 rather than $5,000, these potential sources of recovery would be thoroughly investigated. Investigation might well disclose that the sheet metal cap was damaged after the building was finished by Allen's own maintenance personnel.

Given the relatively minor number of dollars involved, Allen will probably process a claim with is own insurance carrier and that will be the end of it.

7.3 Latent vs. Patent Defect

Phil Quinn mentions latent and patent defects. Patent defects are those that are noticeable to a person who conducts a reasonable inspection. Latent defects are those that are not so observable. The distinction is important because in most states the statute of limitations starts to run on patent defects upon the completion of the project, while the statute of limitations on latent defects runs only after the defect has been discovered by the property owner.

8

The Joint Venture
A Bonded Roofing Subcontract

The frantic scene of organized chaos in the offices of Hanley Constructors was common on big bidding days. They were still receiving last minute telephonic and fax bids and incorporating them into the overall breakdown. Subcontract prices were incrementally dropping by the minute. The bid form was all typed up, except for the final price. Hanley's courier was staked out in a phone booth across the street from Paradise Gardens Development Company's offices in Beverly Hills. Bids were due at 2 o'clock from a selected list of six bidders, all experienced general building contractors, all eminently capable of doing the job.

Hanley Constructors' courier was Hanna Hanley, Greg Hanley's daughter who was learning the business from the ground up. She'd enjoyed her stint in the estimating department for the past three months. In her hand she had the envelope containing Hanley's incomplete bid form and the required bid bond in the sum of $145,000. At 1:53 she'd phone back to the office and get their up-to-the-minute final bid price. She'd then enter the figure in the bid with a ball point pen, seal the envelope, and deliver it handily and safely to PGD before the 2 o'clock deadline.

Hanna, along with representatives of the other bidders, was invited into PGD's impressive conference room to witness the bid opening. PGD was an enlightened developer and always conducted their bidding procedures in the open. As the bids were opened and the figures read out, Hanna kept a running tally. This was intensely interesting to her, as she'd worked hard analyzing the plans, specifications, and take-offs. She'd talked to dozens of subcontractors and suppliers. She knew exactly how the bid went together. As each of the bid prices was revealed, she knew that Hanley would be lower and finally, as Hanley's bid was read out, everyone present knew it as well. The other contractors present begrudgingly and insincerely congratulated her. Their proposals had cost each of them thousands of dollars in time and effort and they were now up in smoke, useless. PGD's representative announced that they would study the bids and in the meantime Hanley should start lining up their surety bond in the sum of the bid, $14,400,440.

In the next couple of days Hanley Constructors contacted all of their principal subcontractors and suppliers and started getting their contracts and purchase orders organized for release as soon as they had PGD's signed contract in hand. Their bonding company required them to have any subcontract over $300,000 to be separately bonded.

Bobble Roofing of Norwalk, California was the low bidder on roofing. Their bid of $573,000 was $3,400 under the second lowest, Evans & Fogg Roofing. When Anson Bobble got word that he had the contract, he set about getting his bond lined up. He'd never before had to produce a bond. This was the biggest job he'd ever tried for. His insurance broker wasn't much help as Bobble was seriously in arrears in premium payments. He called several other brokers and bonding company agents, but they were all unreasonable. They wanted to see operating figures, financial statements, and tax returns. They also wanted details of his experience, personnel, and equipment. Finally, Anson realistically concluded that there was little or no chance he could ever get a bond in this lifetime. But the contract was enticing, and he couldn't just let it slip away. So he decided to call the second low bidder and see if he could cobble together a joint venture with them.

* * *

Evans & Fogg were reputable, well-established roofing contractors in Burbank, California. Eric Evans and Fenton Fogg, were the grandsons of the founders. They were raised in the business. Eric ran the office and did all the estimating and bidding, while Fenton ran the work in the field. They were hugely disappointed when they weren't low on the Paradise Gardens job. They really wanted that one. Eric was sure he had the right price and they were both counting on this job to keep their work volume up.

It was a pleasant surprise when Anson Bobble called. He wanted to talk about Bobble Roofing sharing the PGD job with Evans & Fogg. Anson didn't want Eric to see his so-called offices, so he offered to drive the 25 miles to the Evans and Fogg offices in Burbank. Bobble Roofing did business from a grimy desk and telephone in the corner of a small corrugated iron warehouse they shared with a painter.

Anson Bobble took off in his decrepit pickup and arrived in Burbank nearly on time, only 40 minutes late, and introduced himself to Eric Evans. He told Eric that he'd like to work out a joint venture with Evans & Fogg. He said he wasn't able to get a bond because they had too much bonded work in progress and their regular bonding company felt they were spread too thin. They would soon be finishing up most of the work and their final payments and retentions would be flooding in. However, they needed a bond right now for the PGD job. They couldn't wait for their bonding capacity to open up in a few weeks.

He offered Eric a sweet deal. They would form a joint venture and Bobble Roofing would run the job, do all the work, purchase all the materials, pay all the bills, and take care of the billings and collections. All Evans & Fogg would have to do was furnish the bond. For this negligible amount of effort Evans & Fogg would receive a hefty ten- percent of the profits. He sat smiling waiting for Eric's answer.

Eric was annoyed by Anson's spraying cigar ashes all over the carpet and desk. He went to another room and brought back a clean ashtray. "Put out the cigar and we can talk about the deal." A deflated Anson squashed out the soggy cigar butt and waited expectantly.

Eric referred to his notes and his own estimate for the job. "I think we can do business but we'll have to make a few changes in the deal. Evans & Fogg will get the bond, keep the checkbook, take care of the billings and collections, and Bobble will do everything else. We'll share the profits equally. On a 50/50 basis."

Anson was incredulous. He sputtered, "But that's ridiculous! Bobble Roofing landed the job."

Eric pointed out, "You've got no job without a surety bond! No bond, no job." Eric's jaw was set. He was dug in.

Bobble, recognizing the bottom line, reluctantly acquiesced, "Okay, Partner. We got a deal. We'll make a lotta money on this. 50/50 partners." They shook hands, both smiling.

Eric, thinking ahead, said, "Well, Anson, I'll get our lawyers to write a joint venture agreement for us so we can get it signed up and then we can sign the subcontract with Hanley Constructors. We'll also get the bond under way with our insurance broker."

Anson, also thinking ahead, said, "Maybe we ought to skip the joint venture idea and just do it under Evans & Fogg. It'll simplify the deal." He was concerned about having to furnish financial statements and who knows what all to Evans & Fogg's bonding company. Bobble then left and lit a fresh cigar on the way out to his truck. Things were looking up! He suddenly remembered that he'd have to get his state contractor's license reinstated. He'd forgotten to pay the renewal fee and post the license bond a few months back.

Eric wondered to himself if he was teaming up with the right outfit. But then, he reasoned, a job's a job. A 50/50 profit split wasn't too bad when all he'd have to do is get a bond. Sometimes you have to take a few risks to make a buck.

When Anson got back to his office, he phoned Hanley Constructors and asked for Greg Hanley. He explained about the bond and the deal with Evans & Fogg and said that they'd go through with the subcontract for the low bid price. Hanley, at first, wasn't too keen on the arrangement but felt that it was probably okay as long as they had the bond and the price was firm. So, he agreed to send out the subcontract agreement in the name of Evans & Fogg in the sum of $573,000.

* * *

Hanley Constructors had the Paradise Gardens project well under way, starting a new foundation every other week. The development consisted of 360 condominium units in 20 buildings of 18 units each. The buildings were each a little different but were generally two story, frame and stucco, UBC Type V construction. Each structure was U-shaped in plan with a one-story section at each end of the U. The buildings were interestingly sited on the land. Lush landscaping would surround the buildings and fill the courtyards. Upon completion, it would be a veritable tropical paradise.

According to Hanley's scientifically created computer-generated CPM construction schedule, the roofing would start on the first building in the 13th week of construction on May 24, 1993. Thereafter, a new roof would be started every two weeks. The principals of the two roofing companies figured that they would need only eight days to roof a building so they had an extra two days every other week for picking up loose ends or doing other small jobs. This schedule would give them 40 weeks of work and, at $28,650 per building, they would have a cash flow of $14,325 a week. "Fat city!" is the way Bobble described it. "We'll make a killing!"

Eric told Bobble that they would each have to put up about $7,500 into the new joint checking account to take care of payroll and other costs pending billing and receipt of the regular payments from Hanley. Bobble said, "Okay, Partner."

Ian Jason, Hanley's no-nonsense job superintendent, was constantly on top of the work schedule. He had the office notify Evans & Fogg of their starting date right after the subcontract was signed and the bond submitted. He notified them again in the 21st week of construction that they had two weeks to start the first roof.

Jason was on the job early on Monday, May 24th, as he wanted to get the roofers started properly and let them know where they could park and store materials. But the roofers weren't there. At around 10:30, an ungainly truck, listing ponderously to the port side, pulled slowly onto the property towing a two-wheeled asphalt pot. In among the rolls of roofing felt and ladders on the truck were lumps of bulk asphalt, an open carton of nails, and a crew of four roofers. Following in his decrepit pickup was Anson and two more workers. Jason directed them to the first building to be roofed. He walked over to where Bobble was getting out of the truck, and greeted him, "Where the hell have you been, Bobble? It's almost noon!"

"I was lining up the crew and arranging for the materials. Quality Roofing Suppliers had some problems with our line of credit but it's all straightened out now. Materials'll start arriving later this morning," Bobble explained.

"Well, you better get moving on this roofing. We've got a schedule to meet. This building's gotta be roofed by a week from Friday. If it isn't, I'll get it out of your hide. Y'understand what I'm saying?" He wasn't kidding.

Anson, broadly smiling, placated him with, "Sure, Boss. No problem. Nothing to worry about. Leave it to me. Bobble Roofing is here."

Bobble's crew set up a ladder and a materials conveyor and started getting their limited supply of materials up onto the roof. They were making great time applying felt over the plywood sheathing when they ran out of materials about an hour after lunch. Anson went over to Jason's job shack and asked to use the phone. Jason said, "Sure, help yourself. The pay phone's right over there."

Anson said, "Lemme have a cuppla quarters, willya?"

"Okay, this time. But never again. And you better pay me back tomorrow or you'll regret it." He left, muttering under his breath about flakes and morons.

Anson dialed Quality Roofing Suppliers and was referred to the credit manager. "Look Bobble, we called your bank and they said the check you gave us this morning is worthless. We'll deliver your materials when you give us a good check or cash. And no sooner." He hung up as if he didn't care if he lost this customer.

Anson looked up Jason and hit him up for a couple more quarters. "I'll pay you tomorrow. All I've got on me is big bills, twenties and fifties, no small change. I'll have some quarters tomorrow." Jason couldn't believe this guy. Anson then called Evans & Fogg.

The Joint Venture

When Eric answered the phone, Bobble said, "Look, Eric, we got the job started but we have a small problem. We need materials and Quality Roofing Suppliers wants to be paid in advance."
"Oh? Why is that?"

Anson, thinking fast, improvised, "Well, uh, Evans & Fogg don't have an account there, so they want cash. I'll drop by your office and pick up a check."

"No dice, Bobble. You haven't put up your $7,500 yet."

"Haven't you received that check yet? It must be held up in the mail. You know how lousy the postal service is. I'll be there in an hour to pick up the check for Quality. We gotta keep the job going."

* * *

Bobble told Quality to apply the check in payment of his long overdue account and told them that he would soon be receiving substantial regular payments from Hanley Constructors. Quality agreed to send the first shipment of materials to the job.

Nearing the end of the second week, the first building was ready for application of a flood coat of asphalt and the roofing gravel.

Anson reckoned he better work out some way of getting his costs down if they were going to make any money. He found a source of roofing gravel that would knock off about thirty percent from the normal price and they would deliver to the job directly from the quarry in Colton. Additional savings came from eliminating the sacks. It would be delivered in bulk and dumped on the jobsite. Anson had planned on renting a high lift loader so they could dump the gravel right onto the roofs. They'd save a bundle.

The loader dumped the first few loads of gravel onto one of the one-story sections of roof. Bobble was watching the unique operation and congratulating himself on his innovative logistical brilliance. He didn't notice that one of the loads of loose gravel landed squarely over the toilet vent. About half the load poured into the gaping 4-inch diameter vent. The rest of the gravelling operation went smoothly and the first building was completed within the allotted time. Bobble phoned Eric and told him to send in the bill for the first building. Eric responded, "Swell, we'll get some cash flowing. Incidentally, Bobble, your $7,500 check never arrived."

"Oh. I may have forgotten to mail it. I sent out so many checks last week and I've got so much on my mind. I'll take care of it when I get back to my office today. No problem. You can count on it."

* * *

About the third or fourth day into the third building, Anson was on the roof barking orders when he backed into an asphalt bucket, fell, and rolled to the edge of the roof. He grabbed a roll of roofing felt but it shifted and he plummeted to the ground like a rock. The roll of felt and the asphalt bucket landed on top of him. The crew heard him yell but couldn't reach him in time. The ambulance took him to Riverside Memorial Hospital where he was patched up and was bedbound for nearly three weeks. The doctors ordered additional home recuperation. Anson didn't object, as he couldn't straighten up or walk without excruciating pain.

It was six more weeks before he returned to the job. By then his crew was roofing the seventh building. They seemed to make as good progress without him as with him. During his home stay he phoned Eric once a week to ask for a check to be mailed to him. Eric objected but Anson complained, "How am I going to live without money?" Eric didn't want to kick him while he was down, so he sent the checks. Eric also had to keep Bobble's payroll up so the job wouldn't stop. Meanwhile Quality kept sending materials to the job as Anson got Eric to send enough money to keep Quality quiet.

On Anson's second day back on the job, he couldn't get his pickup started when it was time to go home. One of his men who knew a lot about cars looked it over and advised, "This old heap is finished. You've driven the life out of it. It's thirteen years old. How long do you think it ought to last?" So Bobble rode home with his crew on the company truck. That night he called Eric to get some money to buy a new truck. Eric said, "Hell, no, Bobble. We can't buy a truck out of this contract. Are you crazy?" Then, relenting a little from his harshness, he offered, "Look, Anson, we've got an old truck in our yard. It's a 1979 Ford pickup with almost 400,000 miles on it. But it still runs. You can have it." Anson picked it up the next day. It wasn't so bad. Not much worse than his old truck.

* * *

A couple of weeks later when Bobble's crew was finishing up the eighth building, they had the asphalt pot parked fairly close to the building and one of the apprentices had it fired up full blast to get the operation moving faster. Jason came by and shouted to Anson up on the roof to move the pot away from the building and not take any chances with a fire. Anson shouted down, "Okay, Boss, right away." He was going to tell the apprentice when he came down the ladder. But, rotten luck intervened. The asphalt burner, having lost its thermostatic control years ago, was hotter than the fires of hell and ignited the plaster backing paper and studs and the building started to smoulder. The wood framing, now drier than a chip, burst into flames. The apprentice, utterly overwhelmed with the enormity of the situation, was rushing about frantically shouting at Bobble and the crew to come down off the roof. They scurried down the ladder and tried to get the flames controlled but there was no ready supply of water and it soon got away from them. By the time Jason got the Riverside Fire Department on the scene it was a lost cause. Building #8 was gone except for the lower parts of the first floor studs and the concrete floors and foundations.

Eric Evans prepared the bi-weekly invoice as usual, billing for the roofing of Building #8. Upon receipt of the invoice, Greg Hanley phoned Evans & Fogg and got Eric on the line. "You guys must be out of your minds. There's no way we're going to pay you clowns for roofing the building you burnt down. You can forget it." He hung up.

* * *

In the 30th week of construction Bobble and his crew were finishing work on Building #9. They had to borrow an asphalt pot from Evans & Fogg as their own was a complete write-off after the fire. Bobble wanted to buy a new one but Eric wouldn't hear of it. Bobble was driving the high-lift loader depositing gravel on the roof, when Jason showed up shouting at him to shut off the motor. When it was quiet, he said, "Bobble, your supplier has filed a lien on the job. You better take care of it. And damn quick. The office says no more payments until you do."

Anson called Eric and said, "Eric, there's some sorta mix-up at Quality Roofing Suppliers and they've liened the job. Will you please phone 'em and find out what's eating 'em? I can't understand those guys. After all the business I've thrown their way. We've always got along just fine and now they're getting greedy. They've got no appreciation."

Eric called Quality and spoke with the credit manager, who complained that Quality had shipped over $102,000 worth of materials to the Paradise job and had been paid only $40,000. So they filed a lien in the sum of approximately $62,000. Eric replied, "Hey, wait a minute. We've sent you over $100,000. There must be something wrong with your records."

"Our records are okay, but the first $62,000 was applied to the long outstanding balance on Bobble's account. From now on, we won't ship any more materials except on a C.O.D. basis. The account's frozen until it's brought up to date."

* * *

Bobble called Eric with the demand that his weekly check be increased. "Look, Eric. I can't live on it any more. My expenses have gone up."

"Bobble, we've got to have an understanding here. You've got the job all screwed up in the field and our costs have risen out of sight. We're going to lose money." Eric sounded desperate.

"How can we lose money on a $573,000 contract? You must be mishandling the checkbook. I never should've agreed to a 50/50 split. You can't be trusted." Bobble hung up.

* * *

Jason's general construction was following the schedule fairly well and they were just finishing up Building #1. The plumbers, after installing all the fixtures and hooking up the water supply, were testing the plumbing in the one story section. Suddenly there was water all over the floors. One of the toilets was overflowing. There was obviously an obstruction in the soil line. They tried to clear it but found it immovable. They removed the toilet, tore out some gypsum wallboard, removed some flooring, and broke into the piping. There they found the plug of roofing gravel that had been dropped into the vent by the roofers. By the time the piping was cleared and repaired and the wall and floor finishes restored, the total cost was over $3000. Jason told Bobble that it would be deducted from their next payment. Anson didn't say anything about it to Eric, as he'd only fly off the handle. He was getting a little touchy lately.
Bobble wasn't too happy with Eric as the gift truck had thrown a rod and had to be abandoned. He left it where it fell, next to Building #7. Eric told Bobble to find his own transportation.

* * *

Eric was sitting at his desk taking stock of the situation. He had produced reams of notes and schedules and yards of adding machine tape. He was unkempt and distraught as his partner Fenton Fogg came in and sat next to the desk. Fenton inquired, "How's the Paradise job coming along? Are we making any money?"

Eric winced and started a realistic update, explaining, "Fenton, we're in big trouble on this job. We've completed about half the contract and collected almost half the money. The checkbook is depleted and we still owe our suppliers nearly $100,000. I've advanced over $20,000 of Evans & Fogg money to keep the job going. Bobble has paid out nothing. Hanley's fire insurers have paid for the burnt out building but they plan to sue us for $275,000 to recover their losses. Hanley has backcharged us $3200 for the gravel that got in the plumbing vent and won't pay us for the roof on Building # 8. The architect is complaining about the quality of the roof rock and it'll have to be replaced on ten buildings. Now Hanley is complaining about the two trucks abandoned on the site and threatens to have them removed and backcharged to our next payment."

Fenton suggested, "Why don't we try to get out of the deal with Bobble and finish it up ourselves? If we work hard we may be able to cut our losses."

"Well, it may be too late for that. As it stands, if we finish the job and collect all the money we'll still have a deficit of nearly $400,000. Bobble will be of no help to us. He's got nothing. We'd be better off to let our bonding company finish up the job and we'll have to go into bankruptcy. We've lost our business."

Fenton sadly concluded, "Maybe the bonding companies knew what they were doing when Bobble couldn't get a bond. Tell me, why did we ever go in with that idiot Bobble in the first place?"

"Well, at the time it seemed like a good idea," observed Eric.

* * *

Points of Law 8

THE JOINT VENTURE
Choose Your Joint Venturer More Carefully Than Your Spouse

8.1 Bonding Capacity

In our story, Bobble Roofing is the low bidder on a $573,000 roofing contract, but is not qualified for an award of the contract because it can't make bond. The bond requirement is imposed by prime contractor Hanley Constructors in order to guarantee that Bobble Roofing has the working capital, equipment, and qualified personnel to perform the contract. Hanley Constructors' own bonding company insists that the major subcontractors be bonded so as to protect Hanley Constructors from subcontractor defaults that could jeopardize the ability of Hanley Constructors to perform the work. After all, a breach of contract by a subcontractor turns into a breach of contract by the prime contractor. If such a breach occurs, and both the prime contractor and the subcontractor have supplied performance bonds, the breach of contract gives the owner a claim against the prime contractor and the prime contractor's surety, and the subcontractor's breach gives the prime contractor and the prime contractor's surety a claim against the subcontractor and the subcontractor's surety. By this mechanism, the prime contractor's surety lays off some of the risk on the sureties of the principal subcontractors.

Anson Bobble, a flagrant bad actor, would never be able to get a performance bond for his company because sureties make a thorough investigation of the licensure and credit ratings of any contractor who seeks to establish bonding capacity. Bobble Roofing is not properly licensed and does not pay its bills. That's why Anson Bobble conceives the idea of joint venturing the job with the second low bidder. When Anson proposes his deal to Eric he pretends that Bobble Roofing has plenty of bonding capacity but that the bonding capacity is tied up in a few jobs that are nearing completion. So he needs Evans & Fogg to tide him over. Anson is not going to admit that his insurance agent would not supply a bond since Anson has not been paying his insurance premiums. Eric falls for the story and agrees to joint venture the job for half the profit. This is a serious mistake. By agreeing to joint venture the job with Bobble Roofing, Evans & Fogg will become legally liable for any default by Bobble Roofing. Of all the different forms of business organization available, the joint venture relationship is perhaps the most dangerous.

8.2 Joint Venture

Most contractors are organized as either sole proprietorships, corporations, partnerships, or joint ventures. The corporate form is preferred because the officers, directors, and shareholders of a corporation are not usually responsible for the debts of the corporation. The construction business is recognized as a risky one. By organizing as a corporation, a contractor can avoid personal liability. If the worst should happen, and the corporation should have to file bankruptcy, the assets and the credit of the officers, directors, and shareholders may be unaffected.

Partnership is a different story. With some exceptions, partners are personally liable for the debts of a partnership. (The exceptions are for limited partners.) The distinguishing feature of a partnership is that partners jointly manage their business and split the profits and losses.

A joint venture is a special kind of partnership. In the construction industry, a joint venture is a partnership to do a single job. When the job is over, the business of a partnership is over. (Many joint ventures in the construction industry undertake more than one job. The joint venture confines its business to those particular jobs, and when the jobs are finished, the business of the joint venture is over.)

Construction industry joint venture agreements can be simple – only a few pages long. The joint venture agreement assigns the operating responsibilities between the joint venturers, identifies the job to be performed, and includes a formula for a split of profits and losses.

Although a joint venture is usually established by a written document, it is possible to enter into an enforceable joint venture agreement on a handshake. This is the kind of simple joint venture that we find in our story: a joint venture between two corporations each of which is to contribute $7500 working capital with the understanding that the corporations will split the profits 50/50. Evans & Fogg is to handle collections, the bank account, and bonding and Bobble is to purchase the materials, supply equipment, provide workers, and run the job. If the venture makes a profit, Evans & Fogg will get half of it. If the joint venture should default, however, the injured party can collect 100% of its claim from either corporation, and the corporation that pays the claim would have the right to even the books by collecting half of the loss from the other venturer.

In our story, we have another wrinkle because the joint venturers have decided to do business in the name of Evans & Fogg. However, the fact that the existence of the joint venture may be a secret from Hanley Constructors does not affect any of the liability relationships discussed above.

8.3 The Account to Which a Payment Is Credited

Eric has paid more than $100,000 to Quality Roofing Suppliers for felt, asphalt, and other roofing supplies. Unfortunately for, and unknown to, Eric, Bobble was already a customer of Quality before the joint venture came into existence. At the time Eric started making payments to Quality, Bobble already owed Quality $62,000. Without making any inquiry about where the payments came from, Quality just applied the first $62,000 that came in to the old account. As a result, even though Eric may have paid for all the roofing materials that were delivered to the job, Quality's accounts receivable ledger shows a deficit of $62,000. This was the basis upon which Quality recorded a $62,000 mechanics lien claim against the Paradise Gardens project.

8.4 Crediting Payments

Subcontractors often utilize materials from the same supplier on scores or even hundreds of different jobs.

Furniture stores have no reason to keep track of sales on a job-by-job basis. The ultimate destination of the furniture is pretty much irrelevant. But things are different in the building material business. Building material dealers have a vital interest in keeping track of the different jobs to which to their materials are supplied, because it is only by knowing the balance due for materials supplied to a particular project that a building material dealer can take advantage of the mechanics lien law. The mechanics lien system gives a material supplier a lien against the property where its materials were installed. Material dealers keep separate accounts on a job-by-job basis in order to protect their mechanics lien rights. Since Quality is selling roofing materials to Bobble on credit, it should keep separate accounts for each job.

When a subcontractor makes a payment, the supplier must credit the payment as directed by the subcontractor. If the subcontractor gives no direction, then the supplier can credit the payment to any account.

In our story Eric failed to designate the job to which the payment should be credited and therefore Quality simply credited the payment to Bobble's most delinquent old account. Quality might have a motive to credit the delinquent old account, because that would be the account on which Quality's mechanics lien rights most likely would have expired because of the lapse of time.

8.5 Mechanics Liens

There is an issue here that wants further investigation. In many states, construction funds are held in trust to pay for work or materials supplied to the job that provides the funds. In these states, a material supplier is required to credit payments to the proper job if the supplier had enough information to know what job produced the funds.

To resolve this issue, we have to find out what Quality knew, and when Quality knew it. Even though Eric did not give Quality any specific directions as to how the payments should be applied, maybe Quality had enough information from other sources to impress the funds with a trust. For example, maybe Quality delivered roofing materials to the job site and maybe the amount of the payment matched a bill which would have enabled Quality's accounts receivable clerk to match the payment to the job.

Such information could be a good defense to the mechanics lien claim, but it does not do Eric much good right now because the owner is going to withhold the amount of the mechanics lien claim (plus attorneys fees) from Hanley Constructors, the prime contractor, which will in turn withhold that amount from Evans & Fogg until the mechanics lien issue is cleared up. It could take years to work its way through the court system. Meanwhile, the money is tied up.

8.6 The Fire Loss

Bobble carelessly set fire to one of the buildings, and the owner's fire insurer paid for the loss.

Under a legal doctrine known as *subrogation*, an insurance company that pays a property owner's loss steps into the shoes of the property owner to enforce claims against the party that caused the loss. Therefore, the insurance company has a claim against Evans & Fogg because Evans & Fogg is a joint venturer with Bobble, and Bobble caused the loss by locating its kettle too close to the building.

Before he throws in the sponge and puts his company into bankruptcy, Eric should make a careful review of the contract between the owner and Hanley Constructors. The most popular form of construction contract nationwide, known as AIA Document A201 (General Conditions of the Contract), includes a *waiver of subrogation* clause. This provides that the owner waives its subrogation rights against the contractor and subcontractors for losses that are covered by its fire insurance policy. As we said, the insurance company steps into the shoes of the owner. If the owner has waived its rights against the subcontractor then the insurance company has no claim against Bobble or Evans & Fogg.

8.7 Bankruptcy Woes

Eric explains to Fenton that even if they finish the job and collect 100% of the contract price, they will lose $400,000 and bankrupt the company. Bankruptcy is a process by which the assets of an insolvent company are equitably distributed (after attorneys fees) to the creditors of the company and the company goes out of business. The fact that a corporation goes through bankruptcy does not usually mean that the officers, directors, and stockholders also go bankrupt.
There are lots of exceptions to this rule. One exception is when a corporation goes broke on a bonded job.

8.8 Bankruptcy and Bonds

Eric is talking about turning the job over to the bonding company and putting the corporation through bankruptcy. This would be a fateful decision. He may have forgotten that when his company first established bonding capacity, the surety made a thorough investigation of the financial condition not only of the corporation but also of Eric and Fenton.

Before the bonding company would issue any bonds it required Eric and Fenton and their wives to sign long legal papers with complicated language in small print. It was explained to Eric that he would have to personally guarantee that the bonding company would not sustain any loss. If Eric and Fenton turn the job over to the bonding company and the bonding company finishes the job, he and Fenton will have to reimburse all that expense plus attorneys fees. This puts them in the sad position of having to contemplate not just the bankruptcy of the corporation, but also their own personal bankruptcy.

They may come out better if they dig into their own pockets to finance the corporation rather than turning the job over to the bonding company. This would avoid the attorneys fees and expenses of the bonding company. After all, Eric and Fenton are roofers and the bonding company is not. They can probably get the job done cheaper.

Shangri-La Gardens
The Short Life of Eternal Waterproofing

Shangri-La Gardens was the latest and most fashionable condominium development in West Hollywood when it came on stream early in 1994. Reasonably priced and attractive, all 120 units were promptly sold. Because of the sloping site, most of the units had spectacular views out over Beverly Hills and beyond to Santa Monica.

The buildings were skillfully arranged around landscaped courtyards on top of a subterranean garage. Each condo unit was a two-story townhouse. Every unit had a so-called bonus room at the garage level that gave each of them direct access to their own parking spaces. The bonus rooms were not designated by the developer for any particular purpose, leaving that to the condo owners' imaginations. Some were furnished as music or entertainment rooms while others were developed as home offices or hobby rooms. That was the genius of the idea; the bonus room could be used for anything that the prospective purchaser might have in mind. It contributed greatly to the rapid sell-out of all 120 condos in less than 60 days.

* * *

In early February 1995, the short Southern California rainy season was off to a robust start. It had been raining steadily all afternoon and evening and poured off and on all night. The next morning was beautiful, clear, bright, and sunny just as promised by the TV weather broadcaster.

After a hurried breakfast, Ken Lange, in Unit B23, kissed his lovely wife, said "Good-bye, Hon, seeya tonight," hoisted his briefcase, and proceeded spiritedly downstairs to cut through the bonus room and out to his car. Stepping off the last stair tread onto the bonus room carpeting, he could feel it squish. He also heard it. Instinctively, he quickly retreated back to the bottom stair tread. He couldn't believe his eyes. The whole bonus room carpeting glistened as though it was a small shimmering silver lake. He rushed back upstairs and shouted, "Barbie, come down here quick. You won't believe this!"

She followed him apprehensively down the carpeted stairway and they both stared at the indoor swamp. He could see that the carpet was like a huge wick that entrapped hundreds of gallons of water. He sloshed gingerly across the room and opened the door to the garage. Water had been seeping out under the door and now it streamed out more freely. They spent the next hour carrying furniture up the stairs and drying it off with towels. Later they'd call the carpet people to pick up the carpeting and padding. Barbie said, "I hope it's not ruined! And our beautiful new furniture." She was heartsick, close to tears.

Ken, glancing at his watch, said, "Barbie, we better call the management company and report this. It must be a plumbing leak. A water pipe must have burst."

"I'll do it right now, Ken. You better get going. You're already late for work." Their faces were drained and grim-faced. They didn't know yet if this was covered by their household insurance. Ken said, "I'll check it out today. I'll call our broker."

As Ken was driving out of the garage he noticed some of his neighbors looking at the floors just outside their bonus rooms. He saw some water flowing toward the floor drains and said to himself, "There must be more plumbing leaks."

* * *

The Certified Property Managers telephone lines were all tied up. Only one was equipped with an answering machine and it had been busily recording one message after another. The other two lines were ringing steadily when Melanie Newman opened the office at 8 AM. She answered one of the phones, asked the caller to hold, and went on to the second. It took nearly an hour to catch up with the ringing phones. In the first lull she played the answering machine back to pick up the recorded messages. The eleven stored messages were similar to the calls she had just taken. Something terrible was happening at Shangri-La Gardens.

Owen Pierce, one of Certified's senior property managers, running a little late this morning, showed up at 8:30 to find Melanie on the phone and furiously writing service requisitions. There was an untidy pile of the pink service slips on her desk. She looked up at Owen with a concerned expression and shoved the pile his way. He shuffled through them, quickly absorbing the information, then flew out the door headed for Shangri-La Gardens. There he was met by an anxious group of condo owners, some still in their bathrobes, all talking at once. He was ushered to one bonus room after another. They were all water-soaked. Some of the homeowners were angry and aggressive while others were more civil, realizing that Owen was probably not responsible for their problems and would only help solve them.

He immediately started marshalling assistance on his car phone. He called Melanie to send over another property manager, as he correctly perceived that the problem was widespread and too big to handle alone. He called Pronto Plumbing to come as quickly as possible. Two plumbers from Pronto were there within 20 minutes and started tracking down the leak. After two hours of probing and searching they reported to Owen that they couldn't find any plumbing problems. It must be something else. They asked for his signature on a service call bill for $180.

Quincy Roberts, a junior property manager, soon arrived and started organizing labor to help move furniture, pick up carpeting, squeegee floors, and dry off belongings. He really pitched in and kept the operation moving effectively. He helped with carrying furniture and placating homeowners.

Owen was sleuthing around himself and, in several of the bonus rooms, he saw water seeping up through the joint between the floor slab and the concrete block wall. He also found numerous wet spots and water seepage in many of the concrete block wall joints. He saw water seeping through and running down the wall surface.

He spent the day at Shangri-La and looked at dozens of bonus rooms. Around 6 o'clock many of the condo owners who had been at work started arriving home. He looked at their damage and started making a record of the locations of all the water intrusions.

It took most of the week to clean and dry up the mess, but they couldn't risk moving any of the carpeting or furniture back in as it could happen again.

* * *

Half a year later, 94 out of the 120 condo owners were completely disgusted with their purchases. Hundreds of complaints had been lodged with Certified Property Managers and, so far, nothing constructive had been done to alleviate the problems. The developer who had built the complex had workers running around with caulking guns practically every day but it did no good. Every time it rained more water showed up. More caulking would be done. Many of the homeowners had installed wood paneling or wallpaper in their bonus rooms and now most of it was warped, streaked, stained, peeling, and ruined.

The bonus rooms were useless. They were more suitable for mushroom culture or bean sprout growing than for all the wonderful dreams of the owners. Subterranean leaking was also widespread in the project's mechanical equipment rooms, storage areas, and communal exercise rooms. The garages always seemed to have trickles of water flowing out of the bonus rooms and into the floor drains. A dank musty odor was beginning to permeate all of the garage level spaces. Many of the condo owners were complaining of unpleasant odors getting into their upstairs living areas. Unsightly mildew and fungus stains were forming on the walls of their bonus rooms. Several had visited their physicians as they felt that their previously minor respiratory complaints were worsening.

Some had decided to bail out of the investment but quickly changed their minds when told by neighborhood real estate agents that resale prices would remain severely depressed as long as the leaking problem persisted. They pointed out that such defects must be disclosed to prospective purchasers. The situation at Shangri-La Gardens was by then well known in the real estate sales community.

* * *

Recent meetings of the Shangri-La Homeowners Association, usually held monthly, had been consistently running past midnight so they decided to increase the frequency to bi-monthly, and finally, weekly.

Sam Tracy had been President of the Association from inception and now he fervently wished he hadn't been so quick to accept the honor. This was his retirement activity after 45 years of running his own wholesale grocery business. He was businesslike and worked well with people. But it was no longer a pleasure. The meetings were getting downright unpleasant. The neighbors were no longer friendly. Tempers were short. They bickered and wrangled. They treated Sam as though he was the cause of their problems. Every meeting of the association was a horrible and distressful experience for him and the other co-owners.

The developer was cooperative but didn't really know what to do. IJ Developers was headed by Ignatius Jones, who was more of a promoter than a builder. He put the deals together and hired staff people to do the physical things. Iggy was a great salesman and coordinator. He told the Homeowners Association, "Tell me what to do and I'll do it. This type of thing never happened before in any of my developments. It must be an act of God. Probably caused by the earthquake." Iggy was always friendly and cooperative.

The homeowners didn't know what to do either. The meetings didn't accomplish very much, if anything. They functioned mostly as a catharsis, a forum for sounding off, and for huddling together for mutual comfort.

Owen Pierce attended most of their meetings and was extremely supportive. He spent most of his time at Shangri-La, almost daily. He eased the owners' burdens as well as he could. But there wasn't really much he could do beyond being helpful and friendly. He had been keeping records and correlating them on a chart posted in his office. He noticed that practically all of the leaking had been occurring in Buildings B, C, and D. There were no leaks on the north and west sides of Building A. He couldn't account for this strange and interesting anomaly.

Sam Tracy appreciated Owen's support but he realized that all of his costs and expenses were being billed by Certified Property Managers to the Association each month. It was expensive and the costs were mounting. Some of the Board members were starting to complain and questioned the adequacy of the Association's maintenance reserves. Although Iggy wasn't billing them for his frequent caulking efforts, they didn't seem to be doing any good.

At one of the Association meetings, someone, in desperation, asked, "Why don't we sue someone?"

Someone else replied, "Great idea! Who'll we sue?"

Sam said, "Maybe we should find an attorney to represent us. We need professional advice on whether someone is responsible for our problems."

The small committee formed to select a suitable lawyer received nominations from various Board members and talked to the three lawyers who were homeowners in Shangri-La. They accumulated names of a few law firms that specialized in condominium associations and then narrowed it down to one firm that also had considerable experience with defective construction. They invited Ursula Vickers of Vickers and Walsh, Attorneys at Law, to attend the next Association meeting.

* * *

Ursula explained that there is no point in filing lawsuits willy-nilly without some rational objective. She continued, "We could enmesh innocent parties and completely overlook those responsible for your problems. First, we'll need to know all the parties involved in the development of Shangri-La. Second, we need to look into all possible insurance coverage. And finally, we need to find out exactly what is wrong with this building. This Board can help me with the first two categories, but we'll have to hire a competent technical expert to determine the precise nature of the problem. Where is the water coming from? Why is it leaking? How can it be fixed? What caused it? Then we'll have to discover who's responsible."

Sam Tracy asked, "Who can we get?"

Ursula replied, "Our firm has worked with Frank Grimm, FAIA, an experienced forensic architect. I'd like to call him and get him started."

Sam, answering for the Board, said, "Okay, Ursula, you have our approval."

"Good. Now we have to compile the list of individuals and firms who had anything to do with the development, construction, sale, and maintenance of Shangri-La. And I want to see all your insurance policies."

The Board meeting ended in a spirit of optimism. Finally, something constructive was happening.

* * *

During the next couple of months Ursula and her paralegal assistant collected, dissected, analyzed, and organized volumes of relevant information. Frank Grimm had made extensive examinations of the building and construction documents and had rendered his report. Ursula had considerable trouble piecing together the circumstances surrounding the construction, but after numerous informal interviews with the homeowners, Iggy Jones, and many of the subcontractors and suppliers, she finally uncovered the whole picture.

* * *

It all started in early March 1993. Alexander Barnes was now a free man for the first time in twenty-seven months. Freedom felt good. He was just released from Chino Men's Prison where he was sent following conviction for soliciting a $4500 bribe as a county building inspector. Now he had to find a way to make a living. All he knew was the construction industry but he knew he couldn't be a building inspector anymore. He'd look around for a suitable opportunity. He still had the $4500 stashed in a secure place. He'd testified that he'd lost it all on a weekend in Las Vegas.

Alex was at a coffee shop counter in Culver City when he struck up a conversation with the dejected-looking fellow on the next stool. He'd seen him drive up in a pick-up truck with "CD Waterproofing" painted on the side. Alex and Carl Dunn got along well from the start. Alex's natural friendliness, optimism, and upbeat personality were a counterbalance for Carl's constantly gloomy outlook. Carl gradually revealed his hopeless condition to Alex. He was ready to go out of business. His economic reverses were overwhelming. He'd just finished two large waterproofing jobs that were financial disasters. Although all the contract payments had now been received, he not only had nothing left, he owed his suppliers over $8000. His insurance had been cancelled for non-payment of premium. All his men were gone except for his son who was looking for another job where he'll get paid cash money. Carl had no work in the offing. "Now is the time to face facts and throw in the sponge," he concluded.

Alex paid for Carl's coffee and doughnut, and as Carl left, Alex asked him for his business card.

The more Alex thought about it, the more he was convinced that this was the opportunity he was looking for. Why not? So, he phoned Carl.

Alex agreed to go to CD Waterproofing's Office on Venice Boulevard in Mar Vista. He caught the Sepulveda bus, as he hadn't found a car yet. He walked the rest of the way. CD's office wasn't fancy. Actually, it was a double garage on the back of a lot that had a body and fender repair shop on the street frontage. The garage had an old wooden desk and a couple of scruffy chairs, a 4 drawer filing cabinet with only 3 drawers, some shelving, and a work table for mixing waterproofing materials. There were a number of empty cans and buckets that formerly held waterproofing materials and half a dozen unopened cans. A few rolls of black plastic sheeting were stacked in one corner. There was barely room left over to park the truck.

When Alex arrived, Carl was sitting at the desk aimlessly doodling on a memo pad. He looked lifeless and despondent. He hardly looked up.

Alex came in, sat down, and said, "Carl, this is your lucky day!" Carl raised his head listlessly.

Alex continued, "I'm gonna buy your business! Whadda ya want for it?"

"It isn't worth anything. All I've got left is the old truck and I owe over $8000 to my suppliers."

"That's okay. You give me the business and the truck and I'll pay your suppliers. I'll even give you a job. You have the waterproofing contractor's license so you can be my Responsible Managing Employee. Whatta ya say?"

"Okay with me. You gotta deal. Whatta we do now?"

"I'll change the name of the business to Eternal Waterproofing. You'll be my Field Superintendent. You can run the work. Let's go. I'll buy you a cuppa coffee." Carl locked up the garage and gave Alex the keys to the garage and the truck. Maybe his bad luck streak had ended. Alex drove the truck.

In the next few days Alex had a bright new sign painted and put it up over the door. Eternal Waterproofing. He was now calling the garage their warehouse. He applied for a city business license and a state contractor's license. He lined up a contractor's license bond and got a minimum insurance program in place. He paid $300 as a deposit on the premium. He had the truck repainted with Eternal Waterproofing on each door and on the tailgate. He ordered business cards for himself and Carl and some padded bid proposal forms. He opened a checking account at a nearby bank and deposited most of the balance of his funds, $2900.

* * *

Alex paid a visit to Economy Builders Supply to make a deal for paying off Carl's delinquent account. "I've just taken over CD Waterproofing and I'm gonna pay off the account," he announced.

The dour sales manager brightened a little and said, "Good. The balance is $8123.47. Just make your check payable to Economy Builders Supply."

Alex promised, "I'll give you $500 on account right now and I'll start making monthly payments to clear the rest off. Now, I want to open an account for the new business, Eternal Waterproofing.

We've got to start making money to pay you off." The sales manager wasn't 100% pleased with the arrangement but it was a lot better than the hopeless situation he had with Carl.

* * *

Alex now ventured out in earnest to scare up some action. He drove the truck around visiting construction sites where large-scale residential work was commencing. He'd look up the site superintendent and engage him in friendly conversation. He'd try to find out if the below-grade waterproofing subcontract had been let. If not, he'd then start probing to find out what their low bid was at the moment. He'd usually offer the superintendent $50 or $75 for the information. Most of them would promptly throw him off the jobsite, but occasionally one would continue talking to him.

Alex felt that this type of work was ideal, as many residential developers didn't engage their design architects and engineers to examine the construction in process. And he knew that no one paid much attention to work going on in an excavation.

After about three weeks of promotional activity, he landed on a site where foundations were being poured for four large condominium apartment buildings. Concrete block walls were going up on the subterranean garage of the first building. A flashy sign proclaimed Shangri-La Gardens, 120 luxury condominiums, IJ Developers. As Alex was walking up to the site office, he could hear the superintendent just winding up a phone call.

"That's $20 on Easymoney to win in the fifth at Santa Anita. Ya got it? Good. Don't worry. I'll pay ya what I owe ya. Don't worry. I'm good for it." He looked up as Alex stood in the doorway of the Portacabin that was his construction office.

"Whatta ya want?" he growled.

Alex came in. Smiling. Friendly. Introducing himself, he casually handed over his business card. He started his usual ingratiating banter and, eventually finding that the below-grade waterproofing subcontract was still up for grabs, he offered a friendly gratuity of $50. The superintendent, Glen Harker, now interested, didn't turn it down. Alex was still standing there and started ostentatiously counting off ten-dollar bills from a fat wad.

Glen said, "This is a big contract. Fifty bucks won't go very far. Five hundred would be more like it."

Alex said, "That's too much. After all, I don't have the contract. I haven't made a cent yet."

"Well, I've got expenses. I'll tell ya what we can do. Give me $300 now and $700 when ya get the contract."

"Ya gotta deal. Here's the other $250. Now, what's the low bid, as of right now?"

"$87,750. Go to the IJ Developers office in Santa Monica. Here's the address." He wrote it on a scrap of paper. "Tell Iggy I sent ya."

As Alex headed back to his truck, elated, he said to himself, "Now we're in business. Ya hafta make your own opportunities."

His next stop was IJ Developers in Santa Monica. The secretary, after receiving the Eternal Waterproofing card, steered him to the plan room. He messed around with the plans for fifteen or twenty minutes. He saw no point in studying the plans and taking off quantities. That had already been done by the low bidder. He began writing out a bid on his pad of standard proposal forms. He considered several figures, raising and lowering them in his mind. It had to be lower than $87,750 but not too low. No point in throwing away money. But it had to be low enough to sign up the deal today. He didn't want IJ shopping his bid around. That would be unethical. He finally decided on a bid of $86,250. He started writing the figure when he decided to lower it to $85,250. That's the right amount, he decided.

He went back out to the secretary and asked if he could present the bid directly to the construction manager. She disappeared for a moment and came back following a well-dressed energetic man. "I'm Iggy Jones. Lemme see your bid." He seemed interested, as he invited Alex into his office.

"This bid is way too high. Most of the waterproofing bids are much lower than this. I've got a bid for $84,750. You're just too high. Outta the ballpark."

Alex, knowing now that he was dealing with a kindred spirit, said. "I'm willing to give you my best bid if you're willing to sign up today, right now, before I walk out of here."

"Well, what's your figure?"

"$83,750. There's practically no profit in it. I'll take it just to keep my crews busy."

Iggy thought it over for about 5 or 6 seconds and said, "Okay, we'll prepare a subcontract." Then, referring to his construction schedule on the wall, continued, "You gotta be ready to start on Building A on May 10th and finish it in two weeks. You gotta start Building B on June 7th, Building C on July 5th and Building D on August 2nd. You can have each building for two weeks, then we backfill. Unnerstand?"

Alex walked out to the truck with a signed contract in his pocket. He was thinking, "$83,750 is still not too bad. Even after paying that crook Harker $1000. We'll just have to find ways to economize."

He tooled his truck over to Economy Builders Supply and made another payment of $500. He now had them softened up for all the credit he needed. The sales manager was his.

* * *

Carl had spent the last couple of weeks cleaning up the warehouse and the miserable collection of tools. He had painted the warehouse and it was now shipshape. He was feeling a lot better even though he had lost his business. At least he wasn't forced into bankruptcy.

He looked up from his broom and saw his old truck heading in for a landing. It pulled into the warehouse. Alex was waving a piece of paper and shouting, "Eternal Waterproofing is off the ground. We're in the big time now!" He described the whole deal to Carl. At least he disclosed most of the deal. He didn't mention the $1000 for Glen Harker and he didn't explain how he figured the price. Carl assumed he'd made a take-off from the plans and specs. He also neglected to mention the economies that would have to be made.

Alex further explained, "We gotta start in about a week and a half, on Monday, May 10th. The first building has to be done in two weeks. Then we'll be off two weeks before we start the second. There are four buildings altogether. We'll need about five workers to help you. Can you get a crew together?"

"Sure. That's easy. I'll get right on it." Carl headed for the desk and the telephone. He flipped through his soiled, dog-eared card file. Carl was happy again.

Alex waved to Carl as he left in the truck. He sped to Economy Builders Supply and went to the order desk. He ordered ten 15-gallon buckets of Superseal Primer Coat and the same quantity of Superseal bituminous, fiber-reinforced waterproofing compound. This was top-grade stuff; exactly what was specified in the construction documents. He also ordered fifty 15-gallon buckets of Econ-O-Coat primer and the same quantity of Econ-O-Coat waterproofing compound. The Superseal products cost $49 per bucket while the Econ-O-Coat equivalents were only $11 per bucket. This preliminary order would be enough to get the job started. He asked them to deliver the 20 buckets of Superseal products to the jobsite and the 100 buckets of Econ-O-Coat to the Eternal Waterproofing warehouse. When they arrived in the warehouse, Alex covered the stack of Econ-O-Coat buckets with black plastic sheeting.

* * *

Early on Monday morning, May 10, 1993, Alex and Carl, with 5 helpers in the back of the truck with an assortment of trowels and other small tools, arrived ready for action. Glen Harker told Alex where the twenty 15 gallon buckets of Superseal were stored. Glen suggested that the $700 balance be paid now before any work is done. Might as well be businesslike. So Alex peeled off 14 fifty-dollar bills and placed them in Glen's greedy hand.

Alex told Carl to get the crew going on Building A. They should start at the southwest corner and proceed clockwise around the building. Alex cautioned him to save the empty Superseal buckets as they were emptied and bring them back to the warehouse. Carl asked, "Why do you want to save them? We'll just have to get rid of them. It'll cost us. Maybe Glen'll let us dispose of them in the general trash heap."

Alex said, "Just save them. No more argument." Carl didn't understand. He wasn't used to anyone giving him orders.

Alex took off in the truck to continue his promotional rounds. Maybe he could find a fill-in job to utilize the two weeks idle period after each building at Shangri-La. He returned to the job at the end of the workday to drive Carl and the crew back to the warehouse. They also loaded up 10 empty Superseal buckets. After the men left the warehouse, Alex told Carl to go on home and get some rest. He would stay on to tie up some loose ends.

Carl put his hand on Alex's shoulder and said, "It's really good to be back to work, Alex. I'm sure glad I met you in the coffee shop. You're a great guy, a true friend. You've saved my life. See you in the morning." He left and walked home.

Alex closed the warehouse door and got to work. He filled the 10 empty Superseal buckets from the Econ-O-Coat buckets and then went home.

The next morning, he told the crew to load the refilled Superseal buckets onto the truck. After work that day, Carl went home leaving Alex to finish up whatever he had to do. About 20 minutes after Carl left he returned to pick up his forgotten lunch box. He found Alex pouring Econ-O-Coat into Superseal buckets. He was astounded. He demanded, "What're you doing, Alex?"

"What the hell do you think I'm doing? I'm making a profit for you."

"But, Alex, you can't do that. It's a specification violation."

"Now I know why you went broke."

"But, Alex, that Econ-O-Coat is no damn good for waterproofing. It's only dampproofing. The buildings will leak. Sure as hell!"

"So what? By then we'll be long gone."

"I don't like it."

"Who asked you to like it?"

Alex continued pouring Econ-O-Coat into Superseal buckets. Carl went home.

The next day the crew loaded the newly filled cans of Superseal onto the truck and they went to work, continuing on Building A. The west and north walls were completed with Superseal products and then Econ-O-Seal was used on the east and south walls. By the end of the week, Carl was so disgusted with the situation that he told Alex he could have the business and he left. Carl never returned. Alex promoted one of the crew to Foreman and gave him a 50 cents an hour raise. Alex told him that they were applying the waterproofing compound too thick. He explained that it was a much better job if it was spread out thinner and would be less likely to crack.

Eternal Waterproofing finished the work and all of it was promptly covered up with earth backfilling.

Alex was very efficient in his billing procedures and had collected all the contract payments from IJ Developers by 30 days after completion of Building D. He'd been paying Economy Builders Supply a little at a time to keep the deliveries coming but still owed them a little over $28,000 including the unpaid balance of Carl's delinquent account.

When Economy's mechanics lien arrived at IJ Development, Alexander Barnes was long gone. Eternal's warehouse was locked and the bank account closed. Alex was never located. Carl Dunn was not found financially responsible but lost his contractor's license. IJ Development found little consolation in Eternal's $5000 license bond or their written guarantee.

* * *

Ursula Vickers wound up the incredible story by revealing the forensic architect's recommendations. "In his report Frank Grimm concludes that the sub-grade walls will have to be excavated and completely exposed so they can be properly waterproofed. Earth backfill will have to be replaced and the surface features such as walks, landscaping, and irrigation sprinklers will have to be restored. His estimates vary from $280,000 to $300,000. This doesn't include the cost of repairing interior damage to homeowners' property and that in communal areas."

One of the Board members asked, anxiously, "But, Ursula, who's responsible? Who can we sue?"

"Fortunately, IJ Development is a solvent company and they're insured. They're responsible and they'll pay." Sighs of relief were heard all around the table.

* * *

Points of Law 9

Shangri-La Gardens
Intentional Misconduct

9.1 Who Shall Bear the Loss?

The condo owners in this story have taken losses. Water has come into their bonus rooms and spoiled their carpets, wall coverings, and furniture. Mildew is attacking. Unpleasant odors persist. Though many would like to, they can't sell their units now because the source of water intrusion has not been determined and problems have not been corrected. The market value of a unit may be less than the amount owed on the mortgage. The owners want compensation. They and their lawyers want to transfer the loss so that it will fall on the persons who are responsible for the problem. One of the functions of the law is to determine under what circumstances it is appropriate to transfer losses and the procedures by which this may be achieved.

The person who should bear the loss is the one who dishonestly and surreptitiously substituted Econ-O-Coat for Superseal. Since Econ-O-Coat is damp-proofing material and not waterproofing material, once the substitution was made, the water intrusion and the damage to condo owners was inevitable. Alex, the truly responsible person, has absconded with his loot, leaving behind his company, Eternal Waterproofing.

The owners must therefore seek to transfer their loss to parties who are more or less innocent of wrongdoing. The first target is IJ Developers. Iggy Jones accepted a low bid (perhaps suspiciously low) from a waterproofing subcontractor that, although it was properly licensed, was not known to Iggy either personally or by reputation. IJ Developers did not inspect the work closely enough to detect the switch from Superseal to Econ-O-Coat.

Carl Dunn, who sold out to Alex and then acted as his superintendent, bears responsibility because he briefly participated in the deception, and didn't blow the whistle.

To determine the potential liability of these parties we look to three branches of law: *torts*, *contracts*, and *sales*.

9.2 Branches of Law

The function of *tort law* is to deter wrong conduct by requiring wrongdoers to compensate innocent people who are damaged by the wrongdoing.

The law of the *contracts* seeks to deter breaches of contract by imposing loss caused by breach upon the breaching party.

Sellers of defective goods are responsible under the law of *sales*.

9.3 Torts vs. Contracts

Although the condo buyers do not have contracts with Alex or Carl, they have rights against them under tort law, which provides compensation for damages caused by conduct that the defendant knows, or should know, will cause unjustified injury. Alex and Carl are liable to the condo owners even though they had no way of knowing who they would be as personal individuals. The whole purpose of the development was to build condos for sale so Alex and Carl knew that the buyers would be damaged even though they might not have known their individual identities. The condo owners should try to find out whether Eternal Waterproofing carried liability insurance. Even though Alex has disappeared the owners could make a claim against his liability insurance carrier. They will find, to their disappointment, that most liability insurance policies do not cover damages caused by intentional misconduct.

The condo owners may also have a tort claim against Iggy and his company, IJ Developers. This claim would depend on showing that IJ negligently failed to provide proper waterproofing.

9.4 Negligence

Negligence is defined as failure to exercise due care to avoid causing foreseeable harm. It would come down to a question of whether Iggy exercised due care to prevent Alex from substituting Econ-O-Coat for Superseal. The case against IJ would run as follows: The circumstances of the low bid submitted by Alex should have made Iggy suspicious enough to investigate Alex's background and to carefully inspect the "waterproofing" materials and their application. Careful inspection would have revealed the fraud and prevented the loss. Iggy knew that the waterproofing work would be covered up when the walls were backfilled and that the potential condo purchasers would have no opportunity to inspect the waterproofing for themselves. So he had a heightened duty to make rigorous inspections in order to make sure that the waterproofing was properly performed.

9.5 Sales

The owners would also have a case against IJ Developers under the law of sales. When a developer sells a residence the law imposes in favor of the buyer an *implied warranty of habitability*. Water intrusion breaches that warranty and entitles the buyers to compensation.

9.6 Liability Insurance

When the condo owners make their claims against IJ Developers, Iggy will tender the claims to his liability insurance carrier. Whether the claims are covered will depend on the type of policy that Iggy purchased. The standard form policy excludes damage to work performed by IJ Developers or its subcontractors. The broad form covers damage to work performed for IJ Developers by subcontractors. Therefore the standard form would not cover but the broad form would.

9.7 Property Insurance

The condo owners should look at their homeowners policies. The damage to their new units might be covered, although it must be said that many homeowners policies exclude coverage for damage caused by water intrusion.

9.8 Mechanics Liens

Economy Builders Supply recorded a mechanics lien for $28,000, which included the unpaid balance for the Econ-O-Coat material plus the unpaid balance of Carl's delinquent account.

Carl's old delinquent account is "not lienable" because it is for materials that were not used on the Shangri-La Gardens project. The underlying purpose of the mechanics lien law is to prevent a property owner from being unjustly enriched. Shangri-La Gardens owners could not be enriched by materials that were not consumed on the Shangri-La Gardens job!

There are also a couple other defenses to the lien. Economy Builders Supply may not be able to fulfill the requirement that it sold its materials with the specific intention of improving the Shangri-La Gardens job.

If Economy Builders Supply does prove that it delivered the Econ-O-Coat with the intention that it be used on the Shangri-La Gardens job, that could raise another defense. Economy Builders Supply had sold Superseal for the Shangri-La Gardens job. Therefore Economy Builders Supply should have suspected that Econ-O-Coat was destined to be misused as waterproofing rather than damp-proofing. This would bring into play the doctrine of "unclean hands". The mechanics lien remedy is known as an "equitable remedy" that can be defeated if the property owner can show that the mechanics lien claimant has unclean hands. Economy Builders Supply would have unclean hands if it knew the materials for which it claims mechanics lien rights were to be misused to the detriment of the condo owners.

10

ABC Warehouse VI
The Pressure Regulator Ordered by the Building Inspector

Architect Ivor Judge, AIA, of Judge & King, was feeling pretty good. It was a beautiful invigorating spring morning in mid-May 1990. The ABC Warehouse construction was nearing completion. His incessant barely audible humming was an indication of optimism and peaceful contentment with his lot in life. He was just finishing off his second cup of coffee when his conscientious secretary placed the stack of morning mail on his desk. She'd already opened and emptied the envelopes, clipped the contents of each together with its envelope, and had branded everything with her trusty purple-inked adjustable date stamp. He'd warned her years ago to look before stamping when a client's original grant deed was inadvertently date stamped in.

He was well into the pile, when he suddenly stopped humming, held his breath, and took in a lung full of air. "What the hell is this?" he said, to no one in particular. "Damn!" He read incredulously from the change order request from Hyde Construction:

> "Dear Mr Judge, The enclosed costs were incurred in connection with our furnishing and installing a pressure regulator and relief valve ordered by the building inspector as required by the plumbing code. Please issue a change order at your earliest convenience. Thank you. Sincerely, George Hyde, President."

Enclosed was an itemized bill from W.C. Plumbers for $3523.30 and, after Hyde's customary add-ons and mark-ups for supervision, profit, overhead, and bond premium, the total was $4703.00. Ivor's great day had just plummeted down the soil pipe.

Hyde Construction Company was building the warehouse and offices for ABC Warehouse Company. The job had been going along fairly smoothly, with only the occasional glitch, and was due for completion in about two weeks. Ivor knew that ABC's president, Allen Brady, would undoubtedly blow his stack when he became aware of this totally unanticipated extra. He was still smarting from the cracked floor incident.

Ivor's immediate knee jerk response to this change order request was to get George Hyde on the phone and kick some butt. "George, what the hell's this Mickey Mouse change order?" Hyde accurately sensed that the architect was up-tight and expected he'd be unreasonable. He'd have to employ some sweet talk.

George calmly elucidated, "The plumbing inspector insisted we had to install a pressure regulator and a relief valve because the water pressure is 135 pounds per square inch. The regulator's expensive because we have a two inch water service. The plumber had to revise the piping to get it and the relief valve installed. It was required by code. We had no choice."

"You mean the plumbing work is all done?" The architect couldn't believe that Hyde had gone ahead with such a costly change without prior approval.

"Yeah, the plumber was lucky to find a 2-inch pressure regulator so quickly, and his men have installed it. The plumbing inspector has already reinspected and approved it. We're in great shape for completion on time." George was upbeat, hoping to overcome Ivor's pique.

"Dammit, George, why didn't you get my or the owner's approval first?" Ivor was obviously not impressed.

George's patience was rapidly waning. He chose his words carefully. "Look, Ivor, the plumbing inspector had red tagged the job and we had to get right on it. We couldn't get our final inspections without doing it. If the building isn't completed in less than two weeks, we'll have hefty liquidated damages to pay. If we sat around waiting for paperwork and approvals we'd never get it done."

Ivor dug his heels in, "Well, I can't approve this change order." He was adamant. "The specifications require all work to be in accordance with the building code and if the code requires a pressure regulator and relief valve then the plumber should have included them in the bid. Tough luck, George!"

George was starting to lose his cool. "Look, Ivor, you know damn well that regulator wasn't specified and it's a proper extra. The owner hasn't paid for this work. It wasn't in the bid. We've kept the costs to an absolute minimum and we gotta be paid."

Ivor, intemperately, snapped back, "Well, dammit, George, I'm not going to approve it. That's all there is to it. You and the plumber can go hustle someone else!"

George angrily slammed down the phone, his blood pressure way over the top. He strode into his chief estimator's office looking for sympathy. He was livid. "Dammit all, we're caught in the middle again. We gotta pay the plumber and the damn architect won't approve our bill to the owner. Someday Ivor's gonna get his and he'll have earned it! He expects us to pay for his mistakes! Well, I'm not going to do it! I'll fight this one all the way!"

* * *

Ivor picked up Hyde's change order request and walked into his partner Leo King's office. He threw himself angrily into a chair and said, "Leo, look at this crap. Hyde's going off the deep end again. He's conspiring with his plumbing sub to milk the job. I told him where to get off!"

Leo said, "Cool down, Ivor. Are you sure we're right? Maybe we should have specified the pressure regulator and relief valve. If we've made a mistake it isn't fair to stick the contractor with it."

Ivor rationalized his position. "The specs require the plumber to do all the work necessary for a complete job, all in compliance with the building code. He should have known what the code required."

Leo replied, "Have you called Ell about this?" Ell Street of Street & Rhodes, Mechanical Engineers, was their plumbing design consultant on the ABC job.

"No, I haven't talked to him yet. I plan to call him."

"Maybe he can explain why he didn't specify the pressure regulator and relief valve in the first place. Perhaps the plumbing inspector's wrong and the work's not even required."

That wouldn't be too helpful, thought Ivor, then Hyde would want to be paid again for removing it. Ivor returned to his office to phone Ell Street.

Getting the mechanical engineer on the line, he briefed him. "Ell, the plumbing inspector red tagged the ABC Warehouse job and required the plumber to install a pressure regulator and a relief valve. The general contractor wants an extra of almost five grand! How come it wasn't on the drawings or in the specs?"

Ell explained, "The inspector must think that the pressure is over 100 pounds per square inch. We were under the impression that the pressure was less than 100. Let me check our engineering notes." He left the phone for a minute to dig out his ABC Warehouse file. "I'm back. Here are my notes on water pressure. You instructed me in a letter I have here that the client had informed you that the pressure was 85 psi. We used that in our design. The plumbing code requires a pressure regulator whenever the pressure exceeds 100 psi. We'd also have to install a relief valve to protect the water heater."

"Well, Ell, shouldn't the plumber have included these items anyway if they're required by code? Shouldn't they have included all incidental parts to comprise a complete and integrated system in accordance with the Uniform Plumbing Code?"

"No, Ivor. The plumbing contractor wouldn't have any way of knowing the water pressure when they were bidding the job. They would normally assume that we would know whether or not a pressure regulator was needed. Besides, these are a lot more than mere incidental parts. You can tell by the size of the bill. If we knew that the pressure was over 100 psi we would have shown the regulator and relief valve on the drawings and specified them."

Ivor asked, "Then you think the contractor is justified in charging for this work?"

"Hell yes!"

"What do you think about the price? It seems pretty high to me!"

"It seems reasonable to me."

"Oh."

* * *

Ivor then called ABC Warehouse and got Allen Brady on the line. "Allen, I've got to come over and talk to you about something. How about this afternoon around two o'clock?"

Allen said, "Okay, two o'clock it is. See you then." Allen was afraid to ask the topic of Ivor's concern. He was apprehensive. Ivor sounded worried. This is going to cost me money, thought Allen. I just know it. Damn!

* * *

The ABC Warehouse Company was in a seedy part of Pacoima's unplanned industrial district. The office was in a front corner of the cramped corrugated iron building that served as ABC's warehouse. The small cluttered office was walled off from the warehouse storage area with some old garage doors. They had a suspended electric space heater but no air conditioning. A window provided light and ventilation. Four old desks were pushed together and Allen shared it with two of his clerical employees and his sales manager. Steel files and cardboard transfer cases filled the rest of the space. They had long since outgrown the building. In spite of their cramped conditions they were fairly efficient and their inter-office communication was immediate.

Carl Daly, Allen's loading dock manager, had a stand-up desk adjacent to the main loading doors. His telephone shared the wall space with several clipboards hung on brass hooks and approximately 7,000 scribbled notes impaled on nails.

Business was brisk and ABC Warehouse Company was more than ready for the new premises, properly designed for a modern warehouse business. They were all dreaming about the big move into the new building in less than two weeks.

Ivor arrived at ABC with time to spare but used it all up looking for a place to park. By the time he walked from his car to ABC's office, he was 10 minutes late. When he came into the office, Allen was consulting his watch for the seventeenth time. After the usual exchange of civilities, Allen asked, "Well, Ivor, what's up?"

Ivor, solemnly removing papers from his slim leather case, cleared his throat and said, "Allen, we have a slight problem on the job."

Allen, preparing for the worst, stiffened in his chair and held his breath. He exhaled slowly. His nearby employees, sensing their boss' apprehension, buried their noses deeper in their work. They clearly wanted to avoid any involvement. Nevertheless they had their ears cocked.

Ivor got it all out in a rush. "Hyde's plumber had to install a pressure regulator and relief valve. Requested by the plumbing inspector. Required by code. Plumbing work's all done and approved. It won't hold up completion."

Allen, obviously relieved, replied, "Well, that sounds great. I don't know what it is, but if we need it, it should be put in. What's the problem, Ivor?"

Ivor blurted out, "Hyde Construction has sent a change order request for $4703.00."

Allen, shocked, repeated the amount slowly and deliberately, "Four Thousand Seven Hundred and Three Dollars ? That's ridiculous!" Then, recovering, "Why should we have to pay extra for something required by code? I thought the whole job had to be built according to code! What the hell's going on here? Can't you control Hyde?"

"I've talked to George and he feels pretty strongly that he has the money coming."

"I thought that extras had to be approved before the work was done. I never approved this!"

"The plumber should have gotten approval first but he didn't. Now we have to live with it. I've consulted with our plumbing design consultant, Ell Street, and he said the work had to be done and the price sounded about right to him. The work is clearly not in Hyde's contract, so he must be paid."

Allen then asked, "Why is the work an extra? Wasn't the work shown in the plans and specs?"

Ivor was surprised that Allen had zeroed in so quickly to the crux of the matter. He replied, "No, Allen. The pressure regulator wasn't in the documents."

"Why not? Why should I have to pay extra for something you didn't put in the documents?"

"When I was talking to Ell Street this morning he reminded me that I had written him a letter last January reporting the water pressure as 85 pounds per square inch. That amount of pressure wouldn't have required the pressure regulator. A regulator is needed only when the pressure is over 100. After talking to him I looked into my own file notes to find out where I got the 85 psi figure. That's the value you gave me last January when we were working on the design. Where did you get the 85 pound figure?"

Allen, listening attentively, answered, "Oh, I remember now. You asked me for the water pressure at the new site so I asked Carl Daly to look into it. That's the number he came up with. I'll call him in here and he can tell us about it." He went to the office door to the warehouse and asked one of the dockhands to send Carl in, pronto.

Carl showed up less than a minute later and stood waiting expectantly. Allen told him to pull up a chair. "Carl, do you remember checking on the water pressure at the new site? It would have been last January."

Carl was glad to help, "Oh, sure. I talked to the water company's field engineer. He was very helpful and cooperative. He said that the water pressure varied from 80 to 90 pounds per square inch and that it should be sufficient for anything we'd want to do."

Allen was smiling. "That's fine, Carl. That's what we needed to know."

Carl then helpfully added, "And he said that the water company would be making some improvements in their mains sometime in the future. And he said they expected our pressure would go up to around 130 to 140 pounds. I figured, so much the better. Then we would never have any water pressure problems." Carl was beaming. He liked to be helpful.

Ivor and Allen sat and looked at each other incredulously. They both knew there was nothing else to discuss.

* * *

Points of Law 10

ABC Warehouse II
Who Pays for the Pressure Regulator?

10.1 The Building Code Requirement

Neither the drawings nor the specifications specifically call for installation of a pressure regulator on the ABC Warehouse job. WC Plumbers installed it because it was required by the city inspector. The inspector required the pressure regulator because the building code requires such devices at projects where water pressure will exceed 100 psi.

The contract documents require Hyde Construction to comply with all building code requirements. That provision is also incorporated into the subcontract between Hyde and WC Plumbers. Therefore the contract requires the installation of the pressure regulator even though it's not specifically called out in the drawings or specifications.

Such being the case, neither WC Plumbers nor Hyde Construction would be entitled to receive extra compensation for installing the pressure regulator except for the fact that Carl Daly, Allen's manager, told architect Ivor Judge that the water pressure was 85 psi. Thus, the failure to spec the pressure regulator was caused by misinformation given by the owner to the architect. Accordingly, the architect was justified in omitting the pressure regulator from the contract documents.

10.2 Implied Promises

Every contract includes an implied promise (legalese: implied covenant) that a contracting party will not do anything to interfere with the ability of the other party to perform and obtain the benefits of the contract.

> (An implied promise is a promise that is not expressly written down or stated in the contract but is implied from the very existence of the contract.)

ABC Warehouse, although unintentionally, violated this implied promise when it gave incorrect information that the water pressure would be less than 100 pounds. Giving the false information was therefore a breach of contract that caused Hyde (through its subcontractor CD Plumbing) to omit installation of the pressure regulator required by the building code. This breach of contract caused Hyde (through its subcontractor) to incur the extra expense of installing the pressure regulator. As a result, Hyde is entitled to recover, as damage for breach of contract, the reasonable cost of installing the pressure regulator.

10.3 The Subcontractor

What of CD Plumbing? Since CD Plumbing has no contract with ABC Warehouse, its claim to be paid for the pressure regulator cannot be based on a breach of contract.

10.4 Unjust Enrichment

Here we must examine an ancient legal doctrine known as quantum meruit ("that which he deserves"). This forbidding Latin term embodies the concept that a person who provides valuable goods and services should normally be paid for it, to avoid unjust enrichment.

When Hyde Construction gets paid for the pressure regulator it will be unjustly enriched unless the value of that performance is passed on to CD Plumbing. (Hyde is entitled to deduct its reasonable overhead and profit.)

10.5 Straight Thinking vs. Legal Thinking

The reader may well ask, "Why do we have to go through implied covenants and quantum meruit? It's a simple problem. It's the owner's fault that the pressure regulator was not in the plans, and why should the owner get something without paying for it anyway?"

It's true that we know intuitively how this case should come out before we put the law to it. Why not rely on a rough sense of justice to resolve such legal problems?

The idea is attractive, but it just doesn't work in practice. If adverse parties come to agreement by applying the rough sense of justice, as they often do, then there's no dispute to be submitted to lawyers and courts for resolution. When the rough sense of justice breaks down we resort to formal rules of law that can be applied in a uniform and predictable manner. Under the rule of law, like cases have like results. If a judge does make an error in law, it can be corrected by a Court of Appeal. As the experience of nations with emerging economies makes clear, the existence of a flourishing and vigorous economy depends on rules of law.

11

Davren Effingwell
dba
Economy Home Improvement Company I
Who Protects the Owner? Conflicts of Interest

Anita and Brad Coyle dearly treasured their comfortable contemporary style home nestled in the eucalyptus trees of the La Canada foothills. They'd worked hard for it. The nearby Glendale Freeway brought them conveniently to their jobs. Her law office, specializing in family law matters, was in Old Pasadena. His work place was in a Downtown Los Angeles office tower. Brad was a certified public accountant serving the music recording industry.

At first the house had seemed immense to them, but now, after 8 years, it was closing in on them. Their social and professional activities required more space. After much consideration of selling out and buying something more spacious they decided instead to build an addition. There was a suitable area at the rear between the swimming pool and their east property line. That part of the property sloped off to the south and had a spectacular view of the valley with the city lights beyond.

Anita and Brad visualized all kinds of wonderful uses for the addition. The wish list grew by the day until finally it encompassed a spacious multi-use pavilion with a bar, a powder room, and a guest dressing room for the pool. The large space would have a stone fireplace, bookcases, and plenty of other storage. There'd be room for music and television and desk space for both of them in an exhilarating, comfortable, and uncrowded environment. They'd use the new facilities for entertaining friends as well as doing professional work lugged home from their offices. There'd be a landscaped covered patio overlooking the swimming pool and the distant city view.

They relished talking and dreaming about it. They also thought of adding a solar heating system for the swimming pool. They'd keep the old gas fired heater merely as a back-up system. Brad figured that the reduction of their monstrous gas bills would help pay for the improvements.

* * *

By chance, they discovered that one of their nearby neighbors, Davren Effingwell, was a home remodeling contractor. What extraordinary luck! They surmised that he must be pretty successful judging by his impressive house and his late model Mercedes. One of the other neighbors told them that Dave headed his own business, Economy Home Improvement Company.

So they invited Dave to come over one evening for cocktails and to pick his brain about construction of the proposed addition. They hit it off quite well, the Coyles never before having been involved in such fascinating discussions with a building professional. Dave was charming and witty and they felt that he was extraordinarily conversant in the arcane lore of construction, building costs, planning, building regulations, and finance. He was a real expert in luxurious home improvements. They were absolutely enchanted with him.

Dave was similarly intrigued with his prospective customers, a lawyer and an accountant, their cultured tastes, and their obvious ability to pay for a substantial, high quality, home improvement. After that first meeting, the threesome went out to dinner several times to continue discussing the project. Brad and Anita always paid the bill.

Anita and Brad spent hours devising and embellishing their vision of the ideal expansion of their beloved home. They compiled reams of notes and magazine photographs. They described to Dave all the marvelous features they wanted included in the addition. Dave encouraged them by his wondrously tantalizing descriptions of remarkable new materials and trendy decorating ideas.

Having their rapt attention, Dave pontificated, "Based on my extensive past experience, the addition will have to be about 800 square feet to accommodate all of your requirements." Anita and Brad hung onto his every word.

Brad, suddenly concerned with the economics of the situation, cleared his throat, and was almost afraid to ask, "Approximately how much do you think this is going to cost, Dave?"

Dave, hastily scribbling notes and figures, performing complex operations with his pocket calculator, and, in his most sincere manner, replied, "Well, this quality of construction would cost about $105.00 per square foot. A total of about $80,000 or $85,000."

Not detecting any resistance, he added, "But we should allow another $10,000 or $15,000 for the patio and landscaping. We can refine the figures later."

Brad, exchanging concerned glances with Anita, replied, "Well, that's pretty steep, Dave, we'll have to think about it. We'll get back to you."

Dave warned, "Well, don't take too long. Costs are going up every day, you know. I heard today that lumber is going up."

"Thanks for telling us, Dave."

* * *

In the ensuing week Anita and Brad concluded that you only live once so they might as well go for it. That's what they really wanted and they'd dig up the money one way or another. They called Dave for another dinner and told him they'd like to go ahead. Brad asked Dave if the sloping site would cause any problems.

Dave said, "No problem, Brad. Although your site is extremely difficult, we can easily overcome it by using cantilevered wood floor construction. It's very simple when you know what you're doing."

Brad said, "That sounds like a great idea." Anita nodded her agreement.

Dave added, "We'll use flat roof construction with a built-up roofing membrane. That's the logical place for the solar panel array for the swimming pool heating system."

"Dave, you think of everything," said Anita.

Brad and Anita were both mightily impressed with Dave's knowledge, ingenuity, and sincerity.

Brad then inquired, "Dave, do you know a good architect we could work with?" He'd been assiduously taking notes and was ready to write down Dave's nomination of a suitable architect.

"You don't need an architect," said Dave. "Architects charge too much. Probably would charge 15% of the construction cost. That'd add about 15 grand to the job. A complete waste. I have a talented free-lance draftsman that can draw up the plans just as good. I send him a lot of work so he gives me a good price. It'll cost only about $500."

Anita asked, "But, Dave, don't architects examine the construction as it's being built to see if everything's being built properly?"

Dave said, "Don't worry, Anita, I'll be there for that. I'll be looking out for your interests. Architects just cause trouble, run the costs up, and slow the job down. Every time I worked with an architect, it cost me money."

Anita replied, "Whatever you recommend, Dave. We just want to make sure our addition is built properly. It's very important to us."

Dave insisted, "There's no problem, Anita. Nothing to worry about. This is a simple residential addition. I'm a state licensed general contractor. Bonded and insured. I've built hundreds of 'em. Never have any trouble. I deal with the same reliable licensed subcontractors year after year. I use only the very best of materials and workers. All my people are skilled artisans, Old World craftsmen. If there's an architect in on this, count me out."

Brad quickly assured him, "We want you, Dave. We trust you. Your draftsman sounds great. Whatever you recommend is all right with us."

So they asked Dave to go ahead with the plans. Dave requested a $500 advance to pay the draftsman. Dave told them they might as well start getting their financing lined up.

Brad explained, "We're planning to refinance the property. We have a pretty good equity built up. I'm sure we can borrow all we need."

After Dave left, Brad told Anita they'd try for about $15,000 extra for furniture and a great vacation. Brad felt that they could easily borrow $300,000. This would yield $185,000 to pay off the present loan balance and the new loan costs, $100,000 for the addition, and $15,000 for furniture and the vacation. They were both elated.

* * *

Dave, in his office next morning, was certain that he had the Coyle deal practically sewed up. He dialed Homer Farkleberry who draws plans at home afternoons. Nights he drives a taxi. Homer had taken drafting courses in high school and enjoyed working with his hands. It was like a hobby and the extra money was handy.

Dave told him to run over to the Coyle house and get a few measurements of the end of the house and the pool location. He read Homer the list of the Coyles' requirements and instructed him to go ahead with a floor plan sketch.

Dave said, "I got you $300 for this one, Homer, so do a good job for me. No more stupid errors like the last time."

"Don't worry about me, Mister Effingwell. I'll be careful. That was the carpenter's fault. He should've known better. Remember, the building department engineers will do a thorough job of checking my drawings."

* * *

Dave went to the Coyles' home for dinner that Saturday night and he brought the sketch. After a considerable amount of discussion and numerous minor adjustments, enhancements, and refinements, Anita and Brad were satisfied and told Dave to go ahead.

Dave explained, "Okay, we'll finish the construction drawings, start taking sub-bids, and apply for the building permit right away. I'll need a check for $1200 to pay the plan check and permit fees. By the way, Brad, how're you coming with the financing?"

Brad proudly reported, "The loan's been approved and we can have access to the funds any time we want." He wrote out and handed the $1200 check to Dave.

* * *

Two weeks later, Dave met the Coyles right after they got home from work. During cocktails Dave announced, "I have fantastic news. We have the building permit and we're ready to break ground next Monday."

Anita and Brad were overjoyed. They toasted Dave and their deepening friendship.

Dave took two copies of a contract out of his portfolio and, smiling warmly, laid them before the Coyles who were seated together on the sofa. Their quick reading revealed the contract price, $116,700. Anita and Brad looked up simultaneously, both shocked by the stark reality of the unexpectedly large number.

Brad, with a hurt and concerned expression, said, "This is considerably higher than we've talked about, Dave. What gives?"

Dave, still smiling, earnestly explained, "This is a final all-inclusive price, Brad, including the patio and landscaping, the solar pool heating, and the building permit. Everything's included. The price is up a little bit on account of some ridiculous unexpected building department requirements and, as you already know, lumber's going up."

Anita was perfunctorily reviewing the contract as Brad and Dave continued their discussion. She was reading and half-listening. She noted that it was a standard form contract published by the Home Addition Contractors Association. It was a single sheet of paper printed on both sides with several blanks that were filled in.

Brad said, "Dave, if we go ahead with this, we've got to have some financial safeguards. Both for your benefit and ours."

Dave wholeheartedly supported this sentiment, "I couldn't agree with you more, Brad. You're so right."

Brad continued, "In the first place, I think we ought to have a surety bond for faithful performance of the contract and for payment of labor and materials."

Dave patiently explained, "Brad, I can see where you're coming from but I'm going to have to disagree with you on this. Bonds are expensive and they do no good. The only way you can collect any money from a surety is by filing a lawsuit. A bond is nothing but a ticket to a courtroom. Your best bet is to have a top-notch contractor like me and save the cost of a bond. I'll be there to protect your interests. What could go wrong? And you'll save around $1500."

"Well, Dave, if you feel that strongly about it, we can do without it. Right, Anita?"

"If you think it's okay, Brad. After all, we'll have Dave protecting us. And we'll save $1500."

Brad brought up another idea. "I think we ought to have a liquidated damages clause that would reduce the contract price some amount, like $200 per day, for late completion after a reasonable completion time such as, say, 120 days."

"Well, that won't be necessary, Brad. My subcontractors would be insulted. We've never been late with completion of any of our jobs. I'll be here every day expediting the job. We'll easily finish in less than 120 days. I'll keep the materials flowing and the subcontractors moving."

Brad turned to Anita and asked, "What do you think, dear?"

"Dave seems capable of keeping the job moving, Brad. I think it'll be all right. After all, he is an experienced home improvement contractor, licensed and bonded by the state."

Brad gave in on the liquidated damages idea, but felt that there should be some control on the disbursement of funds. He continued, "Assuming that we want to proceed on this contract, we must have a businesslike payment system. Do you agree, Dave?"

Dave immediately agreed, "Oh, yes, Brad. Right. By all means. That's an absolute necessity. I always insist on it."

Brad continued, "How about a system whereby we pay you 90 percent of the value of the work in place each month and the final 10 percent after the lien period has expired? I could audit the bills each month." Brad had gotten this idea from a book on construction administration that he found in the firm's library.

Dave had a ready answer, "That system's been used in the past, Brad. But, it's not flexible and doesn't reflect the needs of the modern construction industry. It's archaic. We don't use it any more."

Brad, crestfallen, asked, "Then, what system do you recommend, Dave?"

"We prefer to use a joint control system. We generally utilize the services of Fidelity Building Control. Your lender will advance the funds directly to FBC who will disburse all funds only upon receipt of my signed vouchers. I'll issue vouchers to pay all my subs and suppliers and FBC will issue them a check upon receipt of proper labor and material releases. I won't receive a cent of my profit and overhead until the very end when the job is completed."

"That sounds all right to me, Dave. What do you think, Anita?"

"I don't see how it could go wrong."

Dave continued, "FBC has a field inspector that visits the job periodically to make sure that the work's progressing proportionate to disbursements. And don't forget, Brad, the lender should be instructed to advance the funds directly to your control account at FBC. I never touch the money. The usual system is to transfer funds as construction progresses, 20% when construction starts, 20% when concrete foundations are completed, 20% when wood framing's completed, 20% when the roof's completed, and the remaining 20% when the drywall's completed."

Brad brought the session to a close by saying, "This is about as far as we can go tonight, Dave. Anita and I have to decide if we can afford to spend $116,700. That's a lot more than we thought it would be and our lender hasn't approved that large a loan. With the $500 we've already paid you for the plans the total is actually $117,200. Isn't there any way you can lower the price? That's a helluva lot of money."

"Look, Brad, I've worked this price down to the bare bones. I've beaten all the subs down to rock bottom. I've chopped my overhead and profit to practically nothing and you don't seem to appreciate it. If you and Anita can't afford it, just let me know and I won't waste any more time on it." He picked up his portfolio, arose, and quickly stalked off, letting himself out the front door. When he got outside, he said to himself, "You're brilliant, Dave. They're as good as signed up."

Brad and Anita were still sitting there on the sofa. She said, "What'll we do now, Brad? I hope we haven't offended Dave. I'd dearly love to build the addition. Perhaps we'll just have to cut something out."

"Maybe we won't have to. Tomorrow, I'll go back to the bank and see if we can increase the loan by $17,000. Then we could sign the contract without cutting anything out."

"Oh, Brad, wouldn't that be wonderful?"

The following morning Brad went back to the Real Estate Loan Department of Certified Savings Bank to ask for an increase in the loan. The loan officer said okay but that interest rates were rising and if they changed the amount of the approved loan the interest rate would have to be adjusted to the new rate. Brad went back to his office to absorb this new information. He concluded that it would be better to scrounge the funds out of some other source than to sign up for 20 years at the higher rate.

He telephoned Anita and told her about this disheartening new development. He detected her profound disappointment and suggested that they could go ahead with the addition anyway by using the funds to be set aside for furniture and the vacation. They would just have to find other money for those items. Maybe put it on the credit cards. She sounded a little more optimistic when she replied, "Okay, honey."

Brad then dialed Dave and said, "Can you come over tonight, Dave? We'd like to go ahead with the contract."

"Sure. I'll be there about 7 o'clock." Dave felt things were looking up. He again congratulated himself. His sales technique was deadly.

That evening, before Dave arrived, Anita told Brad that she'd gone over the contract. She'd never seen a construction contract before but she didn't see anything objectionable in it. Brad said that it looked okay to him. The blanks had been filled in to reflect the joint control procedure described by Dave. None of Brad's suggestions had been accepted so the contract was all right just the way Dave had proposed it in the first place. When Dave arrived, Brad suggested they get right down to business.

Brad started by saying, "Anita and I have signed both copies of your contract. Now, you can sign one copy and give it to us."

"Okay." Dave signed his name with a flourish. "We'll start work on Monday. You'll move in less than 120 days later."

Brad pointed out, "The 120 days completion time isn't mentioned in the contract, Dave."

"No need to mention it. You have my word. We'll finish on time. We always do." They shook hands and Dave left with his copy of the signed contract.

So the work started the following Monday. Anita and Brad were delighted, elated, and excited. The project went ahead and Dave visited the job several times a week. Anita had told Dave to help himself to beer or cokes in the refrigerator and to use the telephone and toilet whenever needed.

Whenever the lender advanced a 20 percent progress disbursement to the joint control, they sent a notice to the Coyles. Brad had started a set of books on the project. He asked the FBC for a notice whenever a voucher was paid but they refused, explaining they would send a complete accounting at the end.
Although they were both at work every day, the Coyles would come home at night and marvel at their dream materializing into tangible reality. Some days nothing happened and they'd call Dave and he'd tell them everything was on schedule. Not to worry.

Although the job dragged, seemingly interminably, it was finished, finally, only 63 days late. When Brad complained, Dave explained that the 120 days were working days, "You don't expect us to work on Saturdays, Sundays, and holidays, do you?" The Coyles were disappointed but the result looked okay to them so they let it slide. They didn't want to disturb their cordial relationship with Dave.

* * *

Anita and Brad moved into the addition although they didn't have much in the way of furnishings. They bought a few basic necessities on their credit cards and would acquire the rest gradually as they were able to afford it.

They would really only use the addition nights and weekends on account of their busy work schedules. On the first Saturday night after completion they invited a number of their friends for cocktails in front of the new fireplace. They also invited Dave so he could share in the glory. Thus they helped him in promoting similar projects for the homes of some of their guests. Brad and Anita were proud and euphoric with their new entertainment pavillion.

But then they started noticing odds and ends of peculiar conditions. For example, they noticed that whenever someone walked down the middle of the main room, about 8 feet in front of the fireplace, a glass candy dish on a coffee table would bounce up and down, it's lid noisily rattling. The floor seemed uncommonly springy. It was embarrassing when they had guests. When they mentioned it to Dave he explained that there's always a little flexibility in wooden construction, "That's why this type of building does so well in earthquakes. Nothing to worry about."

It was now well into the heating season and they were always cold in the new room. And the existing house was getting harder to heat. Dave said he'd send the heating contractor to look at it. It was probably only a faulty thermostat connection or something minor. They'd take care of it. Not to worry.

The monumental catastrophe didn't occur until about two weeks after completion. It was raining heavily all afternoon and evening and on into the night. Anita had retired early and Brad was working late at his desk on an important financial statement. A huge drop of water plopped on the papers before him. He jumped back and saw another drop splash onto the papers. He quickly grabbed the wastebasket, put it on his desk to catch the drips, and mopped up his desk and papers with his handkerchief. Then he noticed another dribble of a few drops on the floor about 3 feet away from his desk. He put a large ceramic ashtray under it. He thought, I'll call Dave. Then, another drip site appeared. He ran to the bar, grabbed a martini shaker, and positioned it strategically under the new drip. Two more drips started in front of the fireplace. He shoved two large beer mugs under them. He dashed into the bedroom and aroused Anita.

He shouted, "Anita, the place is leaking like a sieve!" He sounded panicky.

Only half clad and barefooted, she sped back to the annex to see what on earth he was ranting about. By then drips were steadily falling in about six or eight new places. They ran to the kitchen and brought back pots and pans to position under the fresh leaks.

Then Anita shouted, "Brad, those small containers are filling up. You better empty them before they overflow." He was dashing about emptying and replacing vessels under old leaks and finding empty containers for new drips.

They had to move some furniture to make room for the containers. They spent the next two hours putting every sort of water-holding receptacle they could find under leaks. Nevertheless a lot of water landed on the floor between vessels. They had over a hundred containers of all sorts all over the room. They couldn't go to bed because they had to keep emptying the smaller vessels that were overflowing. They moved some of the lighter items of furniture back into the main part of the house to make room for more containers. They draped plastic sheets over their desks and bookcases. Anita was going around with a towel mopping up the floor between pots and pans. She was concerned that the hardwood flooring would be ruined.

Around 2 o'clock, Brad telephoned Dave and told him the place looked like Niagara Falls.

Dave sleepily replied, "I'll have the roofer go over in the morning to fix it. Probably a little fiber reinforced bituminous mastic is all that's needed. Don't worry about it, Brad. We'll fix it." He returned to bed and promptly went back to sleep.

Brad and Anita, discouraged and utterly exhausted, finally got to bed around 4 o'clock. Brad got up again around 7 and returned to the disaster zone to see if any containers needed emptying. Many were full and overflowing and the flooring boards were soaked, twisted, and buckled. He was shocked to see that the flooring had risen about a foot in the middle of the room. He was afraid to walk on the risen portion of the floor in case he'd cause any damage.

He phoned Dave and got him out of bed to come over to take a look. When Dave arrived about an hour later, after having breakfast, he said he'd call the hardwood-flooring subcontractor and have him come over and fix it. Nothing to worry about.

Brad asked him, "Whatever happened to the heating contractor who was going to look at the heating system?"

Dave replied, "Oh, didn't he come yet? I'll call him again. Not to worry. He'll fix it."

Brad persisted, "And what about the springy floor? It doesn't seem right that it should bounce so much just from normal walking across it! And now, look at that huge bulge in the middle!"
Dave said, "It's like I told you, it's perfectly normal. I'll call the subcontractors. Nothing to be concerned about." Dave walked home. He had important prospects to talk to today. Friends he'd met in front of Anita and Brad's fireplace.

After Dave left, Anita and Brad looked very grim and tense as they had their coffee and toast standing in the kitchen. They were so tired from the night's activities and lack of sleep that they both phoned their offices to cancel the day's appointments. The rain had stopped but the dripping lasted for some time before they could start removing and emptying the pots, pans, cans, and dishes. They finished removing the remaining furniture and mopped up the wet flooring. They were discouraged and exhausted.

* * *

Fair weather returned and the hardwood flooring settled down some but the buckling still showed. Brad and Anita were afraid to move anything back into the addition as they knew the roof would surely leak again in the next rainstorm. But they heard no more from Dave or his subs. They couldn't use the area anyway because it was always too cold. Brad called Dave several times a day but he was always out and never returned Brad's calls.

After about a week, Anita protested, "Brad, I'm getting sick of Dave and his procrastination and incredible excuses. We've got to do something about this ridiculous situation. We can't continue living like this. It's clear that Dave and his people aren't going to do anything."

"You're right. What do you suggest, Anita."

"One of my law school friends, Jerry King, went into construction law. I'll call him and see what he thinks."

King agreed to meet Brad and Anita at their home that evening. They showed Jerry the warped floor situation, told him about the faulty heating system, and demonstrated the springy floor phenomenon. They told him about the rainwater invasion and showed him colored snapshots of the scores of vessels catching rainwater. They had also compiled a list of about 30 or 40 other defects and anomalies that they had recently noticed. Mismatched door butts, some with screws missing, paint dings, runs, skips, and scrapes, nails not set and filled, a cracked lighting fixture lens, inoperative electrical outlets, drywall irregularities, chips in the bar sink enamel, fireplace log lighter won't ignite, deep scratches in a window pane, and numerous other similar imperfections.

They also told him about their contractor, Davren Effingwell dba Economy Home Improvement Company.

Jerry, taking notes, asked them, "Did you check out his references?"

Brad looked at Anita, then back to Jerry, "No, he's one of our neighbors. We thought he was a great guy. Salt of the earth. A friend. We didn't even ask him for references. We thought he'd be okay."

Jerry asked, "What does your architect say about these conditions?"

Brad said, "We don't have an architect. Dave said we could save $15,000."

The lawyer asked them, "Let me see your construction contract and any other documentation you have."

Brad gave him the Economy Home Improvement Company contract, the Fidelity Building Control joint control agreement, and the Certified Savings Bank loan disbursement schedule. He also handed him the construction drawings, the joint control final statement, and the building department inspection record. Anita had been taking snapshots every weekend so she handed Jerry a complete album of dated and titled prints.

Jerry suggested, "I should also review your insurance policies. And let me see the contractor's certificates of insurance."

Brad replied, "You can see our homeowners insurance policy, of course, Jerry, but what do you mean by the contractor's certificates of insurance?"

The construction lawyer explained, "The contractor normally submits them as proof of the existence of all forms of insurance required by the contract."

"Well, Dave didn't submit any certificates, but I'd think he'd have all the necessary insurance. I'll ask him."

Jerry then inquired, "Do you have a surety bond?"

Anita said, "Dave said we didn't need it. We saved $1500."

Jerry said, "We better move quickly to protect your interests. First, I'll notify Certified Savings Bank to hold any further disbursements to the joint control account."

Brad replied, "It's too late for that. They disbursed the final 20% to FBC several weeks ago. We've already made the first mortgage payment on the full amount."

Jerry said, "Well, then we'll notify FBC not to pay any more vouchers until we get this whole matter sorted out."

Brad replied, sadly, "All the vouchers have been paid and the balance in the account has been disbursed to Dave. You have FBC's final statement in those papers I just gave you."

Jerry asked, "Didn't you have any system of retainage?"

Brad said, "No. Dave said it was an outmoded system and not necessary with FBC controlling the funds."

Jerry said, "Then, we'll notify Economy Home Improvement Company demanding that they rectify all these defects within the next 10 days. Failing that, we'll avail of all our rights under the contract. Let's see the contract."

Brad dug it out of the pile of papers and handed it to him.

Jerry continued, "This standard HACA form contract isn't much of a contract. It's all one sided in favor of the contractor and the lender and is woefully incomplete. I'll study it later."

Anita then asked, "How about notifying the City Building Department? Their inspector was supposed to have been examining the construction as it went along and he signed his final approval on the inspection card."

Jerry explained, "You have no recourse to the City, Anita. The building inspector's function is to see that the building code is not violated, but there is no guarantee to you. The inspector only spot checks and if a violation was noted, it would be considered a transgression against the City, not against you. Now that they've signed off on the final inspection, there's little chance that they'd do anything for you."

Jerry continued, "If you could get them to reinspect and they found any code violations, they'd issue you a citation to comply with the building code."

Brad said, "But Jerry, I thought we were protecting ourselves by dealing with a state licensed general contractor and he said he dealt with only licensed subcontractors. Also, as I understand it, all contractors must post surety bonds with the state to get their licenses."

Jerry replied, "It would be a lot worse if they weren't licensed, but the state doesn't guarantee that the contractors are all competent nor that they'll abide by their contracts. The Contractors State Board might revoke their licenses but that won't help you any."

Anita interjected, "Can't we claim against Dave's license bond?"

"As far as the license bonds are concerned, they're too small to mean anything, especially if there are numerous claims against a contractor."

Anita persisted, "But what about making a claim against Fidelity Building Control? They had an inspector here about twice a month and they continued to pay the vouchers. They shouldn't have continued paying out the funds if the construction was faulty."

Jerry said, "I'm sure that if you review the joint control agreement you'll find that their inspections were not to assure compliance with the contract but only to monitor construction progress and the disbursements against the cost breakdown."

Brad asked, "What about Certified Savings Bank? I know they had an inspector here at least five times, once for each disbursement to FBC."

Jerry said, "Well, their inspector wasn't really looking at the quality of the construction. They just want to make sure that the funds they are advancing are being used to improve their security for the loan."

Anita then asked the crucial question that was also on top of Brad's mind, "Then who in hell was looking out for our interests?"

Jerry promptly replied, "You were."

* * *

Points of Law 11

DAVREN EFFINGWELL
dba
ECONOMY HOME IMPROVEMENT COMPANY I

The Dangers of Contracting with a Friend

11.1 Protecting the Interests of the Owner

Anita and Brad put their trust, faith, and confidence in their contractor and friend, Davren Effingwell, and looked to him to protect their interests. This could work if the contractor were honest, professional, careful, and qualified.

No owner would deal with a contractor known to be careless, unqualified, or dishonest. But as the story illustrates, sometimes it's hard to know just how reliable a contractor will be. What can an owner do to protect itself against the Effingwells of this world?

11.2 Strong Contract

A well-written contract gives legitimate protections to the owner. In fact, most of the provisions of a construction contract are there for the very purpose of protecting the owner.

11.3 Liquidated Damages

A construction contract should establish the date when the completed project will be available for use by the owner. Unexcused failure of the contractor to finish the project on time would be a breach of contract, which would entitle the owner to recover compensatory damages from the contractor. "Compensatory" damages are the amount of money calculated to compensate the owner for loss sustained because of the delay. Such damages would be measured by the rental value of the improvement during the period of the delay.

Since it might be difficult to establish the rental value of a recreation room, a liquidated damages clause would be appropriate. "Liquidated," in this sense, simply means "established in advance." Thus, Anita and Brad might have established $200 per day as the rate of damages to be paid them for failure to complete the job on time. In our story, the contract did not even include a completion date, much less a liquidated damages clause, so Anita and Brad are completely unprotected.

11.4 Retention

It is customary for owners to pay contractors monthly an amount determined by the percentage of completion. For example, if a $100,000 contract is 40% complete, the progress payment would be

$40,000 reduced by a 10% retention. The retention, $4,000, would be held by the owner as security for the contractor's proper performance of the work. The retention is usually paid about a month after the completion of the project so as to give the owner time to check for possible mechanics liens and defective construction. For example, if the contractor failed to provide a proper paint job, the owner would have $10,000 to resolve the problem.

11.5 Insurance

Owners usually insist that their contractors carry adequate liability insurance with owners named as additional insureds. This insurance protects the owner against claims for jobsite injuries and in some cases may also protect the owner against construction defects.

Water leaked on Brad's desk. The desk was damaged because of negligent construction by Effingwell. Effingwell is therefore liable for the damage and Effingwell's liability insurance company would pay for the damage.

11.6 Performance Bond

It's too bad that Anita and Brad did not insist on a performance bond. The premium for the bond would have cost them about $1,500, but it would have been worth it. When a surety company issues a bond, it guarantees that the contractor will perform properly. The bonding company would be responsible to fix the construction defects and pay for the damages.

11.7 Joint Control

A reliable joint control can be a good form of protection for an owner. Unfortunately, joint controls, like contractors, may or may not be reliable and professional. A qualified joint control will protect an owner against mechanics lien claims and against defective construction by making periodic inspections of the job and by exercising financial control to insure that the contractor does not get paid unless work has been advanced to the appropriate stage and unless all potential mechanics lien claims have been paid off.

11.8 Architect

Architects are professionally trained and qualified not only to design buildings but also to make sure that they are properly constructed. A qualified architect will usually be able to protect an owner against a dishonest or incompetent contractor.

11.9 Contractors Licenses

In most states contractors are required to be licensed and the state establishes minimum financial and experience criteria for licensure. In some states contractors are also required to procure license bonds, but such bonds are small and afford only a minimum of protection. Contractors sometimes advertise that they are "bonded" because they have license bonds. These license bonds, however, do not afford anywhere near the amount of protection that a performance bond would provide.

11.10 Inspections by the Building Department

Local building inspectors only look for building code violations. It is not their function to determine whether the contractor has complied with the requirements of the contract or whether the contractor is building a quality project.

11.11 Quality Contractor

As Anita and Brad learned, the owner's best protection is to employ an experienced, professional, qualified contractor of good reputation. With such a contractor, little other protection is needed. The most effective thing they could have done to protect their interests would have been to have carefully checked the contractor's financial and performance records.

12

ABC Warehouse VII
Faulty Contractor Selection and Contract Administration

Allen Brady, President of ABC Warehouse Company was counting his blessings. Business was steadily improving in the new premises and nothing serious had gone wrong with the new building lately. He had practically forgotten the retaining wall fiasco and the two roof collapses. His warehouse was getting close to capacity and he was now seriously considering embarking on the Phase II expansion program according to the Master Plan prepared by Judge & King, AIA, Architects.

After discussing his thoughts with Carl Daly, his warehouse manager, he decided to go ahead with the project. He phoned Ivor Judge and made an appointment for the next day.

He always felt safe when discussing development plans with Ivor. Ivor was highly conversant with the design, construction, and costs of industrial buildings and the administration of construction contracts. They discussed and carefully considered the pros and cons of various expansion options and finally decided to follow the original master plan. They'd proceed with the first part of Phase II consisting of a warehouse addition of 100 feet x 200 feet, a total of 20,000 square feet. The construction would be concrete tilt-up exterior wall panels with a timber-framed roof to match the existing warehouse. A new loading dock with five truck positions would be an extension of the existing dock. Ivor estimated that it would cost between $600,000 and $700,000 but promised to refine the estimate after preliminary drawings were approved. They discussed the architectural fee and Allen approved a full service contract including all of the usual contract administration services. At the end of their first meeting, they were both optimistic and looking forward to another successful project together.

* * *

A few weeks later, near the end of August, ABC's sales manager, Vernon Williams, told Allen that he was hot on the trail of a profitable deal if the new building could be available by May 1, 1995.

"Allen, these people will take 10,000 square feet on a 5-year contract with an option for another 10,000 square feet in 2 years. It's Takahashi Trading Company. They import and distribute computer components. We'll receive their shipments, reship to their dealers, and keep their inventory. We'll need to hire 5 more people to service their inventory." Vernon was flushed with enthusiasm.

"Hell, yes! We can do it. That sounds great! Is there anything else we'll have to do?"

"Yeah! We'll have to send them a fat bill every month."

"Well, Vern, you've certainly paid your way this month. I'll call Ivor and see if we can make it by May 1st."

Allen picked up the phone and got Ivor on the line. He told Ivor about the Takahashi opportunity and said, "Look, Ivor, we'll need the building completed for sure by May 1st. Can this be done?"

Ivor said he'd call back in a few minutes. He hung up and, referring to his desk calendar, started roughing out a schedule on a pad of paper.

> We started design on August 1, 1994
> Submit to building department, October 7
> Out to bid, November 1
> Bids due, November 18
> Sign contract, December 1
> Start construction, December 5
> Completion, May 1, 1995

He counted the days on the calendar and found that would give 147 calendar days for construction, more than enough time for a competent contractor.

He then called Allen back and reported, "It looks possible to me if everything goes without a hitch."

"Great, that's what we want to do. This is extremely important to us, Ivor. We should have a provision in the building contract that penalizes the contractor if the May 1st completion date isn't met. We'll want a $500,000 penalty."

"No, Allen, hold on. That'd be completely unreasonable. It'd scare away all the bidders. Besides, you'd never be able to collect it anyway. The amount should bear a reasonable relationship to your actual damages. We can include a liquidated damages clause of something like $20,000 plus $1000 a day, and move the completion date up to April 15th. That would give you a cushion of 15 days and yet give the contractor enough time to build the project and sufficient incentive to meet the deadline."

"That sounds okay to me. The main thing is I want the building available to us on May 1st without fail. We can't risk losing the Takahashi Trading Company account."

After hanging up, Allen instructed Vern to go ahead, sew up the Takahashi deal, and get them to sign a contract. Allen and Vern went off to lunch to discuss the details and hoist a few in celebration of their recent business successes.

After lunch Allen started making plans for recruiting the five people he would need to service the new Takahashi business.

* * *

During design, Ivor and his structural engineering consultant, Myles Nolan, had a number of discussions about the main structural components of the building. Each of the site-cast reinforced concrete tilt-up panels would be 20 feet wide by 30 feet high to match the existing warehouse building. The panels would be 9 inches thick. Each concrete panel would weigh approximately 67,500 pounds. Myles pointed out the necessity of specifying temporary bracing to hold the raised panels in place until the connecting columns were poured and the roof framing in place. The large laminated wood roof beams would also have to be temporarily braced as each was hoisted into place. This would be to resist the possibility of collapse during construction in case of horizontal loads imposed by earthquake or winds. Ivor agreed and made a note to make sure that the specifications adequately covered these crucial points.

* * *

Around the beginning of October, Ivor updated Allen on their progress, saying, "We're right on schedule, Allen. We'll be filing with the building department next Friday, October 7th.

"Now we've got to start firming up the building contract. I presume you'll want to go ahead with Hyde Construction and negotiate a contract with them. I could arrange to have George Hyde come in here for a preliminary meeting."

Allen wasn't too sure about this idea. "No, Ivor, I'd feel better about going out to bid with a selected group of contractors. George might get the idea he could charge whatever he wished if there was no competition."

"George Hyde is a perfectly good contractor and he did an excellent job on the original building. And, don't forget, he was low bidder."

"No, Ivor, I want competition. I want the lowest price."

"Okay, Allen, whatever you say. I'll start assembling a list of bidders for your consideration."

The next day, Ivor faxed a list of eight general contractors to Allen with instructions to reduce the list to five. Allen faxed back his selected list:

> Advanced Contractors
> Certified Constructors
> Fidelity Builders, Incorporated
> Hyde Construction Company
> Industrial Erections

Ivor contacted the five firms and confirmed their willingness to bid the job.

The building department approved the plans for issuance of the building permit and the required design corrections were incorporated into the bidding documents. The job went out to bid on November 1st to the five invited contractors, with bids due on Friday, November 18th. Hyde Construction proved to be low at $646,000, $12,000 under the second bidder.

Allen instructed Ivor to prepare the contract. Ivor reminded Allen that he should alert ABC's lawyer, Philip Quinn, that the contract would soon be sent to him for review. Allen replied, "Is that necessary, Ivor? He'll send a bill, won't he?"

Ivor's terse reply, "Yes, it's necessary, and yes, he'll send a bill."

"Oh. Just asking."

* * *

The next day was Saturday. Allen often came in on Saturday mornings to catch up on his deskwork and get prepared for the following week. The office was usually eerily quiet, with no one else present. It gave him a chance to do his creative thinking without interruption. He treasured these quiet times. Allen was thinking about the new building, the fantastic Takahashi contract, and the generally optimistic business prospects. A loud knocking at the front door suddenly disturbed his solitude. He went out through the general office into the lobby to let in the visitor. He didn't immediately recognize his caller until he spoke, "Allen, I can see you don't recognize me. It's been 25 years! I'm Quentin Quirk!"

Allen now recognized his high school classmate, although he'd put on a good fifty pounds of flab and lost most of his hair. But he looked extremely prosperous, with his tanned face and flashily expensive designer threads. He had a gold chain around his neck with a thick gold medallion at his throat. A single gold earring completed the picture. He'd previously glimpsed his impressive wheels through the glass entrance door out in the front parking area. A 1994 Mercedes-Benz 320SL. Allen surmised that he must be doing pretty well for himself.

"Come in, Quentin! Let's go to the coffee room and have a cup of coffee. What're you doing these days? I haven't heard from you since high school graduation."

Quentin, sitting, smiling, waiting for the coffee, reached in his jacket pocket, dug out a flamboyant looking oversized business card, and threw it on the table. Payless Industrial Builders, Quentin Quirk, President. "I'm a general contractor. Specializing in industrial construction. I've done quite well. Made and lost a coupla fortunes. Right now I'm on top of the heap."

Allen said, "I'm glad you dropped by. Good timing! Just yesterday, we received bids for a 20,000 square foot warehouse addition. The low bid was $646,000. Let's go back to my office and I'll show you the drawings and specifications and the bids." They took their coffee to Allen's office.

"Allen, let me take these plans home with me and I'll bring them back to you on Monday morning. I'll give you the advantage of my opinion. No charge."

"Great, Quentin, I'd really appreciate that. I need someone I can trust. Now, let's go out to lunch and discuss the good old days."

* * *

Allen arrived at the office early Monday morning. Quentin's shining mustard Mercedes was in the parking lot and he was sitting in the lobby. He had the plans and specifications sitting on top of his snazzy alligator attaché case. Allen invited him into his office, picked up his phone and quietly ordered coffee. He looked expectantly toward his guest who was busily opening the roll of plans and his brief case.

"What do you think of our new plans, Quentin?"

"Well, Allen, frankly I think your architect must be in cahoots with Hyde Construction. That bid of $646,000 is absolutely ridiculous! I'd be willing to build these plans and specs for $560,000 and I'd still make a fat profit. We could make some minor changes, value engineering we call it, and cut even more bucks out. They're really trying to screw you. Where'd you find these clowns?"

"I was wondering about Ivor and George. I thought they were a little too friendly for my liking. I'm really lucky you came along when you did."

"I've saved a lotta people a helluva lotta money, Allen. There's plenty of money in construction without cheating the customer. I'm not greedy. I've always been satisfied to limit myself to a fair profit."

"Quentin, give me your bid in writing and I'll send it to my architect so he can prepare the contract. Oh, by the way, can you start on December 5th?"

"Easy. No problem. We can easily work it in with our other projects."

"And can you complete by April 15, 1995? It's extremely urgent that we have timely completion because we really need the building. That's why the contract will have the liquidated damages provision that was described in the bidding instructions."

"Nothing to worry about, Allen. We're always working under a tight schedule. We work best under the pressure of a completion date. You've got nothing to worry about with Payless and Quirk on the job."

After Quentin left, Allen called Ivor.

* * *

"Ivor, I've got some incredible news. It's fantastic." He explained Quentin Quirk's fortuitous appearance after so many years. He ended up by telling about the written bid for $560,000, a saving of $86,000. "So, Ivor, I want you to prepare the contract for $560,000 in the name of Payless Industrial Builders, Quentin Quirk, President. Then send it to Phil Quinn for his review. Quirk is ready to start December 5th."

"Now, wait a minute, Allen. We've already had a competitive bidding procedure and we've got a legitimate low bidder."

"Well, Quirk's bid's a lot lower. $86,000, Ivor."

"Besides, I've never heard of Quirk or Payless. Have you investigated his experience and checked his references? Have you confirmed his financial condition? Have you checked him out with the State Contractor's Licensing Board? Is he bondable?"

"Look, Ivor, I don't need to check anything. I've known this guy for 25 years. We went to high school together. He's an extremely successful industrial building contractor. He's practically family."

"Well," Ivor insisted, "You should check him out anyway."

Allen said, "Okay." But he'd do nothing as he put Ivor's skepticism down to professional jealousy, and probably covering up for George Hyde. He thought, how could these pros be so wrong on the cost of a simple building? Quentin's probably right about them being in collusion.

Ivor, seeing that he was getting nowhere with Allen added, "Well, Allen, you should at least require Quirk to submit a performance and payment bond."

Allen asked, "How much would that cost?"

"It'll be around one percent of the contract sum, depending on his financial strength and other factors. It'll be about $5600."

"That's not too bad considering we're saving nearly a hundred grand. So, will you do as I ask? Prepare the contract for Quirk. And call your buddy Hyde. Tell him the deal's off. We're going with an honest contractor."

Ivor was bristling with anger but didn't say anything more. He was thinking, what an ass. He always thinks he can save a buck with his stupid short cuts. Someday it'll backfire on him. I hope he can afford it.

When Allen left, Ivor called Hyde Construction and got George on the line. He explained the whole deal and George was absolutely livid. He treated Ivor to an impressive demonstration of the latest industrial strength profanity in describing Allen's business principles and his questionable parentage. Ivor was sympathetic but it didn't help George's disposition any.

Later that same day Allen called Ivor and asked if the contract was ready for Quinn to review. Ivor asked, "Have you checked out Quirk?"

Allen wearily replied, "Yeah, he's okay". Ivor knew Allen hadn't and wouldn't do any checking.

Phil Quinn reviewed the contract. He made a few suggestions and asked Allen if he'd checked Quirk out. Allen said, "Yeah, he's okay. Don't worry. He's a great contractor. I've known him all my life."

* * *

The following day, Thursday, December 1st, right on Ivor's schedule, Quirk came in to Allen's office and signed the contract. It was for $560,000 plus $5600 for the bond, a total of $565,600. Allen reminded him, "Order the bond right away, Quentin. And get the job started as soon as you can. We hafta be done by April 15 next year. Without fail."

Quirk, radiating an air of confidence, replied, "Don't worry, Allen. You can relax, it's in my hands now."

* * *

As soon as Quentin left his office, Allen called Ivor. "Well, Ivor, the contract's signed now. Quentin starts next Monday, right on schedule. We've saved almost 100 grand and haven't lost a day's time."

Ivor was still apprehensive. He asked, "Before starting work he should submit his subcontractor list, schedule of values, construction schedule, insurance certificates, and surety bond. Will he have time to do all these things before Monday?"

"He can start collecting the paper work after the construction is started. If he waited for all the red tape he couldn't get going for a week or more. We gotta have that building on time, Ivor."

"But, Allen, you shouldn't let him start without your written notice to proceed."

"Don't be ridiculous. You're too damn bureaucratic, Ivor. He'll get started Monday and everything's gonna be okay. You'll see. Quentin knows what he's doing."

"Look, Allen, you're paying me to administer this contract and you won't accept my advice. If you want to do it your way, then I'll have to withdraw."

"It's okay with me, Ivor, if that's the way you want it. With Quentin on the job I won't need you any more anyway. Just bill me to date and we'll call it quits."

Allen hung up. He was relieved to get away from Ivor and his constant nagging about procedure. The project was now in high gear and he didn't want to slow it down. Besides, he'd save some architect's fees.

Ivor was disgusted with Allen's ignorance and his refusal to listen to reason.

* * *

When Allen arrived at the ABC premises on Monday morning, the Quirk forces were in conspicuous evidence. Quirk was in full charge, scurrying about industriously organizing the job. Quirk seemingly had everything started at once. Hardhatted operatives were fencing off the work area, installing a job office, installing chemical toilets and temporary power, installing a telephone and fax, raising a job sign, measuring, and excavating for footings. Quirk was the superintendent and had everyone on the run. Allen was impressed with Quirk's obvious command of the situation.

When Carl Daly arrived, he joined Allen on the sidelines watching the frenetic pace of activity. Allen said, "I know we've got the right contractor. See how speedily they work and Quentin watches everything himself. Hyde's people never worked this fast." Quentin saw them and came over to join them for a few minutes. He explained, "We gotta get the footings in and the floors poured so we can use them for casting the concrete wall panels." This gave Allen additional reassurance that he had made a wise decision to go with Quirk.

So, Quirk got the construction started without the preconstruction submittals. He started on December 5, 1994 and with completion required on April 15, 1995, this gave him 131 calendar days to complete the project.

Allen visited the job first thing every morning and talked with Quentin to get the latest updates on progress. He asked several times about the insurance certificates and the bond. Quentin always replied that his broker was working on it.

Allen was surprised at the frequent turnover of personnel on the job. He seldom saw anyone there over a few days. He assumed that Quentin must have a sizeable operation and that he had moved them to other jobs. Quentin said, "I've gotta put my people where they'll do the most good. I've gotta lotta irons in the fire. Don't worry, I keep my best people here on your job."

On January 2nd of the new year, 1995, Quentin came into Allen's office with his first progress billing. He handed it to Allen and said, "I'd like to pick up the check today if you don't mind."

Allen motioned to a chair and invited Quentin to sit. He looked at the invoice. It wasn't itemized. It stated, economically, "First Progress Billing......$170,000."

"Quentin, don't you think this is a little high? After all, with the Christmas and New Year's holidays and the weekends, you've only worked less than two weeks and this represents almost 20% of the contract."

Quentin patiently explained, "We've spent a lotta money getting the job started and we'll be laying out a fortune for concrete the next few days. I've got huge payrolls to meet."

"But you're not supposed to bill for anything that's not already incorporated into the construction. And what about the retention?"

"Are you going to pay me or not?"

"All of Hyde's bills showed a detailed itemization of what each dollar was paying for. Where's the itemization by trade?"

"Look, Allen, If you want itemized bills, then you should've hired Hyde. But he was gonna charge you $86,000 more for it. Is that what you want?"

"No, but..."

"We've been expediting the job to meet your completion date and you don't seem to appreciate it. It's obvious you don't trust me."

"I'm only asking, Quentin. You know I trust you."

"Allen, if you don't pay this bill today, then I'll just have to move my forces to one of my paying jobs. We'll be back when you've paid your delinquent bill. The subs'll prob'ly file liens." He stood up ready to leave.

"Wait a minute, Quentin. There's no point in getting upset about this. Don't leave the job. I'll pay the bill. I'll have the bookkeeper cut the check right now."

He punched a few numbers on his desk phone and asked his bookkeeper to write the check and bring it to him.

Quentin, now on a winning streak, pressed on, "Allen, any time you're not completely satisfied with my work, why just let me know. You can get someone else to take my place."

"Quentin, you know I'm satisfied with your work. I'm lucky to have you," said Allen placatingly. The bookkeeper brought in the check and Allen signed it and handed it to Quentin.

Allen thought about the bond and the other submittals that he'd still not received but decided to mention them later.

* * *

Quentin's second payment request was submitted on February 1st. Allen was shocked to see the amount requested. $235,000. And still no itemization. This would bring the total payment up to $405,000, over 70% of the contract. Even he could see that the construction progress was nowhere near 50%. Quentin again threatened to quit work and pull his crews off the job. Allen, terrified of missing the completion date, agreed to pay him. He asked Quentin about the bond. Quentin replied, "We'll have it any day now. I'll be talking to my broker later today. Don't worry about it." Allen thought about calling Ivor or Phil Quinn but decided not to.

* * *

During the next few weeks they finally got all the wall panels poured and cured and were now ready to lift them into place. The building inspector, visiting the job when the crane was lifting the panels, reminded Quirk to properly brace the panels as they weigh nearly 70,000 pounds each. Quirk replied, "Don't you think I know what I'm doing?"

A few days later on Tuesday, February 28th, the building inspector told Quirk that the bracing looked inadequate. Quirk said, "Okay, we'll take care of it." Quirk was getting sick of this inspector's constant inferences that his work wasn't up to standard.

That night a heavy wind blew down one of the 20 precast concrete panels. The fallen panel was badly broken, cracked, and chipped. Upon falling it smashed into the existing warehouse wall and broke through the concrete panel into the warehouse. It demolished a number of cartons of television sets stacked against the wall.

Carl Daly was the first to arrive on the premises on Wednesday morning. He opened the warehouse as usual and soon noticed the gaping hole in the concrete wall. He stepped through the hole onto the floor of the addition. He could easily see what had happened. Fortunately only one panel had fallen but he was very apprehensive about standing around where one of the others could come crashing down at any moment as far as he knew. Soon thereafter, Quentin Quirk and his people started arriving and were standing around looking at the damage and advancing their individual theories on what had caused and how to rectify the problem.

Then Allen showed up. He took one look at the chaos and practically burst into tears as he started in on Quentin, "What's this going to do to our completion date? I thought you knew what you were doing. You said to trust you. I trusted you and now look at the mess we're in."

Quentin's lame excuses didn't impress Allen. He left and went to his office towing Carl along with him. He said to Carl, "What'll we do now? We need some help."

Carl suggested, "Why not call Ivor Judge and have him come over and take a look? He might have some practical suggestions."

Allen was embarrassed to call Ivor after their last meeting but he was in real trouble this time. So he called him. He described the situation to Ivor and then humbly asked, "Ivor, I know you're probably upset with me, but will you please come over here and let me know what you think? I really value your opinion."

Ivor said okay but he'd have to have Myles Nolan, his structural engineering consultant with him. Allen okayed it and thanked him.

* * *

A few minutes later Allen's secretary dropped a large envelope on his desk. She explained, "A motorcycle courier just delivered this. It must be important."

The envelope contained a sheaf of bills from Matrix Ready Mix Company totaling a little over $98,000 with the threat of a lien to be filed if not paid today.

Allen went out to talk to Quirk. No men on the job. Nothing happening. It seemed unnaturally quiet. Quirk told him he's having some minor problems with the crew but there's nothing to worry about. "I'll get it straightened out today. Nothing for you to worry about."

"What about these concrete ready-mix bills? If you don't pay them $98,000 today they'll lien the job. What the hell are you doing with the money, Quentin?"

"Don't worry about it, Allen. I'm going over to talk to them in a little while. I've got my third payment request here and I'll use some of the money to get the ready mix people off my back."

Allen replied, "No more payments until we find out where the job stands. How are you going to repair all this damage? Where are your crews? How much more money do you owe?"

Quentin again threatened to pull off the job. Allen then asked, "What about the bond? And the certificates of insurance?"

"I was talking to my broker this morning. He says that none of the bonding companies are interested in my business. I've got too much work going right now. When I've finished some of the other jobs they'll issue the bond."

Allen was aghast. He said, "But you've charged me $5600 for the bond in the contract price."

Quentin replied, "I'll issue a change order and give you credit for that. I always give proper credits. I'll call my accountant right now and get him to issue the change order." As Quentin was making the call, Allen walked back to his office.

* * *

Allen called Phil Quinn, ABC's lawyer, and filled him in on the current situation. He asked Phil to come over for a discussion. Phil, sensing Allen's desperation, said, "Sure, Allen, I'll be there shortly. In the meantime, get the insurance certificates and surety bonds out so I can review them." He hung up and dashed out to his car.

* * *

While waiting for Phil to show up, Allen walked back out to the construction site. Quentin was just getting into his car. He waited for Allen to come to him. "Sorry, Allen, I just talked to my accountant and he says I'm gonna hafta go into bankruptcy. You'll hafta get someone else to finish the job." Before Allen could answer, Quentin had sped off in a cloud of dust.

* * *

Just after Allen returned to his office, Ivor and Myles showed up on the worksite. They took a good look around and went into the ABC offices. They walked into Allen's office just as Phil arrived. After mutual greetings and placing of coffee orders, they all took seats. The atmosphere was somber and thick with pessimism and foreboding. Allen started the discussion, "Quentin Quirk turned out to be a flake. I've paid him $405,000 and the job isn't half done. Matrix Ready Mix claims they have $98,000 coming and threaten to lien the job. There are probably other unpaid bills. I don't know how much it'll cost to replace the fallen panel and repair the existing warehouse wall. And I still have to have the building ready for Takahashi Trading Company on May 1st. That's only 8 weeks away. This'll ruin me. What're we going to do?"

Phil was the first to offer help, "Things may not be as bad as you think, Allen. We'll notify the bonding company and the insurance carriers and they'll take the brunt of the financial impact. Then you'll have to get a reputable contractor like George Hyde to take over the job under Judge & King's supervision." Allen had his elbows on the desk and his head in his hands. The weight of the world was on his shoulders as Phil explained how easy it would be.

Allen didn't know how to start so he just blurted out, "There's no bond and no insurance, Phil. Quentin just told me today."
Phil looked at Ivor and said, "How could you let this happen, Ivor? I thought you would administer the contract and monitor the construction as a normal part of your usual duties as an architect."

Ivor, not too happy to be here in the first place, didn't take kindly to Phil's questions and inferences. He succinctly brought Phil up to speed on the status of the architectural services contract, "Allen fired us when Quentin Quirk showed up. He's doing his own contract administration. We're out of it."

Phil looked at Allen and brutally laid it on the line, "Well, Allen, you've shot yourself in the foot again. This time it won't be so easy to get out of it. I hope you've got plenty of money because that's what it'll take."

* * *

Allen had considered himself lucky when Quentin Quirk first showed up, but his genuine good fortune was when Ivor, Phil, and George pitched in to help control losses.

Ivor and Myles promptly designed the repair of the damaged warehouse and got the building department's approval. Ivor agreed to administer the contract and observe the construction.

George agreed to take over and complete the job on a cost plus basis with no guaranteed maximum price. He flatly rejected the liquidated damages provision in the contract. He ferreted out the payment status of all the subcontractors and got the job reorganized and back into production. He produced a realistic expedited schedule that recovered some of the lost time.. They accomplished the damage repair simultaneously with normal construction progress. George estimated that they could achieve completion by May 1st.

After all the bills were settled and damage repairs paid for, the net overrun on Hyde's original bid was $178,000. Phil explained to Allen, "You're fortunate that your business is good enough to take this financial setback. We'll put this claim into Quirk's bankruptcy but there's little chance of recovering anything. I hope you've learned your lesson."

* * *

Moral: Shortcuts in the contractor selection and contract administration processes can yield devastating results.

Points of Law 12

ABC WAREHOUSE VII
Allen Hires His High School Buddy

12.1 Penalty Clause vs. Liquidated Damages Clause

Allen has an opportunity to supply warehousing space to Takahashi Trading Company if the space can be available by May 1, 1995. He wants to put a penalty clause in the contract to motivate the contractor to finish the job on time. He suggests a $500,000 penalty for a $646,000 contract. He doesn't know that most courts would not enforce such a severe penalty.

A penalty clause imposes a penalty that bears no reasonable relationship to actual damages. The $500,000 number, when compared to the actual damages that Allen might sustain, is wildly excessive.

The measure of damages for breach of contract is the amount that would put the injured party in the economic position that it would have occupied if the contract had been fulfilled. If Allen is unable to supply warehouse space on May 1 he risks losing a profitable contract with Takahashi Trading Company. In order to avoid that undesirable result, Allen could hire warehousing space elsewhere. The amount of compensatory damages would be determined by computing the cost to rent the alternative space plus the expense of double-moving the Takahashi inventory, and then subtracting the amount to be paid by Takahashi. Since only 10,000 square feet of storage space are at stake, even at $1.00 per square foot it is obvious that the actual damages would be much less than $500,000. The $500,000 penalty would therefore be unenforceable.

12.2 Consultants

Most architects don't claim to have the specialized knowledge needed to design structural, electrical, mechanical, and geotechnical elements of a construction project. Specialized jobs also require more specialized consultants, such as for lighting, acoustical, and irrigation systems.

If consultants are employed and paid by the architect, the architect becomes legally responsible for their errors and omissions. For this reason an architect may suggest that the owner employ its own consultants. The owner thus may gain more control over the activities of the consultant but at the same time will have to assume the responsibility of coordinating the work of the consultants with the work of the architect.

12.3 Negotiated Contract vs. Competitive Bidding

Construction contracts are usually written contracts. A written contract is formed when both parties indicate their intention to be bound, usually by placing their signatures on the contract document. But there are many other ways of entering into a legally enforceable contract.

In legal terms, a contract is defined as a promise that the law will enforce. The requirements for an enforceable contract are 1) competent parties, 2) a legal object, 3) a mutual manifestation of assent, and 4) consideration.

1) Parties are competent to enter into a enforceable contract if they are adult, sober, and sane.

2) The object of a contract is legal if its performance would not involve the parties in the commission of a crime. (For example, a contract to ship marijuana would be illegal and unenforceable.)

3) Mutual assent is determined by objective criteria. It is not necessary to the formation of a contract that there be a "meeting of the minds", since there is no way of proving subjective thoughts. Objective manifestation of assent may be in written words, spoken words, or actions. A contract is often formed by offer and acceptance. At an auction, a buyer may submit an offer by making a hand signal. The auctioneer may accept the offer by bringing down the hammer. Thus, a contract may be formed without words. Even without words, though, the formation of a contract requires an unequivocal objective manifestation of an intention to be bound.

4) Consideration is present if each party to the contract promises to give something up. If Allen promises to take his grandchildren to the movies, he has promised to give something up but they haven't. Therefore there is no contract. But if Allen offers to pay $646,000 for a warehouse, and Hyde agrees to build it for that price, each promise is consideration for the other's promise and an enforceable contract comes into effect.

A negotiated contract is formed when an owner and a contractor bargain the time schedule, the price, and other terms. One advantage to a negotiated contract is that it can come into existence a step at a time, before all the drawings and specifications are 100% complete. Construction projects in which the work starts when the drawings are only partially completed are known as "fast track" projects.

When a contract is formed by the competitive bidding process, on the other hand, the contract documents must be complete except for the bid itself. If Allen's contract documents are ready to go out for bid, the only thing they lack is the name of the contractor and the contract price. These are supplied by Hyde Construction with its bid of $646,000. The bid is an offer to perform the work specified by the contract documents for that price. The offer will be accepted if Allen awards the contract to Hyde Construction. After the contract is awarded, it becomes the duty of the parties to sign the contract documents.

What assurance does Allen have that Hyde will sign the contract? If he is not willing to rely on Hyde's reputation, Allen can require that the bid be accompanied by a cash deposit that Hyde will forfeit if it fails to sign the contract. In practice, bidders usually supply bid bonds rather than cash deposits. The bid bond is a document by which Hyde agrees to forfeit a stipulated sum if it fails to sign the contract, and a surety company guarantees the payment of that amount.

12.4 Value Engineering

To Allen's pleased astonishment, Quentin says he can build the job for $560,000 and could even cut more bucks out by value engineering. The value engineering concept is incorporated in many federal construction contracts. The contractor is invited to propose changes that would reduce cost. If the government accepts the change and the cost is reduced, the contractor and the government split the savings. The idea is to encourage government contractors to come up with ingenious ways to reduce cost without impairing quality.

Value engineering is an important component of negotiated contracts. Contractors, and especially subcontractors, have specialized knowledge unavailable to architects that enable them to suggest design changes that would cut costs without reducing quality. Drawings and specifications may be changed to incorporate value engineering proposals and the cost of the project thereby reduced. Knowing Quentin as we do, we would advise Allen to scrutinize value engineering proposals carefully, because his proposals might cheapen the quality of the job while reserving most of the savings to Quentin.

12.5 Bonds

A bond is a document issued by an insurance company that guarantees that a contractor will perform. In construction, we deal with bid bonds, performance bonds, and payment bonds.

A bid bond is a guarantee that a contractor will sign a contract. A performance bond is a guarantee that after signing it, a contractor will perform the contract. A payment bond is a guarantee that a contractor will pay its bills for subcontract work and materials and that the subcontractors and their subcontractors will also pay their bills. The purpose of a payment bond is to protect an owner's property against mechanics lien claims that could be asserted by unpaid workers, subcontractors, sub subcontractors, and material suppliers.

12.6 Document Submissions

Ivor, as architect, is apprehensive that Allen will allow Quirk to start work before submitting his subcontractor list, schedule of values, construction schedule, insurance certificates, and surety bonds. Quirk, on the other hand, would like to spike the job by getting started. After all, the sooner he gets started he sooner he can bill for his first progress payment.

12.7 Subcontractor Lists

Allen should require Quirk to submit a subcontractor list so that Allen can review the qualifications of the subcontractors who will perform most of the work on his project.

12.8 Schedule of Values

It is the function of the schedule of values to assign a dollar value to each phase of the work. The schedule of values may then be used to compute the contractor's progress payments. Thus, by determining the percentage of completion of the various parts the job, applying those percentages to the schedule of values, and deducting the retention, the architect can determine the amount of each monthly progress payment.

12.9 Insurance Certificates

Allen should be aware that when he starts to build his warehouse he will become exposed to risks of liability that go far beyond the risks of conducting a warehousing business. As the owner of the construction project, Allen will become potentially liable for jobsite injuries. He should purchase his own commercial general liability insurance policy to cover these potential liabilities, and, as a backup, he should also require Quirk to be fully insured. He should require Quirk to name Allen as an additional insured on his (Quirk's) insurance policies, and see to it that Quirk requires his subcontractors likewise to name Allen as an additional insured on their policies. He should not allow Quirk or any subcontractor to start work until the appropriate insurance certificate has been delivered and reviewed for compliance with the contract documents.

12.10 Bonds

Nor should Allen allow Quirk to start work until he has received performance and payment bonds. Allen should be aware that not all contractors are bondable. Before a surety company will issue a bond, it makes an intensive investigation of the qualifications and financial resources of the contractor. Most contractors, Allen may be surprised to discover, are not bondable because they lack the financial resources to convince a bonding company that it is safe to guarantee the contractor's performance. If Allen allows Quirk to start work without making bond, he could have a big problem on his hands.

12.11 Quentin Threatens to Walk Off the Job

Allen was flabbergasted when he received Quentin's first progress billing of $170,000. This represented 20 percent of the total contract price, yet Quentin had only worked less than two weeks, so Allen thought the bill was unreasonably high. When he questioned the bill, though, Quentin threatened to walk off the job, adding that he would move his forces back on to the jobsite when the bill had been paid. Would Quentin be legally justified in stopping work if the bill wasn't paid? The answer to this question is determined under the law of contracts by examining the following three questions:

1) Would the failure to pay the bill be a breach of contract;
2) If a breach of contract, would it be material;
3) If a material breach, would be unexcused?

It would be unfair for a party to insist on the punctilious performance of a contract while at the same time blatantly refusing to perform one's own obligations under the contract. The worker is excused for quitting work if the employer does not make payroll. A tenant is excused from paying rent if the building burns down! If a driver does not make car payments, the leasing company repossesses the car. These are all instances of a material breach of contract that excuses further performance.

Getting back to Allen and Quentin, failure to make a progress payment is a material breach of contract. The only reason the contractor works is to get paid. Therefore the owner can't argue that nonpayment would be immaterial. Since a material breach of contract excuses further performance, Allen's failure to make the progress payment would legally excuse Quentin for walking off the job. Allen, however, suspects that Quentin is overbilling. Let's assume that the $170,000 represents 20 percent of the contract price, but Quentin has only accomplished 7 percent of the work. If that were the case, then Quentin would only have earned $59,500. Were that the case, Allen's failure to pay the $170,000 bill would not be a breach of contract at all and Quentin would not be justified in walking off the job.

Allen made the payment in order to keep the job going, and the charade was repeated when Quentin presented a second payment request for $235,000 which would have brought the total payment up to $405,000, over 70 percent of the contract price even though progress was no where near 50 percent. Quentin persuaded Allen to pay the bill by threatening, once again, to walk off the job.

12.12 Liability Insurance

Heavy wind blew down a precast concrete panel, smashing the existing warehouse wall and demolishing some television sets that were stored in the warehouse. The panel itself was also badly broken and cracked. This event should remind Allen of Quentin's contractual obligation to supply certificates of insurance. If Quentin is carrying proper liability insurance, the insurance company will pay to replace the TV sets and repair the damaged warehouse, but will not pay to fix the broken, cracked, and chipped panel. This is because Quentin's liability insurance policies would not cover property damages to Quentin's own products. Since Quentin built the concrete tilt-up panel that was damaged, that panel was his own product and therefore damage to that product would not be covered by the liability insurance policy. Damage to the TV sets in the warehouse, however, would be covered.

12.13 Mechanics Liens

Allen receives a sheaf of bills from Matrix Ready Mix Company with a threat to lien the job for more than $98,000 if the bills are not paid. How could this be? Allen has no contractual relationship with Matrix. Matrix was hired by Quentin, and it's Quentin, not Allen, that owes Matrix more than $98,000.

Nevertheless, the mechanics lien law, which has been adopted in different versions in all fifty states, would allow Matrix to put a lien on Allen's property and foreclose the lien. Foreclosure would mean that the sheriff would sell Allen's property and use the proceeds of the sale to pay the lien. In order to prevent this foreclosure, Allen would have to pay off Quentin's bill. After doing so, though, he has the right to sue Quentin for the $98,000, since Quentin, as the contractor, is required to protect Allen's property against mechanics liens.

12.14 Quentin Takes Bankruptcy

The bankruptcy law is intended to allow a debtor to get a new start. By taking bankruptcy, Quentin can avoid his contractual obligation to finish building the warehouse and also avoid his contractual obligation to reimburse Allen for the $98,000 mechanics lien. When he hears this, Allen thinks back to the beginning of the job, when he decided to let Quentin start work without submitting his performance and payment bonds. Under the performance bond, the bonding company would have guaranteed that Quentin would finish the job. Under the payment bond, the bonding company would have guaranteed that Quentin would pay off the $98,000 mechanics lien!

13

Davren Effingwell dba Economy Home Improvement Company II

Cleaning Up the Mess

"Honestly, Brad, something's got to be done about this damn house!" Anita looked haunted and exhausted.

"I know, Honey, but what can we do?" Brad was disconsolate.

Anita and Brad Coyle were feeling utterly discouraged over the miserable outcome of the dream addition to their beloved home. Everything had gone wrong. They were reminded of it daily as their previously comfortable house was now in a shambles. Furniture and books from the addition were lying about in their dining and living rooms. Pictures from the walls and their video tapes and CDs were in boxes stacked in their bedroom. The house was always uncomfortably cold. The recreation room office addition was draped in plastic as a feeble protection against more unwelcome rain intrusion and the whole place had an all-pervasive musty smell.

They were also reminded forcefully at the first of each month when they had to pay their $2604 mortgage payment. Before the addition it was only $1497 a month. The higher payment wouldn't be too bad if they had anything to show for it.

They had tried repeatedly to get their contractor, Davren Effingwell, dba Economy Home Improvement Company, their trusted neighbor, to do something about the intolerable conditions. On the few times they ever saw him coming from or going to his home across the street, he'd smile and wave at them but he'd never stop to talk. He didn't answer the numerous messages they left on his answering machines at home and office . Once, Brad walked over to Dave's house and rang the doorbell. Dave's wife answered the door. She claimed ignorance of Dave's business and abruptly closed the door. This visit triggered the only call Dave ever voluntarily made to the Coyles. He delivered a firm message, "Brad, don't you ever come to my house again intimidating and harassing my wife. She has nothing to do with the business." He hung up before Brad could get in a word.

Brad had sent him faxes that were totally ignored and wrote several letters that weren't answered. One day, Dave answered his office phone and was very friendly with Brad. He said, "There's no point in sending me letters and faxes, Brad. I'll take care of all your problems. They're all the subs' fault. I've notified them and they'll do whatever has to be done. Letters and faxes won't do any good."

"What about the insurance, Dave? When are you going to send me the certificates?"

"Don't worry about that, Brad. My insurance broker will be sending them to you direct. It's in his hands now."

"What's the broker's name and phone number? I'll call him."

"I've misplaced his name right at the moment, Brad. As soon as I find it I'll send it on to you."

* * *

Brad and Anita had lost their usual upbeat optimism and were in a constantly depressed state. It was seriously affecting her family law practice and his accounting practice. One morning, about three months after the rainstorm that finished the short useful life of the addition, Anita and Brad were finishing a quick cup of coffee and bagel before leaving for work. They were standing silently in the kitchen, each engrossed in private thought, when Anita quietly started, "Brad. This is ridiculous. We can't go on like this. We've got to do something about this damn house. It's ruining our lives. It's outrageous." Her voice was rising and becoming strident.

Brad, gulping down the last of his coffee, replied, "I'd do something if I knew exactly what to do. I've never heard of anyone with this kind of problem. What can we do?" Anita's displeasure only added to Brad's misery.

Anita, becoming calmer, suggested, "Let's call Jerry King again. Maybe he'll know what we should do. In his construction law practice he must certainly have seen this type of situation before. What do you think, Brad, Dear?"

"Well, you remember when he was here before, he said we had no practical recourse to the lender, the joint control, or the building department. The contractor's cooperative but won't do anything. He can't seem to get his subcontractors in gear. We don't know if Dave had any insurance. The only thing we're sure of is that he had no surety bond. What can Jerry do for us now?"

"We won't know if we don't ask. I'll call him today and see if we can talk to him tonight after work. Okay, Brad?"

"Okay, Dear. Anything." They left for their respective offices.

They were now grasping at straws. Neither really felt they had any practical courses of action that would rescue their home without bankrupting them.

* * *

That evening Anita arrived home first. In the mail was another mechanics lien notice. Three others had come in the past month. The roofing contractor had demanded $2540, the plasterer $1760, and the lumberyard £2380. Today's demand was from the electrician wanting $1350.

When Brad came in a few minutes later, Anita handed him the new lien claim. Brad had been keeping an account of them. The total was now $8030. He'd have to write another letter to Dave. Anita asked, "How many more of these are we going to get?"

"How the hell would I know," Brad snapped. He was getting defensive as though Anita was blaming him for their plight. As a matter of fact, she was. She thought that a certified public accountant ought to know more about contracting and money management. He, on the other hand, felt that Anita, being an attorney, should have been less gullible in reading Dave's contract.

A few minutes later, Jerry King showed up and they all went out to dinner to discuss their problems and try to relax if possible. During the course of a stressful dinner, they told Jerry about life with a crushing debt, monstrous building defects, mechanics liens, and the total paralysis of Dave and his subs.

Jerry made several insightful observations and summed up his advice by saying, "The best thing to do at this point is to engage a forensic architect to look over the addition, inventory the defects, and recommend the most practical repair strategies. We'll need his written report to record the present condition of the building. I can recommend an excellent one. In the meanwhile, I'll follow through on the mechanics liens and do some research to see if there is any possible source of liability and funds. I'll get back to you when I have any useful information."

Brad and Anita felt much better even though nothing tangible had yet happened. At least Jerry didn't seem as negative and confused as they were.

* * *

Jerry had told them about Frank Grimm, FAIA, an experienced forensic architect. They had worked together previously on several construction defect cases and the attorneys in Jerry's firm respected his forthrightness and ability. Brad telephoned Mr Grimm and made arrangements for him to visit their home the following evening at 5 o'clock.

Frank Grimm arrived promptly on time and took a quick guided tour of the premises. Brad and Anita gave him a list of the defects they were aware of: the springy floor, the inadequate heating system, the roof leaking, the warped floor damage, and the 20 or 30 miscellaneous less serious defects. They also told him of their experiences with Davren Effingwell dba Economy Home Improvement Company.

Frank told them, "I'll need a couple of hours to examine the construction more carefully. Also, I'll need to see the drawings and specifications."

"Certainly, you can have the drawings, but we don't have any specifications. Just the drawings and the contract. Here they are." Brad handed Frank the two sheets of drawings and the one page contract. Frank put them in his briefcase and promised to return at 10 o'clock Saturday to perform his examination.

After he left, Anita asked, "Brad, what did he mean about specifications?"

"I don't know."

* * *

On Saturday morning Frank arrived ready for action. He had his briefcase in hand and a camera hanging from his shoulder. He spent almost three hours examining, photographing, sketching, and measuring. He climbed, stooped, and crawled. He picked, peered, poked, and probed, duly recording all his findings neatly on a note pad. Upon leaving he told Brad and Anita that he'd have to study the documents, his photographs and notes, the building code, and other relevant standards.

Ten days later his preliminary report arrived in the mail. In it he provided detailed descriptions of the known defects. In addition, he listed a number of other serious defects that had not been previously noted. He furnished generic repair method recommendations and outline specifications. The highlights of his report were:

A. Original Defects

    1. Springy floor. Caused by the omission of a concrete pier under the floor. Recommendation: Install the pier.

    2. Inadequate heating system. Caused by adding onto the existing house heating system. Recommendation: Install a separate heating and air conditioning system for the addition.

    3. Extensive roof leaking. Caused by installation of solar heating panels by nailing through the roofing membrane. Recommendation: Remove the solar panels. Repair the roof by removing gravel, adding an asphalt layer, a layer of 40# felt, a flood coat of asphalt, and reinstalling the gravel. Reattach the solar panel array on a pipe rack with 8 legs set in pitch pockets.

    4. Hardwood flooring. Recommendation: Repair the warped flooring, resand, and refinish.

    5. Detailed descriptions of 25 lesser defects with repair recommendations for each.

B. Newly Discovered Defects

    1. Absence of roof insulation. This will forever affect the operating cost of the heating and cooling systems. Recommendation: Add insulation board when reroofing.

    2. The hardwood flooring is installed tight to the walls around the edges. There should be at least a half-inch gap all around the edges to provide room for expansion. Recommendation: Provide the gap when repairing the flooring.

    3. The foundation pier posts are not diagonally braced and would probably fail in an earthquake. Recommendation: Provide diagonal bracing in both directions.

4. Underfloor ventilation has not been provided for. A code violation that would probably lead to mold, mildew, and dry rot of the wood construction. Recommendation: Install underfloor vents.

5. Inadequate underfloor access. Code violation. Recommendation: Provide an additional underfloor access door.

6. The ceramic tile grout in the powder room counter top is cracking and flaking out due to too much water used in mixing. Recommendation: Remove and replace.

7. The roof slopes to the side property line and drips to the ground thereby causing erosion and undermining of neighbor's concrete block retaining wall. Recommendation: Install gutter, downspout, and underground pipe drain to front street and repair neighbor's retaining wall.

8. Detailed descriptions of 15 lesser defects with repair recommendations.

After reviewing the report with Anita, Brad called Frank Grimm for an appointment to discuss the findings and recommendations. Also, they didn't really know what to do next. Maybe Frank could help.

At their meeting, Frank recommended that they engage an architect to design the repairs, prepare the drawings and specifications, obtain general contract bids, observe the construction, and administer the construction contract.

Brad asked, "Couldn't we hire the drafter Dave got to prepare the original drawings? His fee was quite reasonable."

Frank said, "These original drawings are incompetent, incomplete, and inadequate. There are no sections or details and only sketchy generic specifications. The dimensions don't add up. It's a miracle the addition is no worse than it is. The contractor must have had a lot of trouble with these plans during construction."

Brad said, "They did have a number of problems but Dave worked them all out on the job. He always told us, 'Don't worry. We'll fix it.'"

Frank said, "You need a qualified architect to design the repair work, prepare drawings and specifications, and see to it that the work is properly carried out."

Anita asked, "What about all those other things you mentioned? The observation of construction and administration of the contract. We didn't need it before, why have it now?"

Brad said, "Yeah. It sounds expensive."

Frank explained, "Well, the reason this project is in such deplorable condition now is that most of the people you entrusted it with were incompetent. Now it has to be rectified and it must be done by people who know what they are doing."

Brad said, "Okay. Who should we get to be our architect?"

Frank said, "There are several local architects quite capable of doing this type of work. I'll give you three or four names and you can select one." He wrote down several names, addresses, and phone numbers and left.

Anita and Brad, having no rational basis for judging an architect's competence, selected Angela Brooks, AIA, since her office was nearest to their home. After meeting and talking with her they found that she had considerable experience in the design of attractive residential additions. They liked her and felt they could work with her. They gave her Frank's report and engaged her to prepare the repair drawings and specifications. She made arrangements to visit the Coyle home.

In the process of making field measurements and recording existing conditions, Angela noticed numerous minor defects not mentioned in Frank's report. She and her office staff prepared the drawings and specifications needed for the building permit application and for obtaining bids from contractors. She reported to the Coyles that she had filed for the city building permit.

Brad asked, "Why do we need a permit? We already have one and now we're just making repairs." He sensed more fees to be paid and more red tape with the building inspector.

"The work is too extensive to leave the building department out of it. I'll have the permit in about two weeks. Incidentally, we'll have to pay a plan check and permit fee." Brad knew it before he heard it.

When Angela picked up the plans and Plan Check Correction Sheet from the building department, she was surprised to find that the engineer had questioned the adequacy of the existing roof beams. Angela protested, "We're not changing the beams or the loads. They were checked and approved under the original permit."

"Well, I checked them and they don't figure. The original plan checker must've missed it. I can't approve these beams." The plan check engineer was adamant. He was edging away from the public counter, as he was already late for lunch.

Angela rolled up the plans and other documents and left for her office. She called her structural engineering consultant and asked him to drop by as soon as possible. That afternoon they discussed the beams.

"The existing beams are 6-inch by 8-inch, spaced at 48 inches on center," Angela pointed out to the engineer.

Stanley Stout had been doing her engineering for several years and she respected his ability to devise practical and ingenious solutions to structural problems. He was already making rough calculations while Angela was explaining the problem.

Angela continued, "If we have to change the beams, we'd have to remove the entire roof. That'd be disastrous. If we have to insert more beams, it'll look like hell. What can we do?"

Angela waited expectantly for Stan's reaction. He was still engrossed in calculating and sketching. He looked up.

"Originally, the 6 by 8 beams would have worked at a 36-inch spacing, but at 48-inches the beams should have been 6 by 10. I think the most practical solution now will be to attach steel flitch plates to the sides of the existing beams. We can use relatively thin plates and lag bolt them in place. My preliminary figures require a 1/8-inch by 7-inch steel plate on each side of each beam fastened with 1/4-inch by 2 1/2-inch lag bolts. I'll give you a sketch of the bolting pattern." Angela was impressed. This would do the job without much expense or trouble and would look all right.

"I'll need the calculations and details to go back to the building department right away." Angela obtained the building permit the next morning.

* * *

Angela met with the Coyle's a day later at her office. She explained to them that they had to find a competent contractor to do the work.

"This time we want competitive bids," Brad suggested. Anita nodded her approval.

Angela explained, "Unfortunately, that's not practical this time around. No contractor in town would give you a stipulated price proposal for this type of repair work, unless it was sky-high to cover all contingencies."

Anita asked, "Well, then, how do we get a contractor that won't take advantage of us like Dave did?"

Brad added, "And how'll we know how much the work'll cost?"

"I'll make a short list of experienced contractors that do this kind of work and you can check out their references. We'll narrow it down to one and negotiate a contract." Angela started writing a list of contractors with whom she had previous experience. Trustworthy Construction, Gibraltar Home Construction, and Superior Home Improvement Company. She handed the list to Brad.

Angela said, "As soon as you and Anita have decided which contractor you want to deal with, let me know and we can talk to them about doing your work."

During the next few days the Coyles interviewed the principals of the three contracting firms at their home and decided they would like to deal with Gibraltar Construction Company. In furtherance of Angela's previous advice they asked their bank to run a credit check on Gibraltar and now they were ready to talk business. They invited Sam Gibraltar to a meeting at Angela's office. After considerable discussion they agreed on a cost plus fee contract. Sam Gibraltar had studied the plans and specifications and had prepared a detailed breakdown of anticipated costs. This was attached to the contract as an indication of the scope of the work but not as a guaranteed maximum cost. Angela would observe the construction as it progressed and would have to approve all payments to the contractor. Sam Gibraltar had estimated the construction period to be seven weeks. Sam and the Coyles signed the contract. Sam submitted certificates of all required insurance and a surety bond and a week later the work was started. Gibraltar Construction's budget estimate including all construction costs, overhead, and profit, was $39,292.

* * *

Brad and Anita were seriously concerned about how they'd finance the repair work. There didn't seem to be much hope of getting any immediate help from Dave or any of his subs. The Coyles had already paid Frank Grimm's fee and they'd have to pay Angela Brooks as well as Gibraltar Construction. They'd probably have to pay the outstanding mechanics' lien claims of $8030. And Jerry King would have to be paid. There was no way of knowing exactly how much all this would add up to, but Brad had started a schedule where he inserted all known costs and estimated all the rest. It added up to almost $60,000. Brad figured that the fair market value of the property would be approximately $425,000 after the addition was properly finished. His present loan of $300,000 was about 70 percent of the value. If he could talk the bank into raising the loan another $60,000 the ratio would then be just under 85%. If he could talk them into it, that is. It would raise his monthly payment by $521 per month. He and Anita could handle it. They'd just have to curtail some of their activities for a couple of years. However, the bank didn't buy it. They'd only go for a new loan at a higher interest rate on the entire amount. Now what?

One of Brad's acquaintances at the golf club was in the business of buying and selling second trust deeds. Brad didn't really know him, only by sight. People called him Ace. Brad contrived to meet him in the bar and engaged him in some exploratory discussion, ostensibly in behalf of an unnamed client.

Brad got his eyes opened quickly. He found that the loan would be possible all right. But the interest rate would be 12% and the payments on a ten-year amortization schedule. Brad was mentally working out the amount of the monthly payment. He came up with $900. Fairly close -- it would actually be $861. Pretty tough, but he thought he and Anita could handle it if they tightened up on their living expenses a little more. Maybe drop the golf club membership, among other things. But his dream was quickly shattered when Ace told him that there would be a 5-point loan origination fee up front. That would be $3000 more. Ace also mentioned that the loan would have to be paid off in four years. Brad quickly mentally computed that they would have to dig up about $44,000 to meet the balloon payment in four years.

Brad then told Ace that this loan would actually be on his own property. He said, "The terms are steep but it'll only be temporary as we'd pay it off as soon as we make a recovery from our original contractor or some other source."

Ace immediately informed him, "That's okay, Brad, but don't forget the prepayment penalty. Six months' unearned interest."

Brad's quick computation indicated that an early payoff in the first couple of years would be around $3500.

Ace reminded him, "Don't let it go too long, Brad. Interest rates are going up, you know. I don't know how long I can hold these terms."

Brad thanked Ace for the information and told him he'd be back to talk some more about it. He figured he'd find something more reasonable and would never have to talk to Ace again. But, alas, it was not to be. He checked several other financing sources and could find nothing better. So he and Anita decided that they better go ahead with Ace.

* * *

As Sam Gibraltar got the work underway, Brad and Anita could see a great difference in the way Sam's people went about their work. They arrived early and worked diligently, efficiently, and neatly. They never sat around waiting for materials, as Dave's people seemed to be doing a lot of the time. Sam kept in constant touch with his working superintendent, Monty, by mobile phone and visited the job at least once a day.

"Sam," Monty called, "Come over here and take a look at this." He was pointing to a crack between the stone chimney and the wall paneling. They walked around to the outside and noticed that the exterior plaster was cracked and crushed on one side of the heavy stone fireplace and chimney while a gap was developing on the opposite side. The chimney was also leaning away from the wood framing at the roofline. They concluded that the stone structure was founded on inadequate soil. Monty climbed a ladder to get a better look at the joint between the timber construction and the stone. He reported, "There's no steel connection between the chimney and the framing. What should we do?"

"We better talk to the architect. Angela'll be here tomorrow morning on her weekly site visitation."

* * *

The next morning, one of Monty's carpenters was breaking an opening in the exterior wall below the floor line to install an underfloor access door and to start installing the underfloor air vents. As soon as he could see under the floor he was aware of a soggy earth condition around the fireplace footing area. He could smell it before he first saw the green moss-covered moldy earth and puddles of stagnant water. He then saw a copper cold water pipe attached to the bottom of the floor joists. A slow drip was dropping off at a defectively soldered elbow joint. It must have been leaking from the start as the soil around the fireplace was completely saturated.

When Angela arrived and examined the situation with Sam, they agreed that this was undoubtedly the cause of the chimney settling. Sam inquired, "What'll we do now, Angela?"

Angela had been mulling the problem over in her mind. She concluded, "I think we better refer this to Stanley Stout, my structural engineering consultant. He might want to talk it over with a soils engineer. I'll call him this afternoon."

Sam told Angela that the building inspector had visited earlier that morning. He'd approved all of the new work done so far but noticed a few new items, the most serious being too many outlets on two of the electrical circuits and a lack of attic access over the bar and powder room areas. He also pointed out that the fireplace hearth was only 18 inches in depth although the code requires 20 inches. He insisted that these be rectified. In his written code violation report he also included several other minor items.

Angela was pleased with the work otherwise and told Sam to keep up the good work. That afternoon she talked with Stan and included his engineering recommendations in her weekly observation report to the Coyles. A copy went also to Sam Gibraltar so he would know that the architect and structural engineer, in consultation with the soils engineer, had reached a conclusion. The entire fireplace, chimney, and footing would have to be removed, the saturated soil removed, and the footings repoured for a new fireplace and chimney. This would also resolve the problems with the hearth and structural anchorage to the roof framing. The original stones could be cleaned and reused.

The work progressed and Gibraltar Construction completed their work within the promised completion date, which had been extended to cover the extra work. Fortunately Angela had had the forethought to specify repainting the whole addition inside and out as very few of the surfaces were intact after the repair work was done. The building inspector approved the project and signed off. Angela's final inspection revealed a few punch list items which Sam had rectified promptly.

Anita and Brad were pleased with the result and quickly moved all of their belongings out of their living, dining, and bedrooms back into the addition.

Anita commented, "Brad, this is how it should have looked when Dave was done." She was a completely new happy person.

Brad was pleased with the result, but at what cost? His mind was reeling at the bills that were still rolling in. He had compiled a schedule of financial obligations and he couldn't believe what had happened. All of the borrowed money from both loans had been disbursed and they still owed $10,500 on their credit card accounts after buying their furniture and paying Ace's loan origination fee. Their monthly payments were now $3465 on their two mortgage loans and $500 on the credit card accounts. They still didn't know what Jerry King would charge for his services. Brad was wondering how they got into this mess and how could they have avoided it.

* * *

A couple of weeks later Jerry King called Anita at her office and asked for a meeting with her and Brad at Jerry's office. Anita and Brad were impressed with the luxurious reception room and offices of the large construction law firm that Jerry was associated with. It was on the 85th floor of the Imperium Tower, the newest and tallest office building in the city. When the attractive receptionist brought them to Jerry's private office, she offered them coffee, which they declined. They were nervous and tense. Jerry immediately told them he had some interesting news.

He explained, "Our office investigator has been looking into the background of Davren Effingwell dba Economy Home Improvement Company." He was referring to a typewritten inter-office memo.

"First of all, we've discovered that Dave has no assets. His house is mortgaged for more than it's worth and it's hours away from foreclosure. His car is leased and was repossessed this morning. Economy's telephone has been removed and their office is now closed. We can't locate Dave at the moment. His wife left him a week ago."

Brad and Anita exchanged anxious glances.

Jerry continued, "As I told you before, the bank and the joint control are both off the hook. They only did what they promised to do. Their site inspections were for their own benefit."
Brad interposed, "That's what I expected. We're screwed." Brad was at his lowest ebb since the night of the rainstorm. They were financially ruined and they would have to find $44,000 in four years. They'd be lucky if they ever worked their way out of this mess. Anita looked crushed.

Jerry continued, "And you already know there's no recourse to the building department. As you know, their inspections are for the city's benefit, not yours."

"Yeah. We know," said Brad. Anita nodded.

"And the state contractor's license bond is too small to do any good. We could probably get Economy's license revoked but that won't do you any good."

Anita said, "Well, it might save others from what we've had to endure."

Jerry got back to his report. "We checked with the state contractors license board, however, and found that the Economy Home Improvement Company license was actually in two names. Dave's brother-in-law, Herman Grubb, was the Responsible Managing Officer and it was his license. Several months ago Dave and Herman had a falling out and Herman pulled his money out and moved to Fresno. He's now forming a new business, Fiscal Home Improvement Company. In Economy Home Improvement, Herman was the contractor and Dave was the salesman. Dave talks good construction lingo but knows next to nothing about it. He was only in the business because his wife forced her brother Herman to take him in. After Herman left the license was never properly transferred so Herman is still responsible for Economy."

Brad and Anita were now hanging onto every word.

Jerry continued, "Your job was the first one Dave took on after Herman left. Another thing we found out was that although Dave didn't submit certificates of insurance to you, there was actually a builders risk policy in effect. It had a broad form endorsement so your project was covered, since all the work was performed through subcontractors. Economy didn't do any of the work."

Jerry shifted his papers and looked up. "We have a good chance of collecting from Herman Grubb or the insurance carrier, Fairweather Indemnity Company. We're preparing the lawsuits now and will file them tomorrow."

* * *

Points of Law 13

DAVREN EFFINGWELL
dba
ECONOMY HOME IMPROVEMENT COMPANY II
The Aftermath

13.1 Saved by the Law of Agency

Brad and Anita have learned the hard way that an owner who undertakes a remodeling project should carefully examine the financial responsibility and the liability insurance coverage of the prospective contractor and subcontractors. Although they failed to do so, and although Dave didn't provide liability insurance nor was he financially responsible, the situation is saved in this case by the law of agency. Although Brad and Anita didn't know it, Dave didn't act on his own, but was serving as the agent of his brother-in-law Herman who was carried on the records of the contractor's licensing agency as the responsible managing officer and owner of Economy Home Improvement Company. Dave was really nothing more than a salesman who took over the management and supervision of a group of subcontractors.

Under the law of agency, the principal is responsible for the acts of the agent. (Social policy: public safety and welfare would be undermined if a person could avoid liability for dangerous or immoral conduct merely by operating through agents and employees.) Liability is fixed both on the agent and the person who employed the agent to do the dirty work. Thus employers are encouraged to train and supervise employees to perform their work honestly, safely, and competently.

13.2 Liability Insurance

The function of liability insurance is to protect a person or company against claims, lawsuits, and liability for negligence. Liability insurance is not intended to protect against claims, lawsuits, and liabilities for crime, fraud, or breach of contract. Liability for breach of contract is known as a "business risk" voluntarily assumed by a contracting party. It is not the function of liability insurance to guarantee that contractors will perform their contracts!

Liability policies, however, are written to protect the insured against claims and lawsuits for bodily injury and property damage. Property damage is sometimes caused by breaches of contract. So what happens when a contractor's breach of contract (failure to build in a workmanlike manner) causes property damage (flooding and water intrusion)?

Insurance companies try to deal with this by excluding coverage for damage to the contractor's own work. Some policies ("broad form policies") afford coverage for damage to work performed by subcontractors. In our case, all of the work contracted by Economy Home Improvement was done by subcontractors so all damage is covered by the broad form policy.

13.3 Property Insurance

The damage caused by the settlement of the chimney is another matter. The chimney settled because supporting soil was saturated by a plumbing leak caused by inadequate soldering. Let's assume this pipe was not installed by Economy Home Improvement but by the plumbing subcontractor for the original construction. That subcontractor's liability insurance would respond because the leak damaged work performed by other subcontractors (masonry, framing, and plaster, to name a few). Brad and Anita might have a hard time tracking down the original plumbing subcontractor. Perhaps this loss is covered by their homeowners policy.

Most homeowners policies cover property damage that is not specifically excluded from coverage. Most policies exclude coverage for loss caused by soil settlement. So the question becomes, did soil settlement cause the loss?

Here, the moving efficient proximate cause of the loss was a plumbing leak. Losses caused by plumbing leaks are not excluded, so Brad and Anita should be able to call upon their homeowners insurance carrier to fix the chimney problems.

14

ABC Warehouse VIII
Shop Drawings Procedure

It was a crisp dry Monday in fall, November 13, 1989. It was mid-morning and the pre-construction jobsite conference was almost over. The architect, Ivor Judge, of Judge & King, was just ending his usual explanation of the shop drawing process. He finished up with, "George, please submit your shop drawings early. Don't wait until the last minute. Give our engineering consultants and us a reasonable time for checking. And make sure you've checked them carefully yourself before sending them on to us. Also, I want to see all the shop drawing scheduling shown in your overall construction schedule."

He was directing his remarks to George Hyde, President of Hyde Construction Company, low bidder on the ABC Warehouse job. It was a 100,000 square foot concrete tilt-up warehouse with a 5000 square foot office wing. Hyde's bid was $2,100,437, low enough to get the job, and high enough, Hyde hoped, to make a few honest bucks.

George was standing there with a contrived fixed smile on his face trying not to look bored. He was fed up with listening to all the instructions on how to do things he'd been doing for 30 years. He'd heard Ivor's shop drawing lecture so many times he could almost recite it from memory. But he didn't really mind hearing it again, as this was one of the biggest jobs he'd landed in some time. The scheduled profit was substantial and he was looking forward to a successful job. Nothing could dampen his enthusiasm at this point.

When Ivor seemed to have run dry, George said, "You know us, Ivor, you can count on Hyde Construction. We always do a good job and we won't disappoint you on this one."

"Okay, I know, George. Another thing, we plan to follow the shop drawing procedure of the AIA General Conditions to the letter."

The contract had been signed on Friday, November 10, and construction was due to start 10 days later on Monday, November 20, 1989.

* * *

George, back at the offices of Hyde Construction, checked in with his office manager, Ulysses Vance, to make sure all the ABC contracts and purchase orders were being prepared and sent out to the low subcontract bidders. He told Vance to fax reminders to all those who had to submit shop drawings, samples, or material lists.

George explained, "Ivor is getting crotchety about shop drawings. He said he was going to make us toe the line on this job. He's gonna enforce the AIA General Conditions procedures to the letter. So maybe you better read up on it. He said it was Paragraph 3.11 of A201."

Vance listened attentively and opined philosophically, "It'll probably be to our advantage. It might eliminate some of the usual confusion. I'll get right on it."

* * *

Prospects were looking up at Newhall Iron & Steel. This was the third contract they'd bagged in a week. They were low bidder on structural steel and miscellaneous iron on the ABC Warehouse job. Their bid of $9480 included a nice profit. In fact, Abner Bent had been concerned that he might lose the job when he had added a proper mark-up for overhead and profit. But his ace in the hole was his plan to keep his material cost down by using some steel angles left over from a recent job. They weren't quite the right size, but near enough. Who'd notice? When he received the fax reminder from Hyde Construction, he directed his estimator and shop drafter, Chuck Dalton, to get hot on the shop drawings.

"Chuck, on your way in to work tomorrow morning, stop off at Hyde Construction and pick up a set of the plans and specs. And then let's get our shop drawings started."

Chuck replied, "Okay, Chief. Will do."

On Tuesday morning Chuck picked up the plans and specs at Hyde and, referring to Newhall's original take off sheets and bid, started roughing out the shop drawings. The work consisted primarily of a number of steel pipe columns with steel base plates and steel top saddles to support the large wooden glue-laminated beams. Their work also included beam-to-beam connectors, beam-to-purlin connectors, loading door corner angles, miscellaneous angles and bars, steel gratings and frames, and some galvanized iron handrails.

It was all pretty routine except for the 13 steel pipe columns. They were to be 6-inch inside diameter standard steel pipe columns varying in length from 25 to 27 feet to provide for the roof slope. They'd have to be specially ordered, since Newhall didn't carry these lengths in stock.

The shop drawings consisted of only three sheets of drawings so by the end of the day they were completed, reproduced, and ready for submission. Chuck told his boss, Abner, that he'd drop off the drawings at Hyde Construction's office on his way home. Abner suggested, "Why not drop them off at the architect's office instead and we'll save some time. We'll get the approval faster."

"Hey, that's a good idea, Chief. I'll do it."

By the time Chuck arrived at Judge & King's office, everyone had left for the day except Ivor. He was on his way out the door when Chuck showed up. Chuck said, "Here are the shop drawings from Newhall Iron & Steel."

"Well, don't leave them here. They should go to Hyde Construction first."

"Can't you just look at them and let me know if they're okay? Then we can get going on accumulating the materials and getting them fabricated in our shop."

"No. Hyde has to review them first. This is required in subparagraphs 3.12.5 and 3.12.7 of the AIA General Conditions."

Chuck was somewhat surprised at the architect's unyielding and harsh attitude, so he quickly left and headed for Hyde Construction where he flattened and stuffed the roll of drawings through the mail slot.

* * *

Two days later, on Thursday, the iron and steel shop drawings were back in Ivor's hands, having been delivered by Hyde's messenger. They'd been date-stamped in by Ivor's secretary. Ivor then recorded them in his shop drawing log for the ABC job. He assigned them the designation SD-1, since they were the first ones submitted. He first verified that Hyde Construction had stamped the drawings as an indication that they'd reviewed and approved them. Hyde's review would be primarily for verification of realistic job conditions but also for compliance with the contract documents. Ivor secretly harbored the suspicion that general contractors seldom checked the shop drawings before stamping them Approved but he felt he should follow the AIA General Conditions procedure anyway. He was pleased that they'd submitted six copies, since that was what was specified.

Ivor decided that SD-1 should be sent to his structural engineering consultants for review and approval, since the content was primarily, though not exclusively, part of the structural system. He prepared a letter of transmittal sending 5 copies on to Owens & Nolan for structural engineering checking, retaining a copy for architectural review.

He carried the remaining copy to Ed Flair, one of his bright young staff architects. "Ed, will you please review these shop drawings? Owens & Nolan will be checking them for structural design compliance and I'd like you to check them against the required architectural conditions. Please bring them to my office when you finish. If you have any questions, just call me. Okay, Ed?"

"Sure thing, Ivor. I'll get right on it." He wasn't particularly enthusiastic about this assignment, since checking shop drawings is pretty low on the scale of interesting things to do in an architectural office. But it was only three sheets so he felt he could live through it somehow.

He started right in on the task of comparing the shop drawings with requirements of the contract documents. He noticed that Newhall had changed the size of the loading door angle guards from 4 x 4 x 5/16 to 3 x 3 x 1/4. He thought to himself that this was no inadvertent error. It was done either to utilize materials they had on hand or to save a little money. He red inked the correction on the drawing. Everything else looked okay to him. He would hold the drawings until the structural engineering corrections came back from Owens & Nolan.

Later that day, when Myles Nolan was in Judge & King's office on another matter, he picked up the ABC shop drawings and brought them back to his own office. After logging them in, he assigned the checking to the engineer in charge of the original structural design, Grace Hull. "Gracie, can you get on this as soon as possible? Ivor is pressuring us so we don't hold up job progress."

"Sure, Myles. No problem." She waded right in, carefully and cheerfully, as usual.

An hour later she was back to Myles with the drawings marked up and stamped Approved as Noted. Expressing concern, she told him, "It was lucky we had a chance to review these drawings, Myles. Do you remember when we were asked to change the warehouse clear interior height under the roof framing from 20 feet to 24 feet? Well, our revised engineering calculations changed the pipe column size from 6-inches inside diameter to 7-inches, but we forgot to change the pipe size on the drawings. I've marked the shop drawings so they'll build it properly."

"Cripes, what a screw-up!" Myles knew that he would hear more about this. The change would add quite a few bucks to the cost. He'd keep a file copy and return 4 copies of the corrected drawings. He prepared a letter of transmittal and would drop off SD-1 tomorrow morning when he'd be at Judge & King's.

* * *

SD-1, marked up and approved by the structural engineers, landed on Ed Flair's desk. He added the architectural corrections and the Judge & King approval stamp and carried them to Ivor Judge's office. He pointed out, "Ivor, Owens & Nolan changed the pipe columns from 6-inches to 7-inches."

Ivor had a surprised and pained look on his face. "No! Dammit! Why the hell did they do that?"

Ed passed on the explanation he had received from Myles. Ivor knew that there would be a justifiable extra cost impact that was unavoidable. He told Ed, "Well, I just hope that George can hold Newhall Iron & Steel down to a reasonable price for this." Ivor knew that Allen Brady, ABC's president, would be far from pleased. Allen always made such a fuss when unexpected extra costs came to light. Ivor prepared the letter of transmittal to send three copies of SD-1 back to Hyde Construction. He finished logging SD-1 in from Owens & Nolan and out again to Hyde. His part of the paper work on SD-1 was done, except of course for any repercussions.

When the approved SD-1 was back on his desk at Hyde Construction, Vance saw the change in column size and wondered how these things happen. He said to himself, half aloud, "You'd think the engineers would know better than to change sizes of structural members after the job is bid. Newhall will jump all over this. No telling what it'll cost. Well, anyway, whatever it is we'll add on our mark-up too." He kept one copy for the office job file and one would be sent to Ezra Field at Hyde's site office. The remaining copy of SD-1 was messengered to Newhall. By the time Hyde's messenger arrived, the steel fabricator was closed for the weekend. He slid the bulky envelope under the door.

* * *

SD-1 was waiting for Abner when he arrived on Monday morning at Newhall Iron & Steel. He was pleasantly surprised that it took less than a week to travel the circuit from Newhall to Hyde to architect to structural engineer to architect to Hyde and back to Newhall. This could be a world record. When he unfolded the drawings and saw all the red markings, he was relieved that it was finally marked Approved as Noted. At least they wouldn't have to make any more drawings and then have to wait for their eventual circuitous return. He called Chuck on the intercom and asked him to come to his office. Meanwhile he was studying the red markings. The loading door angle guards were red inked. Chuck walked in and Abner exploded, "Damn it all! They caught the damn angle guards. I thought we could use up all those angles we had left over from the Zephyr Industries job!"

"Hell yes. They'd be just as good as the specified size. Those architects are ridiculous!"

"Well, we're stuck now. We'll have to buy all new 4 x 4 angle stock. That'll cost us a few bucks. What a stupid waste."

Chuck was looking at the marked-up pipe column details. "Look, Chief, they've changed the columns from 6-inch to 7-inch! And we've already bought the 6-inch pipe. It just arrived this morning!"

"Damn! Why'd you order them before we got the approved shop drawings?"

"Well, I thought we'd save some time and could start our fabrication sooner. How could I know those idiots would change the size?"

Abner was thinking. A sly smile replaced the scowl on his face. "Hell, Chuck, this may not be all that bad. We can charge a hefty extra for the larger pipe, enough to make up for the extra angle cost as well as the cost of returning the 6-inch pipe. We're in Fat City." He was clearly pleased with the fortuitous discovery and his deadly business acumen.

"Yeah, you're right, Chief. I'll get right on it. I'll order the 7-inch pipe and return the 6-inch. And I'll order the new 4 x 4 angle stock." Chuck was pleased that the Chief was so happy.

"Wait a minute, Chuck. We better get an understanding with Hyde for the extra costs before we spend any more money. Let's prepare an invoice showing the extra cost of the columns and we'll get it over to Hyde Construction. Find out how much the new angles and pipe will cost and how much it'll cost us to return the 6-inch pipe. Okay?"

"Yeah! I'll figure it out and be back to you in a few minutes." Chuck returned to his own desk to work out the prices. He referred to catalogues and recent invoices and phoned their suppliers. In 20 minutes he was back in Abner's office. The Chief was now in an optimistic mood. He lit a new cigar.

Chuck, referring to his papers and notes, reported, "The cost of returning the pipe'll be $250 including trucking and restocking charges. The new angle stock'll be $820. The cost of the total quantity of 7-inch pipe will be $2750 more than the 6-inch pipe would've cost. So, our total extra cost'll be $3820."

Abner was doodling on a note pad. "Okay, Chuck, write this down. We'll charge $4966 for the extra material. Add something to it for extra labor. Then add 10 percent for overhead and 15 percent for profit."

Chuck was punching his pocket calculator and roughing out the cost quotation. The finished calculation looked like this:

Change 13 standard steel pipe columns from 6-inch to 7-inch as required by SD-1	$4966.00
Extra welding and fabrication costs	$650.00
Extra field erection costs	$390.00
	$6006.00
Overhead, 10%	$600.60
	$6606.60
Profit, 15%	$990.99
Total	**$7597.59**

"The total is $7597.59. Not too shabby!" Chuck was proud of his invoice.

"Okay, Chuck, let's get it over to Hyde right away." Chuck left immediately and dropped it off at Hyde's on his way to lunch.

Newhall's extra work cost quotation landed on Vance's desk at Hyde's. He immediately roughed out Hyde's bill so as not to hold up progress. Hyde's cost quotation looked like this:

Newhall Iron & Steel Company cost quotation for changing 13 standard steel pipe columns from 6-inch to 7-inch as required by SD-1	$7597.59
Supervision by Hyde Construction	379.87
	7977.46
Overhead, 10%	797.75
	8775.21
Profit, 10%	877.52
	9652.73
Additional bond premium, 1%	96.53
Total	**$9749.26**

Vance brought the cost quotation in to George's office and placed it on his desk. George took one look and said, "Damn! This'll get Ivor's attention. And when Allen sees it he'll raise holy hell. But what can we do about it? They shouldn't have changed the column size. Well, let's run with it."

Vance had the cost quotation messengered to the Judge and King office. Ivor was getting ready to leave for the day when the Hyde Construction envelope arrived. Ivor was too curious to leave it for next morning. He carefully opened the envelope with an plastic imitation ceremonial Turkish scimitar with a local plumbing contractor's name and phone number on it. When he finished digesting the figures, he sat down. He rapidly punched his telephone key pad and got his structural engineer on the line. "Myles, do you know how much extra those 7-inch columns are gonna cost?"

"No. Why do you ask?"

"Well, it's $9749.26, almost Ten Grand."

The engineer wrote down the number and divided it by 13. "That's roughly $750 extra per column. I didn't think it would cost that much. It's a justifiable extra all right but it seems excessively high."

"Well, that's what it is, Myles. What're we going to do? When I tell Allen about this he'll have a heart attack. He'll collapse. Understandably."

"Ivor, fax me a copy of those cost estimates and I'll see if I can figure out what they're doing. I'll take it home with me. I'll work on it this evening after dinner."

Ivor thanked him and immediately faxed all the information to Myles. Ivor reflected, how ironic that today is Monday, November 20, 1989, the first day of construction on the ABC Warehouse job, and already we have the beginnings of a fiasco on our hands. He was despondent. How could a job go so sour so fast? Why couldn't everyone conduct themselves properly?

* * *

That evening, Myles sat down at his dining room table after dinner with his papers, notes, and calculations to see if he could understand how Newhall and Hyde had compiled their cost estimates. He got his thoughts together but needed more information. He could gather it by telephone the next day at his office.

Tuesday morning, now at his desk, Myles called friends at Acme Steel Fabricators and at Super Steel Suppliers and got the latest authentic information on costs of steel pipe, fabrication labor, and erection costs. He started assembling a cost estimate of his own.

* * *

On Monday morning, November 20, 1989, Ezra Field, Hyde's trusted job superintendent, had been getting the ABC Warehouse job mobilized on the site. Temporary fencing was underway, a power pole was partly constructed, chemical toilets were being placed, and a job office was being installed. It was organized chaos. The key subcontractors had been dropping by to see how things were shaping up. Some had left their shop drawings, product data, and samples with Ezra. As soon as the site telephone was operable, Ezra contacted the Hyde office and Ivor to give them the new number. Ivor told him he would be on the job the next morning and would be there at 8 o'clock every Tuesday from then on.

Ivor then called Allen Brady to let him know that the job was now underway and that he would be on the site the next morning for the first of his weekly site visits.

* * *

Promptly at 8 o'clock on Tuesday, Ivor drove onto the site. Ezra was busy on the telephone making final arrangements for the excavating contractor to start first thing the following morning. He hung up the phone and greeted Ivor with, "Gotta stay on top of the schedule, you know."

Allen Brady and Carl Daly, his warehouse foreman, soon arrived and walked over to the jobsite office where Ezra and Ivor were talking. After the usual mutual exchange of pleasantries, Allen and Carl stood listening to the arcane business discussion between the architect and the job superintendent.

After briefing Ivor on the scheduled activity for the week, Ezra said, "We're gonna need the steel column base plate bolts and templates early next week, for sure. Have you approved Newhall's shop drawings yet?"

"Yes. They were approved and sent back last Friday afternoon."

"I heard that the column sizes were changed and that Newhall was raising their price."

Allen's eyebrows arched up as he shot a quick glance at Carl. He turned to Ivor, stammering, "Wh...what? What's this? What's he saying?"

Ezra added, "Newhall's not going to ship our base plate bolts and templates until the extra's approved. Abner Bent told Vance."

"What? What extra?" Allen was looking back and forth from Ivor to Ezra. "What're you guys talking about?"

Ivor could have kicked Ezra. He knew he would eventually have to break the news to his client, but not this way. He tried to soften the blow to Allen, "There's been a mix-up in Newhall's shop drawings and they want some extra money. We're working on it now. We should have it straightened out in a few days."

"What? What mix-up? What kind of extra money?" pleaded Allen. He sounded panicky.

Ivor was trying to appear calm, radiating warmth and reassurance. "Well, Allen, the column sizes had to be increased. Don't you remember when the warehouse interior height was raised from 20 to 24 feet? When the columns got longer, they had to be increased from 6 to 7 inches in diameter to resist buckling."

Carl Daly, who had convinced Allen of the need for the increased height, chimed in, "But that was months ago, Ivor. The larger column sizes should have been on the drawings."

"I know. That's the problem. Our engineer forgot to change the size on the drawings. So, Newhall is justified in charging extra for the larger columns. The problem is that they're charging too much. We're studying the situation right now and expect to have it resolved soon."

This didn't placate Allen. He pressed on, "How much? How much are they charging? What's it gonna cost me?"

Ezra blurted out, "Ten grand!"

"Ten Thousand Dollars? You've gotta be kidding! I'm paying over two million for this warehouse and now you people want ten thousand more? No way!" He was pale.

Allen and Carl left. Carl had to drive as Allen was too distraught.

Ivor was displeased with Ezra's undiplomatic comments. "Dammit, Ezra, did you have to lay it on him like that? We'll have the whole thing reduced to reasonable proportions in a few days."

"You better have it settled. We'll need those bolts and templates early next week, like I said."

* * *

When Allen got back to his desk in the old warehouse office, he immediately called his lawyer, Philip Quinn, of Quinn & Quinn. He blurted, "Phil, we're being screwed. Those damn contractors are ganging up on me. And the architect is in on it too. I'm dealing with a bunch of money hungry pirates."

"Calm down, Allen, what's going on? Tell me slowly. I'm taking notes."

Allen told him all he knew but Phil felt there must be more to the story. He volunteered to phone Ivor and try to get the complete picture. Allen now felt a little better with Phil on his side, but deep down he knew that part of the ten thousand extra would stick and he'd have to pay his lawyer as well. He felt victimized, trapped, and helpless.

After finishing with Allen, the lawyer called Ivor and made an appointment for him and his client to come to Ivor's office the next afternoon at 2 o'clock. "And Ivor, please have George Hyde there too, if you can."

"Sure, Phil. I'll do my best."

Ivor called George to get him lined up for the meeting. He then phoned his engineer. "Myles, this whole thing's coming to a head quicker than I imagined. You gotta be here tomorrow at 2 o'clock. They'll all be here. Allen, his lawyer, and George."

"Okay. I'll be there. I've gathered some information on steel pricing."

* * *

On Wednesday afternoon a few minutes before 2, they started arriving at the stylish offices of Judge & King. A friendly secretary showed them to the comfortable conference room and invited them to help themselves to coffee.

George arrived first and appeared to be in excellent humor. However this came out today, he'd be in good shape, since the extra would always include his component. It wasn't his fault if costs kept going up.

Allen and Phil came in together and greeted George. Allen was solemn and untalkative. He sat somberly next to his lawyer across the table from the affable contractor.

Myles had arrived about a half hour early and had ever since been closeted with Ivor in Ivor's office. They'd been reviewing all the facts and figures they had available. At precisely 2 they solemnly filed into the conference room, Ivor taking his seat at the head of the table. There seemed to be tacit expectation that he'd preside.

Ivor led off with a complete rundown of the situation underlying SD-1. He quickly came to the bottom line. "Therefore, Newhall is clearly entitled to be paid for increasing the size of the columns. The problem is, they're simply charging too much. Their bill is beyond reason."

Allen rowed in, "That's not the only problem. Why should I have to pay for mistakes made by my architects and their engineers?"

Phil interrupted, "I'd suggest we take one thing at a time. Let's talk about money first, then we can talk about who pays. Okay?"

Ivor invited Myles to reveal his information on pricing. Myles explained, "I've checked around and found that Newhall's bill is way out of line. I've prepared a pro forma cost estimate that's more in keeping with current industry prices." He passed out sheets of paper showing his analysis of the Newhall charges:

Change 13 standard steel pipe columns from 6 inch to 7 inch as required by SD-1	$2800.00
Extra welding and fabrication costs	$250.00
Extra field erection costs	$120.00
	$3170.00
Overhead, 10%	$317.00
	$3487.00
Profit, 15%	$511.05
Total	**$3998.05**

Then he passed out another sheet showing a suggested revision to Hyde's bill:

Newhall Iron & Steel Company cost quotation for changing 13 standard steel pipe columns from 6 inch to 7 inch as required by SD-1	$3998.05
Supervision by Hyde Construction	$159.92
	$4157.97
Overhead, 10%	$415.80
	$4573.77
Profit, 10%	$457.38
	$5031.15
Additional bond premium, 1%	$50.31
Total	**$5081.46**

Allen, looking at the total, observed, "Well, this is a much better figure, but I don't see why I should have to pay it."

Phil interposed, "Allen, let's let Myles explain his figures."

Myles continued, "It's obvious that Newhall has inflated or overestimated their costs of acquiring the materials and they've estimated much too high on their extra fabrication costs. I've adjusted the figures down to realistic industry-wide current costs. All of the subsequent mark-ups and add-ons are reduced accordingly as they are all based on a percentage of the bare costs. The figures are easy to follow."

At this point, George dipped his oar. "Even if Myles' figures prove to be acceptable to Allen, how do we know that Newhall will agree to them?"

Ivor sought to resolve the issue by making a ruling under paragraph 4.4 of the AIA General Conditions. He explained, "We have a claim here by the contractor of $9749.26 which the owner is reluctant to pay. I'm in agreement with the owner that the amount is too high, but my ruling will be that the owner must pay the amount proposed by Myles Nolan, $5081.46. I'll put this in writing and this will be my decision under paragraph 4.4. If either the owner or the contractor wish to appeal the ruling they may do so by demanding arbitration within 30 days. If no appeal is made, the ruling will be final."

Allen turned to Phil, "Can he do this? Do I have to pay?"

"Yes. That's what it says in the AIA General Conditions, which is part of the contract. We could go to arbitration but I wouldn't expect to get very far with it. After all, you'll be receiving the value of the larger columns. After paying my fees you'd have been much better off paying the extra."

George also had his reservations. "I can see the sense of this but I'm not sure that Newhall will agree."

Phil suggested, "George, you might tell Newhall that they can appeal this ruling but I can't visualize any arbitrator buying their story in the face of Myles' analysis. Considering their legal fees and time lost they'd be better off to accept this ruling. This would also cost Hyde Construction a lot of time and legal fees considering that Hyde must furnish the conduit to facilitate the arbitration."

George saw the logic of Phil's remarks. "I can see the advantage of accepting Ivor's decision. Just leave it to me. I'll talk to Abner Bent and get him shaped up."

Ivor summed up, "I'll issue my ruling in writing and mail it out later today. I'll also prepare Change Order # 1 in the sum of $5081.46, with no contract time extension, and sign it. I'll also expect Allen and George to sign it. Is this acceptable to both of you?"

Allen and George looked at each other and at Phil.

Allen hesitated and quietly said, "Yes, I'll agree."

George sighed and said, "Okay, I'll buy it."

Phil said, "Congratulations, Gentlemen. This is a civilized and rational way of avoiding a bitter and costly dispute that no one can win."

* * *

Points of Law 14

ABC WAREHOUSE VIII
Shop Drawings, Changes, Claims, and Dispute Resolution

14.1 The Construction Contract

A well articulated construction contract includes an agreement, general conditions, special conditions, drawings, specifications, and addenda. The documents may run to tens of thousands of words. Most construction contracts, almost of necessity, are based on standard forms. Forms are supplied by commercial publishers, and by construction industry organizations such as the American Institute of Architects (AIA), the Associated General Contractors (AGC), the National Society of Professional Engineers (NSPE), and the National Electrical Contractors Association (NECA). As one would expect, forms are designed to protect the interests of the constituencies of the sponsoring organizations. As a general rule, subcontractors are justifiably reluctant to accept subcontract forms sponsored by AGC, while general contractors likewise view with considerable skepticism the forms of subcontract that are sponsored by NECA and other subcontractor associations.

14.2 The AIA Documents

The most generally accepted forms are those sponsored and distributed by the AIA including the form of general conditions identified as the A201. The architect often exercises controlling influence over the form of contract utilized because the architect is employed by the project owner. Competitive forces usually put the project owner in a position to dictate the contract form.

When a project is competitively bid the contractors essentially either accept the contract documents as prepared by the architect or forget about bidding the job. When a contract is negotiated, however, the contractor is free to bargain over contract provisions.

The 1997 edition of A201 is 44 pages long and is divided into 14 articles, each of which includes many subparagraphs. Every day, across this country, dozens of contractors, subcontractors, owners, architects, and their lawyers (and plenty of judges and arbitrators too), review the AIA general conditions and apply their provisions to the facts of an actual or potential construction dispute.

One young lawyer was driven to poetry when confronted with the A201:

Ode to the AIA General Conditions

I've worked very hard upon your file,
Gathered my materials, stacked 'em in a pile,
Dreamed up theories, researched the law,
Scribbled with a pencil 'til my knuckles got raw,

Construction Nightmares

Toiled countless hours at a furious pace,
And here's my analysis of your case:

A workman employed by a subcontractor,
Was connecting an electrical wire,
A slight mistake in his calculations
Resulted in a terrible fire.
Our client (the insurer) paid the loss,
We're suing to get the money back.
And the issue you asked for research on:
The effect of the AIA contract.

The contract is made up of several parts,
Which you know (deep down, in your heart of hearts),
Means that each must be parsed individually,
To interpret the whole "holistically."
The AIA form is nineteen pages long,
With dozens of difficult terms.
The paragraphs twist and wind their way,
Like a can full of crawling worms.

Yes, a can full of tangled-up paragraphs/worms,
Buried in handfuls of verbiage/earth,
You want me to find how they all fit together,
And tell you what the can is worth.
So I pluck out a worm, stretch it out on the table,
Determine its scope and intent (if I'm able);
Not so fast, Mr. Worm, just a few questions more,
How have the courts interpreted you before?
And how are you related to all the others in the can?
You tell it to me and I'll write it down, man.

When I learn what he knows, where he's been, what he's done,
Throw him back in the can and grab another one.
And when I've examined the whole bloody lot,
Look back in the can, and what have I got?
A big mess o' worms, all tangled and twisted,
And it's not much help that I carefully listed
What each court said about each sonofabitch,
When I look in the can, I can't tell which is which!

Anyway, it looks like the workman is a liable guy,
And so is his company, by the by,
And even against the "prime", liability will lie,
And I'll soon write a memo, telling you why.

--William L. Bryan

We are concerned with the A201's treatment of shop drawings, changes, claims, and dispute resolution. Each subject is covered by several carefully worded paragraphs.

14.3 Drawings

A set of plans and specs represents an effort by an architect and its consultants to establish a perfect representation in two dimensions and small scale of a construction project to be built in four dimensions and at full scale. (The fourth dimension is time: the schedule by which the project will be built.)

The only perfect and unambiguous expression would be the completed project itself. But the architect works not with lumber, bricks, and mortar, but with paper, pencil, and CADD programs. Drawings and specifications utilize numerous conventions that must be understood by the construction professional who is expected to know how to "read plans."

14.4 Shop Drawings

No matter how fully articulated a set of drawings may be, some details are left to the discretion of the trade contractors who make their living working everyday with the systems, materials, and equipment available in the marketplace. Trade contractors are required to submit shop drawings to be reviewed and approved by the general contractor, the architect, and its consultants. Approval is usually signified by a stamp. Architects and engineers exercise impressive ingenuity in selecting the words that are embossed upon the stamp, often eschewing even the word "approved". For example,

> Reviewed for basic conformance with the general intent of the drawings and specifications.

The contract documents usually contain a paragraph providing that the architect's stamp does not constitute approval of any deviation from the contract documents unless that the deviation has been specially called to the attention of architect and has been specifically approved by the architect. The architect's approval of shop drawings is a reverse parody of the girl who, saying "no", means "yes".

14.5 Reading Plans

Many judges learn from trying cases quite a lot about how to "read plans", they are not usually qualified (any more than juries) to convert words, lines, and squiggles on paper into a true representation of three dimensional reality. Lawyers are encouraged by the law to make their points about the interpretation of plans and specifications through the introduction of expert testimony. A credible expert is qualified by education, study, experience, and familiarity with the customs and usages of the trades to understand what the plans mean and to offer a credible opinion as to whether the performance of the contractor or the subcontractor fulfills the requirements of the plans. A "good expert" expresses opinions in a convincing way and has the wit to parry the thrust and slash of cross examination. Ordinary witnesses are permitted to testify only as to facts that they have perceived. Expert witnesses, once their qualifications have been approved by the judge, are permitted to offer opinions on subjects within their areas of competence. As a matter of course, contractors, subcontractors, architects, and engineers are experts and may therefore express their opinions. Opinion testimony is limited to the expert's area of competence. A roofing contractor, for example, would not be permitted to offer opinions on glazing issues.

14.6 Changes

The law allows contracts to be changed. The requirements for the formation of a contract also apply to changes. A change to a contract may be made by competent parties with a legal object who objectively manifest their mutual assent to do or not to do a thing, which assent is supported by consideration on both sides.

Competent Parties. Parties who are intoxicated, insane, or infants are not competent to contract. Suppose the inspector on a city street repair project asks the contractor to increase the thickness of the AC paving and promises that the city will pay an extra dollar a square foot for it. Such a promise to pay extra compensation would not be legally enforceable because the inspector is not authorized to bind the city.

Legal Object. Parties agree to change a construction contract by substituting Romex (not approved by the building code) for rigid conduit. No enforceable contract.

Mutual Assent. Contractor says she would like to omit the ornamental iron from a project. Owner remains mute. The contract is not changed because there has not been a manifestation of assent by the owner.

Consideration. As above, but the owner does not remain mute. The owner agrees. Still the contract is not changed because the owner's agreement is not supported by consideration. An enforceable change would occur if the parties agreed to reduce the contract price.

14.7 CCD's

Just as a contract can only be formed by the consent of both parties, so can it only be changed by the consent of both parties. Allen, on his architect's advice, wants to change from six-inch to seven-inch pipe columns, and Newhall Iron is trying to gouge Allen on the price. Actually, it is a little more complicated than that. Newhall Iron is trying to gouge the prime contractor, Hyde, which would pass the gouge on through to Allen plus markup. If Newhall Iron won't agree to a reasonable price, what can Allen do? He might consider having Hyde delete the columns from the scope of Newhall Iron's work. But that would also be a change to the contract and requires Newhall Iron's consent. Moreover, it would mean that Hyde would have to find another subcontractor to install the columns and that would be expensive. So it looks pretty much as if Newhall Iron has Allen at its mercy unless there is a provision in the contract that would allow Allen to change the scope of the work without Newhall Iron's consent. All well-drafted construction contracts include such a provision. Under the 1997 version of A201, the owner, with the approval of the architect, has the power to issue a Construction Change Directive (CCD) and the contractor must perform the work as modified by the CCD whether it wants to or not.

14.8 Flow Down Clause

The subcontract document between Hyde and Newhall Iron should contain a "flow down" clause by which the subcontractor undertakes toward the prime contractor all of the obligations that the prime contractor undertakes toward the owner. The flow down clause would put Newhall Iron under a contractual obligation to perform the work, as changed, even without any agreement as to the price for the change.

14.9 Pricing the Change

Most contracts provide that if the contractor does not agree to a CCD the work will be performed on a time and materials basis with the contractor keeping track of the costs and submitting payroll, material, and equipment invoices with backup on a daily or weekly basis. This would leave open to dispute the amount of credit to be allowed for the deletion of the six-inch columns.

14.10 Pass Through

In case of dispute about the charges and credits to be assessed and allowed for changes in the work, the owner usually pays the undisputed amount to the prime contractor who passes it on to the subcontractor, and the subcontractor makes a claim against the prime contractor which is, in turn, passed on to the owner by means of a "pass through" clause in the subcontract document. The pass through clause provides that the subcontractor's claims will be passed through to the owner by the prime contractor, and that the subcontractor's recovery against the prime contractor will be limited to the amount that the contractor recovers from the owner.

14.11 Dispute Resolution

In the first instance, claims are presented to the architect who is given few days to decide the claim. Under A201 the decision of the architect becomes final unless a party promptly appeals from the decision by filing a demand for arbitration. Arbitrators are then appointed, and the claim decided by arbitrators under the rules of the American Arbitration Association. The decision of the arbitrators is final. If the parties do not abide by the decision it may be confirmed by the court and thereby converted into a legally enforceable judgment. The confirmation proceedings are summary and can usually be completed in about 60 days.

14.12 Conflict of Interest

The position of the architect in our story is a difficult one because he is expected to impartially resolve a claim against the owner. But the architect is being paid by the owner, and it is the duty of the architect to protect the interests of the owner! How can an architect be expected to be impartial in deciding claims against its own client? This is a job that would seem to require almost superhuman virtue and integrity! The contractor and the subcontractor may well doubt that the architect is a person of superhuman virtue and integrity.

If that isn't troublesome enough, here we have the added complication that extra cost is to be incurred because of a bust in the architect's plans. It's true that the bust was caused by the structural engineer, but as a matter of law a principal is liable for the misconduct of its agent, and the structural engineer, in preparing the structural drawings and specifications, acted as the agent of the architect. Therefore, in effect, the architect will here be asked to act as a judge in its own case. That violates the most elementary concept of due process!

14.13 The Resolution

Under Ivor's decision here, the cost of the change has been cut down from $9,749.26 to $5,081.46, and the owner will have to pay the extra $5,081.46 to the contractor. As between the owner and the contractor this is the proper outcome because just as the architect is responsible for the mistakes of its consultants, the owner is responsible to the contractor for the mistakes of the architect. (There are exceptions to this rule. And exceptions to the exceptions!) But the owner still has a legitimate gripe here. He is going to have to pay another $5,081.46 to his contractor because of a bust in the plans caused by the negligence of the structural engineer. So what happens? The owner backcharges the architect and the architect backcharges the structural engineer. That's the way it would end up if the whole mess were resolved in court.

Only $5,081.46 is in dispute here. It would be foolish to go court with five parties, each represented by counsel. The attorneys fees would soon exceed the amount in dispute.

15

Venture Tower
Responsibility For Window Wall Leaking

It was late on Friday afternoon and Frank Grimm, FAIA, was just finishing up his final report on an extensive wall-cracking problem plaguing the owners of a newly constructed suburban concrete block bank building. He was carefully selecting precise unequivocal language so no one could possibly misconstrue his findings and conclusions.

The insistent ringing of the phone disturbed his concentration. It was Clyde Dixon, Claims Manager at Bulwark Professional Indemnity Company, an insurer of architects' and engineers' professional liability. Frank was retained from time to time by BPIC to look into situations involving claims against architects and engineers for building defects or maladministration of construction contracts.

Clyde briefed him. "Our insured is an architectural firm, Judge & King. They designed a multi-story office building and subterranean parking garage on Wilshire Boulevard in Beverly Hills for Venture Properties."

"I know the building, Clyde. Isn't it the Venture Tower Building on the South side of Wilshire near Beverly Drive?"

"Yes, that's the one. Well, the architects have alerted us that there's severe rainwater leaking through the window walls. There's been considerable interior damage and a substantial claim could be lurking. BPIC wants you to go out and look at the building and review the construction drawings and specifications. We want you to determine, if possible, the cause of the leaking and if the architects are responsible for it in any way. We want to know if it's a design defect or a construction defect. We need to know whether we need to establish a reserve for this potential loss."

"Who should I contact?"

"I'll fax you a list of names, addresses, and phone and fax numbers of everyone involved, the building manager, architect, general contractor, and glazing subcontractor. Talk to anyone you wish."

"Okay, I'll start with the building manager."

Shortly after Clyde left the phone, Frank's fax machine sprung to life. Frank then called Venture's building manager, Emily Fox, and made an appointment for 9 o'clock the following Monday.

* * *

When he was a young idealistic architectural student at USC from '46 to '51, Frank Grimm never dreamed that this is the way some architects make their living. He had now been an architect for over 35 years, the last 10 of which were spent in forensic consulting. He had gotten into it gradually and now spent most of his time examining buildings in trouble, determining the cause of the problems, and trying to devise practical and economical ways of repairing them.

His assignments were always interesting, no two exactly alike, each one requiring independent research and analysis. He was astonished at the variety of problems that could occur even though competent designers and constructors were in diligent attendance. In only rare instances the incompetence or dishonesty of one of the participants proved to be the cause of unsatisfactory results. He was well aware that the easy problems would have been solved by the participants and that only the difficult ones would be referred to him. He also recognized that often the people involved in design and construction defects were not sufficiently objective to recommend fair and proper solutions.

Frank's utilitarian office was sub-leased from a busy, large architectural firm, Bennett & Fielding. His rental included access to their extensive library of architectural and engineering texts, building codes, architectural periodicals, and catalogs of building materials and equipment. A conference room was his upon prior reservation. He also had full use of their reproduction equipment. He had no employees, not even a secretary, and wrote all his own reports on a word processing computer. His combination telephone answering and facsimile machine rounded out his office resources. On those rare occasions when he needed help, he was able to borrow suitable personnel from Bennett & Fielding.

* * *

Frank arrived at the Venture Tower Building a few minutes before nine, turned his car over to the parking attendant, and went on to the lobby. The lobby security attendant directed him to the office of the building on the sixth floor. Emily Fox was waiting at the door when Frank arrived at Suite 600. Emily invited him into her office and promptly started filling him in on the building and its problems.

"Venture Properties owns this building. We're a real estate investment company and we own and operate 13 high quality buildings with a total of over four million square feet of prime office space. Our president is Allen Brandon. Perhaps you've heard of him?"

Frank Grimm nodded and she went on, "The building isn't quite a year old and we've been experiencing severe rainwater leaking through the window wall from day one. The interior damage to finishes and furnishings has been extensive. We've spent a small fortune repairing the tenants' improvements and cleaning and mopping up after each storm. Most of the tenants are extremely upset about this and some are withholding rents. Several have had their lawyers write us letters of protest and putting us on notice that they're considering cancellation of their leases and suing us for damages."

Frank, rapidly taking notes, asked, "Is the leaking occurring in any particular place? For example, any particular side of the building or any particular floor level?"

"No, it leaks on all four sides and on all 18 floors." She added, "We've kept a log on all the leaking, the damage, and the repairs." She'd made a copy of it for him.

Frank asked, "Do you or any of your maintenance people know what causes the leaking?"

"No, but there's one peculiarity that you should know about. We discovered that the leaking has been very minimal during the working day but gets considerably worse at night and on weekends and holidays. Our building engineer figured that one out. He concluded that our air conditioning fans created sufficient internal air pressure to resist the water intrusion. When the air conditioning system was turned off at 6 o'clock the rainwater could penetrate the window wall more readily. So, now we keep the fans on nights and weekends when rain is forecast. It costs quite a bit of money to run those gigantic fans continuously so we don't do it if rain isn't expected. However, we got caught a couple of times when unpredicted rain showed up."

Frank inquired, "Is the building totally occupied?"

"No. We had the building about 70 percent pre-leased before construction started. But now we're having difficulty in leasing the remaining five floors on account of the leaking. Everyone in the office leasing business seems to know about it."

"So far, what have you been able to do about the problem?"

"Two things. First, we have to clean up after every water intrusion. We have to mop up and dry out the tenants' spaces. This has cost us hundreds of hours of extra janitorial labor. We've used thousands of paper towels. We have to lift the carpets and bring in large fans to dry out the carpeting and padding. We usually have to replace the padding. Then we repair the damage. We've had any number of wallcovering installers, drapery cleaners, and painters working here. This all costs money and the tenants are still unhappy.
"And, secondly, we call Hyde Construction, the general contractor, and they call the glaziers who send workers over to inspect the windows and apply caulking. But it never does any good. If anything, it's getting worse."

"Has Hyde Construction been cooperative?"

"Yes, very. Whenever we call them, they always promptly get the glaziers into action. Creative Glaziers send a couple of workers over here right away. Personally, I don't think they know what to do. They borrow our window washing platforms and work on the outside of the building, usually applying some sort of sealant. It generally takes them three or four days to do it."

"Have the architects looked at the problem?"

"Yes, but they don't seem to know what's wrong or what to do about it. They just look around and appear puzzled."

Frank and Emily then toured the building looking at typical offices and representative damage. Many of the tenants were creatively caustic in their comments about the quality of the building.

They entered some of the vacant spaces where the exterior wall construction could be better observed, since interior finishes were not yet installed. Frank was able to remove some of the snap-on extrusions to get a better look at the entire assembly of aluminum, glass, vinyl gaskets, and the screwing and fastening of connections within the window wall and its attachment to the structural frame of the building. He made numerous measurements and sketches of the arrangement of materials. He photographed dozens of joints and intersections.

Frank then made arrangements with Emily Fox to return on Wednesday morning at 9 o'clock to inspect the outside of the building. She promised to have the window washers present to operate the motorized window washing platform so Frank would have access to all parts of the exterior of the window wall.

As Frank was leaving, Emily remarked, "This is the most beautiful building I've ever worked in...and the most unpleasant."

* * *

Upon his arrival Wednesday morning, Emily led Frank up to the roof and introduced him to the two window washers, Max and Joe. They already had the window-washing platform slung into position on the outside of the building and on a level with the top of the parapet wall. They strapped Frank into a webbed harness, then snapped it onto a safety cord that was independently attached to the building. The safety cord was equipped with an automatic braking mechanism that Frank would have to be mindful to adjust as the scaffold descended the outside of the building.

Max explained how it worked, "If, for any reason, the scaffolding should collapse or fall while we're out there, this lifeline will save you. The platform would fall free but we'd be left dangling safely in the air. Then we'll be rescued by hoisting us on up to the roof or lowering us to the ground. Understand?"

"Uh, yeah." Frank had a vivid mental image of himself and Max dangling in the air with the scaffolding plummeting down to the earth 225 feet below.

Max continued, "The platform will only accommodate the two of us, so I'll go with you to operate the motors and keep the cables and lifelines in order. Joe'll wait for us up here. Okay? Unnerstand?"

Frank answered tersely, "Yep." He had his hands full with a sketch pad, camera, and shoulder bag with extra film, paper, pencils, markers, scales, and measuring tapes. He also had a solid grip on the lifeline.

Max asked, "Do ya hafta go to the toilet? If ya do, now's the time 'cause we got no toilet on the scaffold."

Frank said, "No, I'm okay."

Max scampered up and over the parapet and onto the platform. He beckoned to Frank, and said, "Okay, Mister Grimm, c'mon aboard."

As Frank followed Max onto the scaffold, he detected a sudden movement underfoot as the platform shifted with his added weight. He happened to look down the 12-inch space between the parapet wall and the scaffolding. He could see the concrete patio 18 stories down. As his heart skipped a beat, he thought, "What the hell am I doing up here? Why did I ever agree to do this?"

After he was safely on the platform and the entrance gate chain was snapped firmly into place, Max operated the motors for the slow descent. Frank got used to the gentle swaying of the rig and gradually lost his apprehension. He had a solid grip on the guardrail and risked a glance at the distant view of the Santa Monica mountains and the busy traffic below on Wilshire Boulevard. He couldn't resist taking a few photos of the breathtaking views.

Max explained to him that while the vertical cables support the load of the scaffolding, the two cables at the ends of the platform are there to keep the rig from being windblown. The wind cables are attached to trolleys that run down vertical tracks on either side of each window bay. The tracks are integral with the window wall system. Frank felt safer after Max explained all of the rigging and safety devices.

Frank then turned his attention to a detailed examination of the anodized aluminum extrusions that formed the window wall system. He examined the arrangement of the glass and the vinyl gaskets. He paid particular attention to the joints between materials and the way intersections were formed. He observed the weep holes that provide a drainage outlet for any water entering the system. He noticed that the long vertical aluminum members had thermal expansion and contraction joints at every other floor line, as this is a convenient length for the extrusions. He could see extensive evidence of the caulking and sealing repair efforts. He noticed that most of the weep holes were now inoperative as they had been sealed shut. In examining the vertical trolley tracks, he noticed that one could look into the track and see the open joints on the backside. He made numerous sketches and took close-up photographs from all angles.

Frank and Max took about three hours to make the descent down one bay of the window wall from the parapet above the eighteenth floor down to the ground. When they reached the brick patio level, Frank stepped off the platform and left Max to ride it back to the top alone. Frank thanked Max for the safe ride and went back to his office, stopping only to drop off four rolls of 35-mm film at the one-hour photo depot. He then called Judge & King, AIA, Architects, and made an appointment with Leo King for Friday at 9 o'clock.

* * *

As Frank was driving to the Judge & King office, he was aware that the firm had a fine reputation for outstanding design and competent contract documentation. He wondered what could have gone wrong in this situation. He didn't really know Ivor Judge or Leo King although he had worked with both of them on AIA committees through the years.

When he arrived, Leo King was ready with the contract drawings and specifications spread out on the conference room table. Leo King described the project, "It's an 18 story steel frame building, 200 feet x 130 feet, 26,000 square feet per floor, a total of 468,000 square feet, over a four level subterranean parking garage for 1200 cars. It's a first class building with a total construction cost of 33 million dollars."

Leo continued, "The building was about 70 percent pre-leased to highly solvent firms well able to pay the premium rents, and they've all spent big money on their tenant improvements. I've never seen such luxurious interiors, the very best in carpeting, wood paneling, vinyl and fabric wall coverings, the works. The restaurant on the top floor is exquisite. If you ever have a chance to go there, you'll see what I mean. Great interiors, fantastic food, and high prices."

Frank, browsing through the documents, asked, "Can I see the approved shop drawings for the window wall?"

"Sure. I'll go get them." He left and was back shortly with a roll of shop drawings. He unfurled them onto the table.

Leo volunteered, "The first submission of window wall shop drawings was really pathetic. They were based on using Top-Notch's materials. But, don't let their name fool you. We rejected them. Later, when we received these new shop drawings we were very happy to see that they were prepared by Quality Window System Corporation. We've always considered Quality to be the best manufacturer in the business. We wanted the very best for Venture Tower. We, of course, have been very disappointed that the window wall hasn't measured up to our expectations. We don't know what went wrong. We've written a dozen letters to Hyde Construction to get them to rectify the problems and we've withheld money."

Frank stood to leave, picking up all the documents. He said, "Well, I'll take the construction drawings and specifications and the shop drawings with me to study. I'll return them."

"We're really pleased that you are looking into this for us, Frank. We feel certain you'll find the cause of the problem."

* * *

On the way back to his office, Frank stopped off at the film developers to pick up the photos taken at the Venture Tower Building. Back at his desk, Frank spent a couple of hours reviewing the construction documents and the shop drawings. He carefully examined the photographs, and his own field sketches and measurements, comparing them with the shop drawings. He also reviewed the Quality Window System Corporation catalog in Section 8 of the latest Sweet's File.

Frank decided to call Clyde Dixon to let him know how the investigation was progressing. Getting Clyde on the line, Frank told him about his building inspection, his talk with Leo King, and his cursory initial document review.

"It's too early to draw any firm conclusions, Clyde, but it's plain to see that the window wall in the field doesn't match the shop drawings. I don't know, yet, why the leaking occurs. That will require considerably more study and analysis.

"Tomorrow, I'm going to visit Hyde Construction to see what they know about the situation. After that I'll look up Creative Glaziers.

"Okay, Frank, thanks for the update. Keep me informed."

Frank then called Hyde Construction and made an appointment for early Monday morning.

* * *

In the busy Hyde Construction offices, Frank was shown into George Hyde's ample private office-cum-conference room. This was his first opportunity to meet the colorful head of Hyde Construction although he had heard a lot about him and the company. In the local construction industry Hyde enjoyed a fine reputation as competent builders and tough but fair competitors. They landed some of the region's most prestigious jobs, producing quality work at fair prices.

Sam Trent, Hyde's chief estimator was also present. He was sitting quietly waiting for Hyde's signal to speak. George Hyde was a rigid taskmaster and most of his employees were afraid of him, especially when they made mistakes. Hyde held Sam Trent responsible for their signing up with Creative Glaziers. He felt that all the subs should be responsible for correcting their own errors and he didn't want to know why it didn't work. Hyde was still retaining $190,000 of Creative's money, representing the final 10 percent. Venture Properties, similarly, were retaining around $300,000 from Hyde on account of the window wall leaking and a few other less significant matters.

George Hyde appeared to be embarrassed that one of their projects would have such a serious defect. He explained, "We're staying on top of our subcontractor on this. It's up to them to fix it, dammit. Every time we get a service call from Venture Tower, we call Creative Glaziers and they promptly send two or three men out to make the necessary repairs. We're doing everything we can about this. Our lawyers have put their surety on notice." He neglected to mention that Venture's lawyers had also put Hyde's surety on notice.

Frank asked, "How did you happen to pick Creative Glaziers for this job? Frankly, I've never heard of them."

George Hyde looked pointedly at Sam Trent, who answered the question.

"Well, they were low bidder. We had five window wall sub-bids. Creative's bid was $1,900,000. The second bidder was $2,055,000, the next two were $2,060,000 and $2,080,000, and the high bidder was $2,098,000.

So, you see, Creative was almost 10% below the second bidder. We thought their bid was okay because they bid 'plans and specs' and were able to get the required surety bond. We didn't know the company but heard they had moved here from Kansas City about six months previously. They had a lot of Midwestern references."

Frank asked, "Did you check any of their references?"

Sam quickly snapped back, "No, but if we hadn't used their bid we wouldn't have gotten the job. We'd have been second, not lowest."

Frank asked, "George, would you please call Creative Glaziers and tell them I'll be calling them for an appointment?"

"Sure, I'll call them right now. The president is Farley Grimpenmire." George got him on the line and said, "Here, Grimpenmire, talk to Frank Grimm." He handed the phone to Frank.

* * *

That afternoon, after lunch, Frank visited the premises of Creative Glaziers. They had a shop and warehouse combination and a small office. When he arrived there was only one person present. He requested, "I'd like to see Farley Grimpenmire, please."

"That's me. Just call me Farl. C'mon into the office. I'll make us a cuppa coffee."

Frank got right to the point, "I want to talk about the Venture Tower job."

"So do I. To start with I have $190,000 coming and have for over a year. Now my surety company is bugging me. And I'm still spending money on the job. When am I gonna get paid?" He was resentful and angry.

"I wouldn't know about that. You'll have to take it up with Hyde Construction. I want to talk about the window wall and why it leaks so badly."

"Well, whadda you wanna know?" Farley seemed open and cooperative.

"I've been looking at the shop drawings and they don't seem to match the work as it's built. Why is this?"

"That's an easy one. When we bid the job we got sub-bids from three materials suppliers. Superb Window Products was high at $475,000. Quality Window System was next at $465,000. And Top-Notch Aluminum was low at $375,000. We were trying to get the contract from Hyde so we used the Top-Notch price in compiling our bid. If we hadn'ta used it we wouldn'ta gotten the job."

Sounds familiar, thought Frank. He knew something about all three suppliers. Superb and Quality were about on a par. Both companies had designed all their own extrusions. They each had engineering departments and had tested all their components and assemblies. Their research and analysis was ongoing and resulted in constant improvement of their window wall systems. The two companies made high quality products that have stood the test of time. Top-Notch, on the other hand, sold generic extrusions that could be assembled in different ways to make store fronts, windows, office partitions, window walls, skylights, and greenhouses. They didn't make integrated systems, as such, but they sold all the parts needed by glaziers to construct practically anything needed by the trade. They also sold various types of vinyl gaskets and a vast array of connectors, fastenings, sealants, and general glazier's tools and supplies. Some of their components were very similar to Superb's and Quality's but not exactly the same. They kept their prices down and distributed mostly through building material dealers and glazing contractors.

Farl continued, "As soon as we got the signed contract we asked Top-Notch to prepare the shop drawings. We submitted them to Hyde and they sent them on to the architect. A few days later they were back here marked 'Rejected - Disapproved - Revise and Resubmit - Does not comply with the contract drawings and specifications.' So, we're back to square one."

Frank interposed, "What did you do then?" He was taking copious notes.

"I talked it over with my brother, Sherman. He runs our shop and field operations. He said, 'They'll never approve Top-Notch's shop drawings no matter what we change. We better just get Quality to make the shop drawings.' So I contacted Quality and told them to go ahead with the shop drawings. I figured I could get them to come down on their price quotation. They had the shop drawings ready in a week and we submitted them to Hyde. Hyde sends them to the architect and in a few days they're back approved, just like Sherm says."

Frank looked up and asked, "What happened then?"

Farl continued, "Well, I kept working on Quality to lower their price. They started at $465,000, which is $90,000 too high, and the best I could do was get them down to $460,000. So I told them where they could shove it."

"Then what?"

"We sent the approved Quality shop drawings to Top-Notch and asked them if they could furnish these materials. They said they could if we increased their price from $375,000 to $390,000 and if we could accept a few very insignificant deviations.

"I talked it over with Sherman and we decided that even though Top-Notch was screwing us out of another 15 grand, we were still better off than paying Quality all of our profit. So, we beat Top-Notch down 5 grand and told them to go ahead at $385,000.

"Sherm said no one would notice the minor deviations."

Frank couldn't believe what he was hearing. He then asked, "Didn't Quality object to your using their shop drawings?"

"Well, when we broke off negotiations they asked for their shop drawings back. We told them to get lost. We finally agreed to pay them $2500 for the shop drawings, just to shut them up. We found out later that they complained to someone at Hydes. But nothing came of it."

Frank later heard that Quality's complaint to Hyde was taken in a phone call by Kevin Barlow, one of Sam Trent's estimators. Kevin told Quality that Hyde wasn't going to get involved in disputes between subcontractors and their suppliers. Quality said they'd take it up with the architect and Kevin said if they did Hyde Construction would never allow any Quality materials to be used on any of their future jobs. Kevin neglected to tell Sam Trent or anyone else about this conversation.

Frank continued his questioning, "Did Hyde Construction know that Top-Notch furnished the materials rather than Quality?"

"No, because all the materials were delivered here to our shop. We delivered them to the job after we cut everything to the proper lengths and fabricated them into unit modules for the job. Top-Notch's materials are practically the same as Quality's."

"Now, what about the recent repairs?"

"Yeah, what about them? It's costing us nothing but money. Hyde expects us to go out to the job every time it rains and they still owe us the better part of $190,000. I told them we're gonna lien the job and Hyde said go ahead, by the time our lawyers are through with you, you'll wish you never did it. They told us that as soon as we cure the leaking the owner would release the money and we'll suddenly get paid."

"What kind of repairs have you made?"

"Well, mostly, we apply caulking. We've used up hundreds of dollars worth of sealant and the wages are killing us."

"Do you know why it continues to leak?"

"Well, no. Not exactly."

* * *

On returning to his office later that afternoon, Frank punched Clyde Dixon's number into his phone keypad. He brought Clyde up to date on his chats with the Hyde people and Farley Grimpenmire at Creative Glaziers. "Clyde, what do you think I ought to do now?"

"Well, Frank, since I last talked to you, Venture Properties has filed separate arbitration demands against Hyde Construction and Judge & King. Hyde has filed a demand against Creative Glaziers. The insurance carriers for all of the respondents have been put on notice.

"So, for right now, let's just wait and see how the arbitration starts shaping up. Don't do any more until you hear from me.

* * *

A couple of weeks later, Clyde Dixon again called Frank and told him that BPIC had hired the law firm of Grant & Powers to represent the architects, Judge & King. "Oliver Powers will be calling you to discuss the Venture Tower situation."

Later that day Oliver Powers called and Frank gave him a brief run down on his investigation so far. Oliver told Frank that Venture Properties had engaged a firm of structural engineers to study the Venture Tower window wall, to make repair recommendations, and to develop a projected repair cost estimate. They expected to receive the report in a few days and would send a copy over to Frank when it was available.

A few days later the report arrived. It was prepared by Strong & Stout, a structural engineering firm with a national reputation for designing window wall repair systems. Their report was predicated on the removal of the entire window wall system and replacing it with Quality or Superb products. Their reasoning was based on bringing it up to the originally specified standard. It would require the complete scaffolding and tenting enclosure of the entire building so that leased tenant spaces could be environmentally controlled with the least inconvenience to the tenants. This would also preclude any further water intrusion into the tenant spaces. All work would be done from the outside to minimize tenant inconvenience. The assumption was made that 15 percent of the glass would be damaged in removal and would have to be replaced. The entire project was scheduled for completion in one year with an estimated cost of $4,200,000. The report took four months to prepare. Frank found out later that the report cost $125,000.

* * *

Frank Grimm had spent the morning in the Los Angeles Superior Court testifying in a case involving the repair of improperly built clay tile roofs in a tract of 248 homes.

After a hurried late lunch he returned to his office to find the phone ringing. It was Clyde Dixon informing him of a meeting with the defense lawyers in the Venture Properties arbitration. Clyde told him that the meeting would be at 10 o'clock Wednesday in the offices of Hyde's lawyers, Drummond & Eagle. Rex Drummond will be there along with George Hyde, Farley Grimpenmire with his lawyer, and Leo King and his lawyer. He explained, "They would like you to be present to participate in their discussion of the Strong & Stout report. The report's $4,200,000 price tag has focused everyone's attention. Can you be there?"

"Yes. I'll be there."

* * *

Frank drove to The Century City offices of Drummond and Eagle and rode the elevator from the garage to the lobby, then, transferring to one of the high-speed tower elevators, continued on up to the 33rd floor. The firm, specializing in construction and real estate matters represented several of the largest construction firms in the nation. He was shown to a spacious conference room and served a cup of coffee. The room was full and he recognized most of those present.

It was Rex Drummond's turf so he assumed the chair. "I suggest we go around the table and self-introduce. I'll start. I'm Rex Drummond of Drummond & Eagle and we represent Hyde Construction Company." He nodded to the scowling man on his left.

"I'm George Hyde, President of Hyde Construction."

Next to George, "I'm Kurt Marshal, of Marshal & Marshall representing Creative Glaziers. To my left is my client Farley Grimpenmire, President of Creative Glaziers. Farley smiled to all around the table.

Last on the left side of the table, "I'm Frank Grimm, Forensic Architect."

Opposite Frank, "I'm Fred Notch, owner of Top-Notch Glazier's Supply Company."

Next, "I'm Grace O'Malley, with Young & Younger. We represent Top-Notch."

Then, "Oliver Powers, of Grant & Powers, representing Judge & King. Next to me is my client Leo King, Architect." Leo nodded.

Rex Drummond thanked everyone and proceeded, "You've all had a chance to read the Strong & Stout report. We've got Frank Grimm here today to get his reaction to the report. Frank, is it really necessary to remove the entire window wall? Isn't that somewhat extreme?"

Before Frank had a chance to answer, George Hyde interposed, "Damn right it's extreme. Venture wants to screw us."

Farley Grimpenmire pitched in his contribution, "They're not satisfied with my $190,000, they want more. They oughtta change their name to Vulture Properties. The bastards." He was clearly ticked off.

Fred Notch, entering into the spirit of the meeting, "I don't know why I'm here. I've already furnished all the materials and still have over $80,000 coming. The sons of bitches."

Rex stood up, "Calm down, folks. That's why we have Frank Grimm here today. We want to bring some sanity into this operation. What do you think, Frank?"

Frank had the S & S report on the table in front of him. "Yes, it is extreme. The repair method they've recommend would undoubtedly work, but it seems excessive considering that the leaking could probably be controlled for a lot less money."

Rex asked, "Do you think you could devise an effective repair method that would be less expensive?"

"I'm not sure. I'd have to study it. I'd have to analyze the various building conditions and give some further thought to the logistic and scheduling considerations. I'd have to devise a reasonable basic concept and then I'll need some help in pricing it out."

Rex offered, "I'm sure Hyde, Creative, and Top-Notch will give you all the help you need on scheduling and prices. Isn't that right?" He looked around the table and received enthusiastic nods of assent from George Hyde, Farley Grimpenmire, and Fred Notch. "Are you willing to take on the assignment, Frank?"

"Who would I be working for?"

"For now, Hyde will guarantee your fee, but we'll later share it out among the three respondent firms present."

"Okay, I'll give it a try."

* * *

During the next several days, Frank studied the problem, made freehand sketches, and wrote an outline specification. He made a tentative schedule and pro forma cost estimate. He met with Sam Trent of Hyde Construction and refined the schedule and cost estimate. Then he called Rex Drummond and asked him to organize another meeting of the respondents and their attorneys. He sent sufficient copies of his report to Drummond for distribution before the meeting. The repair method was based on the concept of sealing all the joints between materials and opening all the weep holes so they will operate properly. The work would all be done from the window-washing platform, thereby eliminating the scaffolding and tenting costs. The building's air conditioning fans would be operated constantly to avoid any water intrusion during the repair period. No glass would be removed or changed, thereby eliminating the inconvenience, cost and time. The total cost, including all labor, materials, supervision, and architectural fees for drawings, specifications and administration, was estimated at $145,000 and a time period of 60 calendar days.

* * *

At the next meeting, Rex Drummond thanked Frank for the prompt action. "It was only three weeks ago today that we previously met. Everyone's seen Frank's report. Are there any questions?"

George Hyde asked, "Will this be acceptable to Venture Properties?"

Rex replied, "We don't know yet. We'll have to talk to their attorneys. If Grimm's report is satisfactory to everyone here, we'll release it to Venture's attorneys. Okay?"

Each of the parties present indicated assent.

George Hyde said, "I have a question. How are we going to split up this bill after it is agreed to? I don't see why Hyde should have to pay any of it. We didn't do that crappy job. That was Creative Glaziers and their crooked supplier, Top-Notch."

Fred Notch angrily stood up and said, "Look, Hyde, we only furnished what Creative wanted. We have better materials if Creative wanted to pay for them. They beat us down $5000 and we had to cut something out. We're not in the charity business. We shouldn't have to pay anything either." He sat down, still hot under the collar.

"Now, wait a minute, Notch," shouted Farley Grimpenmire, "you guys said you'd comply with the shop drawings, just minor deviations, you said. We just installed your materials. If the materials won't work that's not our fault."

Rex was furiously tapping his pen on his coffee cup and eventually everyone became quiet and looked up at him expectantly. "Let's not start bickering among ourselves. All parties here have some potential liability and that's why we're all here. He turned to the forensic architect.

"Frank, will you please give us the benefit of your analysis of liability? I know you've reviewed the contracts that I sent you, you've seen the building, you've studied all the construction documents, and you've interviewed all the parties. You've served as arbitrator in similar situations many times."

Frank considered the question for a moment. "I do have an opinion."

Rex, addressing the four respondents and the attorneys, said, "Let's hear what Frank has to say. We don't have to agree to it, but his objective informed opinion will give us some idea how we might do in front of an arbitrator. Okay?"

All nodded their acquiescence. Rex turned to Frank, "Okay, Frank, let's have it."

Frank started slowly and thoughtfully. "Rex is right. All present have some responsibility for the existence of the building defect.

"Hyde Construction, as the general contractor, has primary responsibility to the owner. However, Creative has the same duty to Hyde. Creative committed the inexcusable subcontracting sin by cheating on the shop drawings and making unauthorized substitutions, so Creative must be the major responsible party. Top-Notch was in on the fraud and thus should share Creative's burden.

"Hyde and the architects were innocent dupes defrauded by Creative and Top-Notch. However, this doesn't relieve them of their duty to examine the work under construction. Hyde and the architects should have noticed that the work didn't match the approved shop drawings. Hyde, in their day to day superintendence, should have looked at what the Creative people were doing and compared it to the shop drawings.

"Judge & King, in their periodic field observation, should have looked more closely at Creative's work as it progressed. The architect's principal duty in field observation is to look and be observant. The architect and general contractor would be relying on each other to exercise reasonable care in overseeing the work of subcontractors. They let each other down."

Hyde, Grimpenmire, Notch, and King were soberly contemplating a spot on the table in front of each. The four attorneys were all busy with their note taking.

Frank continued, "Taking all factors into account, I'd assess relative liability something like this." He took his pad of paper and listed the names of each of the respondents, Creative Glaziers, Top-Notch, Hyde Construction, and Judge & King. Opposite each he wrote a percentage figure. He passed it around the table. Each party and their attorney soberly looked, some grimaced, and passed the pad along.

Rex looked at the pad, absorbed the information, then said, "Considering the flagrant nature of the building defect there's no point in wasting time in the futile effort of denying liability. So the arbitration will be for the sole purpose of determining the amount of damages payable to Venture and, secondarily, the relative liability of the four respondents."

Rex reminded the grim assemblage, "In addition to the $145,000 repair costs, Venture will want us to pay for their extra janitorial labor and supplies, electrical costs, and interior repairs. They'll also want us to pay for the S & S repair report and their attorney's fees. By my estimate the whole package would be approximately $350,000.

"If we can admit liability and agree to Frank's relative liability figures, we can then approach Venture's lawyers to negotiate the amount of damages. We might be able to avoid the time, trouble, and cost of the arbitration.

"If the negotiation founders we can offer to go to mediation. If that fails, we'll be in arbitration.

"Oh, and don't forget we must share Frank Grimm's fee and you'll each have to pay your own attorney's fees."

There were no happy faces around the table but they all agreed.

* * *

Points of Law 15

VENTURE TOWER
Fraudulent Shop Drawings

15.1 Contracts and Torts

We are concerned in this story with allocation of liability for a construction defect. Of the five great branches of law that are taught in first year law school: contracts, torts, real estate, and crimes, we deal here with the first two.

Contracts concerns itself with the creation and enforcement of binding promises.

Torts enforces the principle that humankind should so conduct its affairs as to cause no harm.

The law of contracts had to be invented to make commerce possible.

The law of torts makes possible the pursuit of happiness.

The law of torts is again divided into two parts: harm caused intentionally and harm caused by negligence. Here, we deal with both.

15.2 Fraud and Negligence

Creative Glaziers submitted for approval a set of shop drawings that it did not intend to follow. The prime contractor and the architect didn't catch it. Creative Glaziers thus committed an intentional tort – fraud. The prime contractor and the architect were careless. This carelessness was a breach of contract and it was also the tort of negligence. (Negligence is defined as failure to exercise due care.)

15.3 Physical Damage and Economic Loss

Judge & King were parties to an architectural services contract with Venture Properties under which Judge & King agreed to inspect the work. Their inspection failed to detect that the window system installed by Creative Glaziers did not match the shop drawings. This negligence was both a tort and a breach of contract, and it caused two different kinds of damages to owner: physical damage and economic loss. The distinction between these two types of damage can be important because one is governed by contract law and the other by tort law.

The physical damage occurred when water soaked the carpets, window treatments, and drywall. Economic loss occurred when the rental value of the building decreased because it did not perform as expected. Venture Properties cannot recover for economic loss under the law of torts. This could provide a defense to the party that did not have a contractual relationship with owner: Top Notch Glaziers Supply.

15.4 Third Party Beneficiary

The architect Judge & King and the prime contractor Hyde Construction both have direct contractual relationships with Venture Properties. Creative Glaziers, however, has no contract with Venture Properties but does have a contract (a subcontract) with Hyde Construction. Since the express purpose of the subcontract is to construct a building for Venture Properties, this subcontract is a third party beneficiary contract that is enforceable not only by the Hyde Construction but also by Venture Properties.

15.5 Claims

The owner has claims for breach of contract against the architect, the prime contractor, and the subcontractor and also has tort claims against all three, plus the supplier. One difference between the contract claims and the negligence claims is that the contract claims include economic damages (reduction in rental value) while the negligence claims include only the physical damage caused by water intrusion.

15.6 Multiple Parties

It is not unusual for four (or even more) different parties to be legally liable to an owner for construction defects. In our example, it is not only Creative Glaziers, but also Farley Grimpenmire, the president of Creative Glaziers, that is legally responsible for the fraud. But the fact that other parties are liable does not provide a defense to any other defendant. The owner is entitled to recover its full damages from any defendant.

15.7 Equitable Contribution

Let's suppose, now, that the owner exercises its right to pick and choose, sues only the architect, and recovers its full damages from the architect. There is an element of injustice here, because the architect's only fault was its failure to discover a fraud perpetrated by a subcontractor. The law handles this situation by giving the architect a right of equitable contribution against other parties who are at fault. The jury will be asked to assign percentages of fault to each defendant, and after "everybody sues everybody" justice is finally done when each defendant is made to pay its fair proportion as determined by the jury. Remember, however, that Top Notch Glaziers Supply, with no contractual relationship to Venture Properties, is only liable to pay its fair proportion of the physical damage.

15.8 Insurance

Our legal analysis will not be finished until we consider the insurance coverage of the defendants. Two types of insurance must be considered: errors and omissions (E&O) coverage and commercial general liability (CGL) coverage. E&O coverage is carried by architects and engineers to cover claims made against the architect or engineer for errors and omissions in the performance of professional services. Under such policies, the insurance carrier agrees to defend the architect or engineer against such claims and to pay any judgment entered against the architect or engineer. It was at the instance of the E&O carrier of Judge & King that Frank Grimm was called upon to make his investigation and recommendations.

The contractor, subcontractor, and supplier will be covered by CGL policies under which the carrier agrees to defend the insured against claims for physical injury to tangible property and pay any judgment that may be entered against the insured.

15.9 Counsel Employed by Insurer

When Grimm attempts to put together a settlement package he will be dealing only secondarily with the contractor, the subcontractor, and the supplier: he will be dealing with the parties who control the checkbook, which is to say, insurance adjusters and claims managers. Each insurance company will employ counsel to represent its insured. Although paid by the insurance carrier, counsel represents the interests of the insured and therefore is placed in the difficult position of being required by the ethics of the profession to bite the hand that feeds by persuading the insurance carriers to come up with enough money to settle the claims.

Most construction defect cases don't go to trial because the legitimate expenses of the attorneys in such "complex, multi party litigation" would be a lot more, in most cases, than settlement costs.

15.10 Settlement

It is in the interest of every insurance carrier to settle claims against its insured as cheaply as possible. Now let's suppose the insurance carrier for Creative Glaziers has an opportunity to settle with Venture Properties on the cheap. Part of the insurance carrier's argument for a cheap settlement is the fact that its insurance policy excludes coverage for fraud and the main claim against Creative Glaziers here is a fraud claim. The trouble is, a settlement with Venture Properties won't take Creative Glaziers out of the case because Creative Glaziers will still have to defend itself against equitable contribution claims of Judge & King, Hyde Construction, and Top Notch.

The courts try to encourage settlements because there aren't enough judges to try more than about ten percent of the civil cases that are filed. Here, the concept of "good faith settlement" comes into play.

15.11 Good Faith Settlement

To make a settlement with a plaintiff that will protect it against equitable contribution claims, a defendant must get the settlement approved by a judge. It is up to the judge to determine, after hearing from all parties, whether the settlement represents a fair appraisal of the defendant's potential liability. If approved by the judge, a good faith settlement will protect the settling defendants against further claims.

To recapitulate:

- Some construction defect claims sound in contract, some in tort, and some in both.
- Tort claims include claims of physical damage but not claims for economic loss.
- The plaintiff may recover its full damages from any defendant.
- Under a doctrine known as equitable contribution, damages are allocated among defendants according to proportionate percentage of fault.

- Defendants are protected against lowball settlements by the good faith settlement procedure, which insures that each defendant pays a fair proportion of the loss.
- Architects and engineers carry E&O insurance, and contractors carry CGL insurance, to protect them against construction defect claims. The insurance does not cover damages for fraud.
- It is the duty of the insurance carrier not only to pay judgments that are covered by the policy but also to provide the insured with a lawyer, paid by the insurance company, to defend the insured.
- Although paid by the insurance company, the lawyer's professional responsibility is to the insured.
- Settlements are hammered out by negotiations between the parties, their attorneys, and the insurance carriers and their attorneys.
- Settlements are based on the expectations of the parties as to the amount of damages that would be assessed against each defendant by a jury.

16

Psi Creativity Center I
The Design and Bidding Phases

At 11:30, Wednesday, March 23, 1994, Ivor Judge, FAIA, of Judge & King, AIA, Architects, was at his desk hurrying to finish his building code research for an exciting new bank project. He had to leave promptly at 11:45 to be on time for a Rotary Club meeting. The weekly luncheons gave him a pleasant break from the pressure of a busy architectural practice. He enjoyed the easy congeniality of old friends every Wednesday and occasionally he happened onto a new professional assignment. Many of the members were influential in the business affairs of their suburban community.

This time he chanced to sit next to Sam Stark, the affable owner of Stark Realty. Sam was bursting with good news, "Say, Ivor, you'd be interested in this. I just closed a sweet deal with a couple of shrinks on a one and a half acre site out by the golf course. They want to build an office building. I gave them your name."

"I truly appreciate your thinking of me, Sam." This was a reciprocal favor as Ivor had often recommended Sam to his clients and friends.

* * *

That afternoon Ivor's secretary received a phone call from Doctor Quagmire's secretary to make an appointment for him and his associate, Doctor Marsh, for the following day, Thursday, at the Judge & King offices at 4 in the afternoon. Ivor already had a dental appointment for that hour but felt it was worth the inconvenience of moving it to accommodate these important new prospects. Fortunately his dentist had had a cancellation, so he was able to move his dental appointment to Friday morning at 10.

Disappointingly, Doctors Quagmire and Marsh didn't show up at 4, but Ivor received a call from their secretary at 4:30 asking if the appointment could be moved to 8 o'clock, after dinner. Although Ivor said okay, he was somewhat irked as he needn't have changed his original dental appointment. Now he'd lost a couple of hours and ruined his entire Friday morning.

He phoned home to tell his wife that he wouldn't be home for dinner, as he'd grab a quick bite at a fast food restaurant, and wouldn't be home until after Doctors Quagmire and Marsh had left. After his hurried and unsumptuous meal at the neighborhood pit stop he was back in the office 10 minutes early for his 8 PM meeting. Eight o'clock arrived but the clients didn't. The phone rang at 8:15. It was Doctor Marsh reporting that Doctor Quagmire had had an important emergency and that they would phone next week for a new appointment.

After a week, and two more cancelled and rescheduled appointments, Ivor finally met his prospective clients. They came to his office at half past 4 on Friday, April 1st, 30 minutes late for a 4 o'clock appointment. He offered coffee. Doctor Marsh accepted. Doctor Quagmire declined, asking for Perrier with three ice cubes and a slice of lemon. Ivor's secretary resentfully walked a block and a half to the supermarket to get the mineral water and a lemon.

Doctor Quagmire introduced himself, "I'm Doctor Karlin Quagmire, Ph.D., and this is my associate, Doctor Marcia Marsh, also Ph.D. We're consulting psychologists. I specialize in psychosexual disorders and Doctor Marsh deals in anxiety, bereavement, and assertiveness counseling."

Ivor was taking careful notes, but wasn't sure he was spelling some of the words correctly.

Doctor Quagmire continued, "Our real estate agent, Sam Stark, recommended you very highly and that's why we're here. We've just purchased a beautiful site near the golf course and want to build an office building. We're thinking of building for our own practice and additional space for 4 or 5 others in the psychological counseling field. The development will create a significant professional counseling center as well as give us an investment to finance our retirement in about 25 years."

Ivor asked if they had any specific space or planning requirements.

Doctor Quagmire answered, "No, we've lots of ideas, but nothing in writing yet. That's why we're here." Doctor Marsh nodded agreement.

Ivor continued his questioning, "Doctor Quagmire and Doctor Marsh, do you have a topographic and boundary survey of the property?"

Doctor Quagmire seemed to be the spokesperson for the duo. "First, please call us Karl and Marsh. No need for formality, Ivor. Secondly, we don't have a survey."

"Well, we'll need one. We'll also need a soil test."

"Where can we get these things? From the City Hall?"

"No." Ivor explained, "I'll contact a reputable land surveyor and a soil testing lab and obtain fee quotations for your approval. Then, if they are acceptable, you can authorize me to instruct them to proceed."

Karl continued, "Can we talk about the project? We know exactly what we want. My office should be large, carpeted, a high ceiling, and plenty of storage. With a private toilet and shower. It should be soundproof and have motorized draperies so I can regulate the light from my desk. And the air conditioning control should be separate for my office. I want a built-in hi-fi stereo music system with provision for automatic video and audio recording of my sessions. The telephone system should have intercommunication to my secretary and to Marsh. I need room for yards of bookshelves. There should be a large walnut paneled wall for my diplomas, licenses, and awards." Then he added, almost as an afterthought, "And Marsh should have an office too." She nodded her assent.

"I've noted all your requirements, Karl." Ivor was taking careful notes. "We should talk about the building in general. Do you have any particular size in mind? Any particular budget?"

"We thought you'd know about these things, building costs and all that."

The architect explained, "The best way to start will be for us to look at the property and make some schematic plans of land use, circulation, parking, and building size and location. We'll have to research the deed restrictions, applicable planning laws, and the building code. When we know the approximate building and parking capacities, then we can estimate the construction costs and study the pro forma operating costs."

"That sounds good. What do you think, Marsh?" Before she had a chance to answer, Karl asked, "When can you get started? The property'll be out of escrow next week and we don't want to waste any time. Right, Marsh?"

"We can get started on the schematic study right away. But, we'll need to have a signed contract first." He handed a blank form of AIA Document B141, Standard Form of Agreement Between Owner and Architect, 1987 Edition, to Karl. Ivor continued, "This is the contract form we use. You can take this with you to study. It describes all of our services and your responsibilities as owner. I'll prepare the contract on this form and have it delivered to your office on Monday. If it's satisfactory, please sign and return it with the required deposit and we'll start work."

Karl handed Marsh the form and she folded it into quarters and slid it into her purse.

Ivor added, "You might want to have your attorney review the contract before you sign it."

* * *

The following Monday Ivor talked to Judd Daid of Daid Land Surveys and Rocky Stone of Geotechnical Services Corporation and asked them to look at the site and send him a written proposal for their services ASAP, preferably today, via fax. They both agreed. He told them that the owner was extremely anxious to get the work under way. Meanwhile, he started preparing the architectural services agreement and by late afternoon he had the contract ready and had received the proposals for surveying and soil testing. He had the architectural contract and engineering proposals delivered to Doctors Quagmire and Marsh before PM.

After about a week, Ivor, not having heard from Karl or Marsh, figured that they had probably submitted the contract to their lawyer and that he was holding up progress. He knew that Karl and Marsh were impatient to get the project moving so he felt he should phone them to see if he could answer any questions or help in some other way. When he called, the secretary said both doctors were "with clients" and could not be disturbed under any circumstances. He left his name and number and a request for a return call.

About a week later Ivor phoned again. The secretary said both doctors were bogged down with appointments. Ivor asked, "When will one of them be free?"

"Doctor Quagmire will have a 10 minute break in 20 minutes."

"Please tell him I'll call back in about 20 minutes."

"Yes, Mister Judge, I'll tell him."

In 20 minutes Ivor dialed again and was left on hold. After about 5 or 6 minutes of Greensleeves, Doctor Quagmire was on the line.

"Hello, Ivor, I've only got a minute. I'm really swamped. How are you coming with the schematic sketches? We thought we'd have heard from you long before this. We can't wait to see your ideas. I was just talking to Marsh about it. It'll be wonderful when we get our own building. We're going to call it Psi Creativity Center. What's holding you up?"

"Sorry, Karl. We've done nothing. We need our signed contract back along with the deposit. As soon as we receive them, we'll start work."

"Oh, the contract. I think it's somewhere here on my desk. Or maybe Marsh has it. I'll take care of it right away. As soon as I find it. But don't let me hold you up. You can get started. Go right ahead. We've already lost three weeks."

"We also need your authorization to order the survey and soil test. We need this information. It's crucial. We can't get very far without it."

"Okay, Ivor. Go ahead with the soil test and survey. It's all right with me. Now, let's get going on the sketches. We're losing valuable time."

"Thanks, Karl. We'll get started as soon as we receive the contract and deposit."

* * *

Although Ivor had Karl's oral authorization to go ahead with the survey and soil test, he decided to wait until the contract and deposit arrived.

The architectural contract arrived in the mail 8 days later, on Thursday, April 28th. Ivor showed it to his partner Leo King. He briefed Leo. "Doctor Quagmire is hot to trot but it took him a month to get the signed contract back. And he forgot to enclose the deposit."

Leo looked concerned. He asked Ivor, "Are you sure we should be working for this guy? He sounds kind of flaky."

"Well, he's signed the contract. He's probably all right. Maybe a little disorganized, perhaps. I'll phone him about the deposit." When he phoned later, Doctor Quagmire said he'd have his secretary mail it right away. "She must have forgotten to mail it with the contract." Ivor thought, I'll bet.

Ivor then asked, "We need a copy of the CC & Rs on the property, Karl. Can you mail them to us with the check?"

"Oh, sure, Ivor." The check arrived four days later but no CC & Rs. Ivor called again and talked to Karl's secretary. She said she'd talk to Doctor Quagmire about it. The CC & Rs arrived four days later on Friday, May 6th.

So, Ivor finally got started. He wrote letters of authorization to the surveyor and the soil testing lab, with copies to the client. He studied the CC & Rs. He went to the City Hall and checked the zoning and planning maps and obtained copies for the office file. Inigo Johnson, one of his staff architects, who'd be in charge of the project, did the building code research. Ivor and Inigo drove out together to look at the site. They walked around the property and took a couple dozen photographs for reference at the office. They could go no further without the client's program.

After three or four attempts to get Karl or Marsh on the phone, Ivor decided to write the program himself and submit it to the client for approval. In less than two hours he came up with an organized word description of a project consisting of five professional suites of 2000 square feet each, a common building entry, toilet rooms, connecting circulation paths, mechanical equipment, electrical, and telephone spaces, and a 40-car parking area. The total building area would be 11,000 square feet with a proposed budget of $1,000,000 including tenant improvements in the client's suite but the building shell only in the rental suites. The building would be one story to comply with the deed restrictions, as disclosed by the CC&Rs. The one and a half acre site would be heavily landscaped and in compliance with the yard requirements of the zoning ordinance. He also wrote a detailed description of the suite to be occupied by Doctors Quagmire and Marsh.

On his way home that evening he delivered the proposed program to Karl and Marsh's office, but he was not allowed to see either of them. He left the program with their secretary/office manager Lois with instructions that the Doctors should study the program and let him know as soon as possible if it was satisfactory or would require any changes.

By the following Monday, he hadn't heard from his clients so Ivor telephoned and got Lois on the line. She said, "Lucky you. Doctor Quagmire's standing here next to my desk. Here he is."

"Hello, Ivor." Doctor Quagmire sounded cheerful. "How're you coming with the design?"

"We're doing nothing, Karl. We're waiting for you to approve the program. We can't proceed without it. How does it look to you?"

"It looks just fine. Don't waste any more time, Ivor. I thought you were calling to show us some sketches."

"Okay, Karl, now we can proceed."

Ivor knew it was pointless to ask for written approval of the program so he wrote a confirming letter and put it in the mail.

* * *

After a few days Ivor and Inigo had developed the program into a responsible schematic site plan, circulation and parking layouts, and building footprint for the site. They had a broad brush budget and a pro forma operating statement. Time to talk to the client.

Ivor called Karl and got Lois on the line. She said the doctor was not available as he was with a client. Ivor requested a return call. After a few more calls, Ivor finally arranged a meeting at Karl's office. It was wedged in between two patients, scheduled for 50 minutes, at 4 on Tuesday, May 19th. Marsh couldn't make it as she had a group therapy session scheduled for the same time. Karl said that she might pop in for a few minutes, but she never showed up.

On the day, Ivor arrived before 4 o'clock but Karl wasn't free until 4:35. The architect showed all his preliminary work to the client and tried to explain it. But Karl wasn't attentive as he was already reviewing his notes in preparation for his next patient. At 4:49 he told Ivor that the program, preliminary sketches, budget, time schedule, and pro forma operating statement all looked great. He added, "It looks fantastic. Go right ahead and get going on the next phase. We're behind schedule now. It's taken you over six weeks to get this far and we can't waste any more time. I've got to get going now. My secretary will show you out." Ivor felt drained and unappreciated.

Driving back to the office, Ivor reviewed the situation. Karl wasted the first four weeks. We've done a colossal job getting the program and the schematic phase completed in two weeks and he thinks we're behind. Well, at least he approved the work to date. When he got back to the office he talked it over with Inigo and they were ready to start the Design Development Phase the next morning. He mailed a letter to Karl confirming his approval of the schematic phase. Karl and his blatant disregard of appointments and his superficial review of Ivor's work was irritating and irksome but at least he approved everything. If Ivor could get used to Karl's bizarre ways, maybe the project would turn out okay after all.

* * *

After about a week and a half the architects had all the planning and design ideas worked out and in another week had the design development drawings complete for presentation to the client. They had also worked up a set of outline specifications of materials and updated the construction budget. They were all ready to talk again to their clients.

Ivor was getting used to calling Karl's secretary for negotiating appointments. When he got her on the line he said, "Look, Lois, we've got to have a decent conference this time. There's a lot of material to present and it has to be explained. It's all ready now and we must meet before any more time is wasted just waiting for a meeting."

"Okay, Ivor, I'll see what I can arrange."

Lois called back the next day and said, "I did the best I could, Ivor. You can come over tomorrow and talk while Doctors Quagmire and Marsh are having their lunch in the group therapy room. I'll order a sandwich and coffee for you too."
"Well, that's not too bad. How much time will we have?"

"Fifty minutes. Starting at noon. So, don't be late."

* * *

Ivor arrived at 11:45 on Tuesday June 7th, a quarter hour early. Lois showed him to the conference table in the group therapy room. He laid out all his drawings and documents so he'd be ready for the clients. He was proud of his and Inigo's output and felt certain the doctors couldn't help but recognize the value of their offerings.

He saw the deli box of fresh sandwiches at the end of the table.

When noon arrived and he saw no sign of Karl or Marsh, he went out to the outer office to talk to Lois. She said the doctors were not in yet but she was sure they would be as they had afternoon appointments starting at one o'clock. "I'll let you know as soon as I hear anything."

At a few minutes to one they arrived, but didn't sit down. Karl said, breathlessly, "We have appointments with clients in a couple of minutes. We can't talk today. Lois will give you another appointment." And they were gone. Ivor, deflated, sat down and slowly ate one of the corned beef sandwiches with mustard on egg bread. As he left, he and Lois arranged a meeting for the following day. Same time, same place. Ivor complained, "Lois, I can't waste my life sitting around waiting for these two prima donnas to show up whenever they feel like it. They've repeatedly made appointments and broken them. From now on I'm going to bill them extra, on an hourly basis, for all of my time that they waste like this. Please tell them that, won't you?"

"Ivor, I can't tell them that. You don't seem to realize that their time is valuable. They're very busy people."

"Okay, Lois. I'll be back tomorrow at noon, and I'll tell them myself."

* * *

The following day Ivor showed up on time and the two doctors were waiting for him. Karl said, "We've been waiting for you, Ivor. Let's get started so we can finish on time. Marsh and I work to a schedule and are extremely busy."

Ivor didn't say anything about billing for his wasted time. He thought, there's no point in complaining when they're on time. Maybe they've changed. Ivor had been laying out the drawings and other documents on the table and had his note pad at the ready. He started, "This is the site plan and..."

Karl interrupted, "Marsh and I've been looking at the program and it'll never do. We can't have shared public toilet rooms. They'll have to be within the individual suites. That way we can charge rent for them. And the suites are way too big. The rent would have to be too high. We want more suites. We've decided on eight suites at 1500 square feet each. But our suite should be a little larger than the others, maybe 300 square feet more. After all, it's our building. Right, Marsh?" Marsh nodded agreement. "Now let's see what else you've got."

Ivor was caught off guard. "I thought you'd approved the program and the schematics. Our design development drawings are based on them. I'll explain what we have here, but we'll have to change the drawings to comply with your new requirements." So, Ivor showed them the drawings and explained how the new requirements would be incorporated. "We'll need another meeting to show you the new drawings and the revised budget. How about a week from today?"

"Okay, Ivor, but let's not waste any more time. Time is money. Right, Marsh?"

* * *

During the next few days, Ivor and Inigo revised the drawings to incorporate the new program requirements and reworked the budget and pro forma operating statement. The building area had grown to 13,000 square feet and the budget estimate had risen to $1,200,000. They were ready for the next meeting with the clients with a couple of days to spare.

The following Wednesday Ivor reappeared at the doctors' group therapy room and had the drawings all laid out when they arrived. They listened attentively while he explained the drawings and other documents.

Ivor said, "After you've approved this phase we'll start on the Construction Documents Phase."

"Okay, Ivor, then start. We approve. Don't we, Marsh." She nodded approval. Karl added, "We can't waste any more time, Ivor."

"I should warn you, Karl, and Marsh, up to now it has been fairly easy to make changes because the schematic and design development drawings are relatively simple and uncomplicated. However, after we start the construction documents, changes are very difficult, time consuming, and expensive. We'll have to charge you for all changes from the approved design development drawings. Understand?"

"Well, get going. We can't have any more delay." Karl and Marsh rose to leave.

"There's one more thing, Karl. We haven't received your check for our June 1st billing. This is the 15th so the bill is 5 days delinquent. We must be paid on time or we can't continue working."

Oh, sorry, Ivor. Lois must not have taken care of it. I'll tell her to send you a check right away."

* * *

Ivor and Inigo commenced the construction phase services. Inigo started planning the layout of the set of drawings and Ivor made appointments with the engineering consultants to start firming up their designs and recommendations.

Doctor Quagmire's check, in payment of the June 1st billing, arrived on July 1st. The July billing was mailed out on July 2nd. Ivor didn't really expect that Karl would pay it on time but was starting to get anxious in late July. He called Quagmire's office on July 28th and could talk only to Lois. Ivor reminded her that the bill should have been paid no later than July 10th.

She complained, "I can't just write checks whenever I feel like it, Ivor. Doctor Quagmire has to ask me. He hasn't instructed me to pay that bill. I haven't even seen it since it arrived in the mail."

"Will you please tell him that the bill is due?"

"No. That's your job. I'll just tell him you called. Maybe he'll call you back."

"Please, Lois, just ask him to call me."

The check arrived a few days later. Meanwhile, the August 1st billing had already been mailed out and Ivor had added interest on the overdue amount. When the bill arrived on Karl's desk, he was incensed at being asked to pay interest. He picked up the intercom and instructed Lois to get Ivor on the line. "Ivor, what's the idea of charging me interest on last month's bill? I've already paid it. You're gouging me. I thought your fees were pretty high in the first place. Now you're piling it on."

"Our contract provides for interest on overdue amounts. See subparagraph 11.5.2. You can avoid interest very easily. Just pay our bills on time. We need a steady cash flow to operate our business."

"Well, I'm not going to pay the interest. I'll pay the bill today and I'll deduct the interest."

"Okay, Karl, if you pay the bill today, you can deduct the interest this time." The check arrived a week later, with the interest amount deducted. Ivor had no further difficulties collecting from Karl, even though none of the checks were less than one or two weeks late. The threat of interest had had a beneficial effect.

* * *

On August 30th, Ivor called Karl to let him and Marsh know that he was going to file the building permit application within the next week. The drawings and specifications were almost completed and would be finished by the time the permit was issued. Ivor talked with Lois, as the doctors were tied up with clients, as usual. Ivor explained, "I'll need Doctor Quagmire's check for $1500, made payable to the City, for the plan check fee. We can't file without it. Please tell Karl that we'll file as soon as we receive his check."

Karl's check didn't arrive within the week so Ivor called again. "Lois, Karl is holding up the job. We can't file without the plan check fee."

Karl's check arrived 10 days later, on Friday, September 16th, and the plans were filed for checking on the following Monday.

Ivor expected to have the building permit in hand in about a month so he felt he better talk to his clients about the contractor selection process. They had told him previously that they wanted the job to be competitively bid and the bidding documents were prepared on that basis. Ivor mailed Karl a copy of the latest edition of the Recommended Guide for Competitive Bidding Procedures and Contract Awards for Building Construction, AIA Document A501 and AGC Document 325. If Karl will only read it he will understand the bidding procedure so we can discuss it intelligently.

He phoned Karl to get an appointment to talk over the bidding process. Lois was able to work him in between client appointments about a week later, from 2:50 to 3:00 on Tuesday, September 27th. Marsh couldn't make it so Ivor and Karl met without her. Ivor had a list of six suitable contractors for this type and size of job. He suggested, "Karl, you can look over this list and reduce it to four. That will provide for adequate competition. If the list is any longer, the contractors won't have sufficient incentive to bid the job seriously."

"Okay, Ivor. But one of our clients is a contractor. I'd like to put him on the list. He owes us quite a bit of money and this would give us a chance to get paid."

"Does he have any experience with commercial construction?"

"I'm sure he does."

"Is he bondable and financially responsible?"

"I'm sure he is."

"What's his name?"

"Hyman Lowe. His business is called Superior Builders."

"I've never heard of him or his company. Do you mind telling me what you were treating him for?"

"No, I can't tell you. That's strictly confidential. A matter of professional ethics. Doctor-patient relationship, you know."

Ivor continued, "I'd suggest your having your bank check him out before putting him on the bid list. Also, ask him for references from architects and engineers he has worked with and the names of some of his customers and projects. It's important to know if he's competent, financially responsible, and bondable."

"I'm too busy to do all that, Ivor. I'll give you his address and phone number and you can do it. Okay?"

"All right, Karl." Ivor wasn't too keen on this unknown bidder and even less enthusiastic about having to check him out.

Ivor continued, "Then, you must shorten my list down to four and tell me which one to drop if Superior Builders is included. We don't want more than four bidders on the list. I'll need it by next Tuesday, October 4th."

"Okay, I'll do it." Doctor Quagmire was already leafing through the next client's folder, only half listening to the architect.

"Don't fail me, Karl. We've got to get this out to bid by October 18th and we need time to notify the bidders in advance."

"You can count on me, Ivor." Karl didn't look up when Ivor left.

* * *

October 4th came and went, with no bidders list from Karl. Two days later, Ivor phoned Karl and Lois said he was tied up. Ivor asked Lois to get the bidders' names from Karl at the next break between clients. She called back a half hour later and told Ivor that Doctor Quagmire said to go with the first four names on the list. Ivor thought, Karl must have given the list a great deal of thought.

Ivor had contacted Hyman Lowe and obtained a resume and references, checked a few of them out, and could find no solid reason for leaving him off the bid list. He checked with all the other bidders on the list to determine that they were also interested in bidding the project. After one drop-out, the list of bidders was firmed up to:

- California Constructors
- Hyde Construction Company
- Preferred Contractors
- Superior Builders

Ivor and Inigo obtained the building permit, finished up the bidding documents, sent a copy to the client for approval, notified the bidders, and had the job ready for bidding on Tuesday, October 18th. The bidding documents were picked up promptly by all four contracting firms. Bids were due on Tuesday, November 8th at 4 PM in the offices of Judge & King.

* * *

The bidding period was uneventful. The bidding documents had been carefully prepared and very few questions arose. Those that did were answered in a two page bulletin that was sent by fax to all bidders and the client on the Friday before bids were due.

A representative of each bidder was present at the bid opening in Ivor's conference room. Inigo was present to record the bids as they were opened by Ivor. Karl and Marsh were invited to witness the bid opening but they didn't show up until after 5:00. By then all the contractors had left. When the doctors arrived, Ivor gave them a copy of the bid summary:

Estimated Construction

Contractor	Bid	Time
California Constructors	$1,219,946	275 days
Hyde Construction Company	$1,189,078	260 days
Preferred Contractors	$1,226,956	270 days
Superior Builders	$997,350	210 days

"Well, Karl, it looks like your friend Hyman Lowe is low bidder. I hope he can do a good job. However, he looks a little too low and he may not have allowed enough time."

"I think he'll do all right, Ivor. What do you think, Marsh? This will give us a chance to collect the money he owes us. And we're saving $200,000 besides." He was elated.

* * *

After the doctors left, Ivor and Inigo picked up the files and papers and went to Leo's office to relax with a cup of coffee before leaving for the day.

Leo asked, "Well, how did the bidding go, Ivor?

Ivor showed him the bid summary, and explained, "The three high bids are all within 1 or 2 percent of our budget estimate. But, the low bidder is almost 10 percent low and 50 days less time than the lowest of the other 3 bidders. I don't know if they can pull it off."

Leo suggested, "Maybe they left something out. They're almost $200,000 under Hyde. And we know Hyde's estimators are pretty good."

"And they're 50 days less than Hyde's estimate of 260 days construction time," Inigo added.

Leo asked, "What do we know about Hyman Lowe and Superior Builders?"

"Not much. Inigo checked some of their references by phone and found nothing negative. Most of their experience is in small towns in Arizona and New Mexico. They've only been in this area for a year or so. Karl says he's a great contractor."

Leo summed up the thinking of the three architects. "Well, Ivor, your and Inigo's work is cut out for you for the next year. A simple project, but a complex, difficult client, and an unfamiliar contractor with a dubious bid."

* * *

Points of Law 16

PSI CREATIVITY CENTER I
A Difficult Client for an Architect

16.1 The Need for Surveys

At the beginning of a construction project, a prudent owner will commission both a boundary survey and a topographic survey of the project site.

Some owners don't seem to realize that some of the important responsibilities related to the construction of a building project must be fulfilled by the owner. Construction professionals like to make the owner feel that the objectives of the owner's program will be easy to fulfill in a prompt and cost effective manner. An architect or a contractor who is intent on selling a job is not likely to emphasize the responsibilities and potential liabilities of the owner.

The owner's impression that between them the architect and the contractor will take care of everything is often strengthened by the contract documents. The construction professions being competitive, as they are, architects and contractors are willing to accept contract documents that are strongly biased toward the protection of the owner's interests, sometimes at the sacrifice of the interests of the design professional and the contractor. A cursory review of most construction contracts reveals that most of the clauses are written for the benefit and protection of the owner.

16.2 The Contractor's Duties

Among the many duties imposed on the contractor in a typical construction contract are the duty to prosecute the work in a workmanlike manner in full compliance with the drawings and all applicable building codes, to perform and complete the work on schedule, to provide a jobsite safety program, to comply with the instructions of the architect, to avoid or resolve strikes or picketing, to carefully supervise the work of the contractor's employees and subcontractors, to protect the job against mechanics liens, to carefully review the drawings and specifications, to call any errors or omissions to the attention of the architect, to provide the owner with certificates of insurance and also require that the subcontractors do so, and to guarantee that all elements of the construction are free from defects and will perform in accordance with the requirements of the contract documents.

16.3 The Owner's Duties

Looking at the other side of the coin, construction contracts require the owner to pay the contract price, and usually to grant extensions of time when the contractor encounters excusable delay, and to pay for extra work. Sometimes the owner is also required by the contract to submit disputes to arbitration and to cooperate with the contractor in timely selection of appliances, floor coverings, and color schemes. In addition, the law imposes upon the owner the duty to fulfill implied contract conditions, such as, for example, to inform the contractor as to the boundaries of the project, warn the contractor about concealed underground conditions that could affect the work, pay taxes and assessments against the property, give the contractor unobstructed access to the jobsite, supply the contractor with information needed by the contractor to proceed with the work, supply the contractor with clear and adequate drawings and specifications, provide clarifications of ambiguities in the drawings and specifications, and avoid interfering with the contractor's prosecution of the work. These duties are implied by law whether they are expressed in the contract documents or not.

The owner also has duties toward the architect which may either be specified in the contract between the owner and the architect or, as in the case of the construction contract, implied by the law. One such duty is to provide the architect with necessary information about the owner's property including a boundary survey in cases where such a survey is relevant to the architect's work. If the property to be improved is other than a flat city lot, a topographic survey may also be required so that the architect may provide for proper grading and drainage of the property. Such a survey will reveal whether the improvements to be designed by the architect might change neighborhood drainage patterns to damage neighboring property. A property owner has a legal obligation to avoid directing storm water on to a neighbor's property in such a manner as to cause damage to that property.

16.4 Need for a Soil Report

In order to design and build proper foundations for the project, the architect and the contractor will need a geotechnical report. This report, prepared by a geotechnical engineer, will include recommendations dealing with such issues as the types of foundations recommended, compaction requirements, and the bearing capacity of the soil. In addition, the owner has an implied contractual obligation to notify the architect and the contractor of any known underground or otherwise concealed conditions, such as old footings, piping, electrical conduit, or abandoned fuel tanks that could interfere with the progress of the work or constitute a danger to jobsite workers.

16.5 The Architect's Estimate

Architects are often called upon to estimate the cost of building the structures they design. In many lawsuits, owners have claimed damages from architects for underestimating the cost of a construction project. Such litigation is usually unsuccessful for the owner, since courts recognize that an estimate is not a guarantee. It would often be unfair to hold an architect responsible for underestimating the cost of a project when the architect is not given control over the size and quality of the project, the bidding, or the negotiation of the construction contract.

It is such commonplace knowledge that architects tend to underestimate the cost of construction that courts seem to allow architects about a 10 percent margin of error without question. The owner who makes an underestimation claim against an architect faces another problem: to prove damages that are recognized by the law. If it costs $1,500,000 to build a warehouse, that warehouse is probably worth $1,500,000 even though the architect may have estimated that it would only cost $1,200,000.

In some cases, if an owner had known the true cost of a building project, the project would have been abandoned as economically unfeasible. For example, assume that the going rental rate for apartments is enough to produce income of $150,000 a year. Such a rental stream would justify an expenditure of

$450,000 to build an apartment building because the $150,000 rental return would then be enough to pay interest on the cost of the land and building, plus taxes, insurance, and operating costs.

Now suppose that the owner enters into a cost plus contract (with no guaranteed maximum price) to build the apartment project based on an architect's estimate that the project will cost $420,000. When the project is complete, the owner has spent $550,000. Had the owner known the true cost, the owner would have abandoned the project. Under such facts as these, an architect could be required by a court to pay damages for underestimating the cost.

16.6 Zoning and Building Codes

An architect who undertakes to design a building for a client has a legal obligation to be familiar with, or become familiar with, the zoning and building codes that apply to the project. Failure to comply with applicable codes in the preparation of drawings and specifications for a construction project is a clear case of architectural malpractice.

16.7 Drawings and Specifications are Part of the Contract

Most agreements, whether purchase orders, leases, buy-sell agreements, or car rental agreements, are embodied in writing. The construction contract for a major project provides an example of a lengthy and complicated contract. The contract is often hundreds of pages long and includes some or all of the following:

> The Agreement: This may be a relatively brief document that provides for such things as the identification of the project, the contractor's obligation to build the project, the contract price to be paid by the owner, terms of payment, bonding requirements, insurance requirements, the rights of the parties in the event of default, extensions of time, scheduling, date for completion of the project, liquidated damages, attorneys fees, alternate dispute resolution, and similar legal provisions.
>
> General Conditions: A usually lengthy exposition of the conditions under which the job will be constructed including a detailed description of the duties of owner, architect, contractor, and subcontractor, indemnity provisions, required notice provisions, and default provisions.
>
> Special Conditions: Provisions dealing with the special characteristics of the project, such as the location of borrow pits, union labor, prevailing wage requirements, access to the jobsite, hours of operation, and types and limits of insurance required.
>
> Specifications: Technical specifications are usually divided into sections that are roughly equivalent to the construction trades, and specify concrete mixtures, lumber, framing, piping, wiring, switchgear, glazing, mirrors, roofing, rebar, structural steel, ornamental iron, and technical requirements for the work, materials, and equipment that are to be incorporated into the project.

The foregoing documents, usually in 8-1/2x11 format, are bound together and called the Project Manual.

> Drawings are also a part of the contract. The drawings include plans, elevations, sections, isometrics, details, and notes. They are usually divided into segments roughly equivalent to construction trades, including foundation, structural, architectural, roofing, electrical, mechanical, excavation, and fireproofing.
>
> Submittals: The drawings and specifications are expanded during the construction process by submittals from subcontractors. The submittals portray the precise manner in which work will be done by subcontractors, and the exact equipment, supplies, and materials to be employed.

All of the documents specified above, along with change orders, addenda, and interpretations issued by the owner and the architect become a part of the contract documents, and thus subject to interpretation by courts. Courts interpret contract language so as to conform with the intention of the parties as revealed by the words and drawings employed in the contract documents. Under a typical contract provision, the meaning of the contract is in the first instance interpreted by the architect but subject to arbitration or litigation.

16.8 Interest on Late Payments

Many construction contracts provide that if the owner fails to make timely payments in accordance with the contract documents (usually by the 10th of the month) the owner will pay interest at a specified rate. If the contract does not specify an interest rate, the law still requires that interest be paid on amounts overdue under a contract at the legal rate of interest prescribed by the state legislature.

This interest requirement reflects the policy that a person who commits a breach of contract should pay damages to the innocent party, and that the damages should be an amount that would put the innocent party in the position it would have occupied if the contract had been performed as specified. A contractor does suffer damages from late payment, measured by the amount of interest it could have obtained on the sum by depositing it with a financial institution, or by the rate of interest the contractor would pay to borrow the money to replace the amount unpaid by the owner.

16.9 The Bidding Process

The objective of the owner in taking competitive bids is to get the lowest price consistent with quality construction. The process customarily followed in bidding a construction contract provides an interesting study of free enterprise.

The first requisite is "contract documents" that clearly depict the supplies, equipment, and materials that will be incorporated into the project along with their proper locations, and methods of attachment and installation.

The owner, or the architect on behalf of the owner, issues invitations to contractors deemed qualified to bid the project. Contractors interested in bidding the work either pick up sets of plans, or go to plan rooms where the plans are made available. Plans are also picked up by or made available to trade contractors and material suppliers who will be invited to figure the job and submit their bids to subcontractors and the prime contractor.

Subcontractors do not normally like to submit their bids to the prime contractor much earlier than an hour before bid opening time because to do so is to invite the prime contractor to "shop" the bid. Prime contractors develop relationships with subcontractors and like to deal with the same subcontractors over and over again. A subcontractor that is not necessarily favored by a prime contractor does not wish to submit its bid so far in advance of bid opening time as to give the prime contractor a comfortable interval within which to invite another, favored subcontractor to match or beat the bid. Prime contractors may receive a hundred or more bids over the telephone within the last half hour before bid-opening time, and the assembly of the sub bids is a fraught process.

Material mistakes are not uncommon. Under the law of contracts, a bid is an offer that ripens into a contract when it is accepted. However, if a contractor's bid is infected with a material mistake, and the mistake is a clerical error rather than an error in judgment, the law regards the contract as not having been formed at all, and therefore a prime contractor with such a materially mistaken bid is not required to perform the work even if the bid has been accepted by the owner.

17

Psi Creativity Center II
The Construction Period

Ivor Judge, FAIA, was deeply concerned about the ability of Superior Builders to construct the Psi Creativity Center professional office building for the contract price and within the estimated construction time. His partner, Leo King, had commented, "Well, Ivor, if they can produce the bond, we won't have too much to worry about."

"That could be just the beginning of our worries," observed Ivor. "Doctor Quagmire wanted him on the bid list because Hy Lowe owed money to the psychology practice for some counseling services. Now, I guess Doctor Quagmire'll get paid."

* * *

Ivor was putting the finishing touches on the construction contract. He was using AIA Document A101, Standard Form of Agreement Between Owner and Contractor where the Basis of Payment is a Stipulated Sum, 1987 Edition and Document A201, General Conditions of the Contract for Construction, 1987 Edition. He had already phoned Hyman Lowe, President of Superior Builders, and was now waiting for him to show up for a brief meeting.

Ivor's secretary signaled him on the intercom to announce that Lowe was waiting in the conference room. Ivor shook hands with him and they took seats at one end of the long walnut table. Ivor was surprised at the youthfulness of the contractor. He didn't look over 25 years old. He was neatly dressed, businesslike, and very personable. Ivor took charge of the conversation by handing three copies of the contract to him, stating, "Here's the contract, in triplicate. If it's satisfactory to you and your attorney, please sign all 3 copies and return them to me. Then, I'll obtain the owners' signatures. Meanwhile, you might start gathering together the preliminary paperwork. We'll need your construction schedule, submittal schedule, subcontractor list, schedule of values, surety bond, and certificates of all required insurance."

The young contractor, taking notes, replied, "I understand, Mister Judge. I'll start working on them."

"As long as we'll be working together for the best part of a year, you might as well call me Ivor."

"Okay, Ivor, call me Hy."

"Are you sure you can build the project for your bid, Hy? It looks pretty low. Maybe you've left out something?"

"After the bid opening we were concerned. We went back to search our estimates and we couldn't find any major errors. I know we're almost $200,000 under the next bidder, but we think we're right. I sure hope so."

"So do I. How about your time estimate of 210 days? You're 50 days under the next bidder. Could that be right?"

"We checked that too. We know it's tight, but we think we're okay. We made a construction schedule when we were bidding."

Hy put the contracts and his notes in his slim brief case, shook hands with Ivor, and left.

* * *

Two days later, on Friday, November 11th, Hy was back in Ivor's office with the signed contracts and a sheaf of preconstruction submittals, in triplicate, as required by the Special Conditions of the contract. A letter of transmittal inventoried the neat bundle. Ivor was surprised to see him so quickly. He leafed through the pile and found no surety bond or certificates of insurance. "This will be the sticking point," he thought, "This guy looked too good to be true."

Hy volunteered, "You'll notice that the surety bond and certificates of insurance are missing. My broker promised to have them delivered to you this afternoon."

Ivor, to himself, "I'll bet. So this guy's a con man. Quagmire was probably treating him for obsessive delusions." He was disappointed. He felt sure they'd soon be talking to Hyde Construction Company, the second low bidder. He'd have to explain to Doctor Quagmire why the price was up $200,000 and the construction period 50 days longer. Where'd Quagmire find this dingbat?

After Hy left, Ivor carelessly flung the pile of documents down on the corner of his desk and tried to figure out what to do next. He was at a low ebb. If he sent the contracts on to his clients, it would be a waste of time if he had to prepare new contracts with the second low bidder. So he and Leo went out to lunch. He tried to forget about Hy and his phony low bid and time estimate. He and Leo felt from the start that there'd be no bond.

When he returned to his office after lunch, a bright red and blue striped courier envelope was waiting on his desk. It was from Builders Insurance Brokers and contained Hy's surety bond and insurance certificates, in triplicate. Now, Ivor was surprised and impressed. His life was back in order. He immediately prepared a letter of transmittal to Doctors Quagmire and Marsh sending along the three signed contracts and one copy of each of the other submittals. He asked his secretary to have them delivered right away.

He then called his clients' office but as usual Karl and Marsh were busy, so he talked to Lois, their secretary/office manager. "I'm having the contracts delivered and you should have them within the hour. Will you please tell the doctors to have their lawyer review the contract? And, if it is satisfactory, sign all three copies, keep a copy, and return the remaining two copies to me. The submittals are for the doctors' information and should be kept for their file."

"Okay, Ivor, I'll take care of it."

"This should be done right away, Lois, because the work can't start without it."

"Okay, Ivor, don't worry about it."

* * *

Early the following Tuesday, Hy Lowe called Ivor to ask some questions. "When'll I get the signed contract?"

"I don't know, Hy. It should be soon. It's in the hands of the owners and I've advised them to have their lawyer review it."

"When can I take over the site and get going with the construction? I'd like to notify my subs and suppliers. I've prepared all my subcontracts and purchase orders and I'm ready to send them out. My superintendent's just standing around doing nothing."

"I don't know. As soon as I get the signed contract back, I'll let you know."

"Do you have the building permit?"

"The plans have been checked, corrected, and approved. All you have to do to get the permit is go to the City Building Department and pay the fee. As you know, that's your responsibility under the contract."

"Good. When can I get some more plans and specs?"

"As soon as we get them back from the other bidders. In a few days."

"What'll I do now?" The contractor was frustrated. He was all revved up and ready to go.

"Don't do anything before the signed contract comes back. Just wait. Be patient." The architect, too, was frustrated. He hated being at the mercy of an eccentric dilatory client.

After hanging up, Ivor dialled Lois and found that Karl wasn't available. "What's happening, Lois? Have they signed the contracts yet?"

"No, Ivor, the envelope's still here on my desk. Karl's too busy to look at it. He's swamped."

"Lois, please tell Karl that he's holding up the job. Tell him to get those contracts signed and returned to me. The building permit is ready to be issued and the contractor's ready to start the job. But remind him to have his lawyer review the contract, the surety bond, and the certificates of insurance. He should also talk to his insurance broker about his own insurance."

"Okay, Ivor, I understand. I'll do the best I can."

* * *

The signed contracts arrived in Ivor's mail two days later, on Thursday, November 17th. Ivor called Hy Lowe and told him the news. "But don't start anything on the site until you receive the written notice to proceed. I'll mail the signed contract to you. I'll also send approved copies of all your submittals."

Hy said, "Don't mail them. I'll pick them up in less than an hour."

Ivor called Karl and got Lois on the line. "Is Karl there?"

"Yes but he can't talk. With clients. He'll have a break in 12 minutes."

"Ask him to call me. It's important. I must talk to him."

"Okay, Ivor."

About 15 minutes later, Lois called Ivor and announced, "Ivor, here's Doctor Quagmire."

Ivor waited expectantly with his ear to the phone. He waited. After 5 or 6 minutes of soft music, Karl came on the line. "What's going on, Ivor? Did the contractor get started? I can't wait to see construction activity on the site. Psi Creativity Center taking shape. It's marvelous. Right, Ivor?"

"No, Karl, nothing's started yet. We have to give the contractor written notice to proceed, as required by Subparagraph 8.2.2 of A201, the AIA General Conditions, a part of the contract."

"We're in a morass of red tape, Ivor. Damn bureaucracy. Why can't you just tell him to start work?"

"The reason I'm calling you is to make sure your financing and insurance are in place. That your loan is recorded. If the contractor starts before your lender records the loan, it could delay your financing and if any insurable loss occurs before the insurance becomes effective, you'd have no coverage."

"I'll have Lois check with the bank and insurance broker and call you back." He hung up and rose to greet the next client.

An hour later, at his next 10 minute break, Karl got Lois on the intercom. "Lois, will you please check with the bank and find out if they've recorded our loan yet? And then call Ivor so he can notify the contractor to proceed. We're up to our elbows in red tape. Rampant bureaucracy. Take care of it, Lois. Please."

"Yes, Doctor Quagmire." Lois called Mr Leech at the Fifth National Bank. "Can you tell me if Doctor Quagmire's loan has been recorded yet?"

"No. The loan papers haven't been signed yet. Doctor Quagmire still has them. I sent them to him at least two weeks ago. Tell him to sign them in the presence of a notary public and get them back to this office. Then, if everything is in order, we'll have them recorded first thing the following morning. Then, I'll let you know."

About a week later, on Monday, November 21st the loan was recorded. Mr Leech phoned Lois and informed her. Lois called Ivor and gave him the news. Ivor asked Lois, "Did Karl say anything about the insurance? Is it in effect?"

"What insurance?" asked Lois.

"The owner should have liability, fire and extended coverage, and boiler and machinery insurance. The owner may also elect to purchase loss of use insurance. His insurance broker can explain all this to him."

"I don't know anything about it, Ivor."

"Well, will you please ask him? The construction cannot start before all insurance is in place."

Lois called back the next day to report that Doctor Quagmire was having his insurance broker take care of all the insurance. She called Ivor the following day, "Okay, Ivor, the insurance is all taken care of."

"Lois, please tell Karl that I'm going to authorize the contractor to proceed with the work."

Ivor phoned Hy and told him it was okay to proceed and that he would fax the notice to proceed and follow up with a copy in the mail.

Hy Lowe was all ready to go to work but replied, "Well, Ivor, that's good news, but tomorrow's Thanksgiving day so we can't do anything on the site before next Monday."

Ivor's patience was running out. The job could easily have been started three weeks ago if Doctor Quagmire had responded promptly to each of his obligations under the contract. Ivor was looking at Subparagraph 2.2.4 of A201 which states, "Information or services under the Owner's control shall be furnished by the owner with reasonable promptness to avoid delay in orderly progress of the Work." Ivor considered taking this up with his client but decided against it considering Doctor Quagmire's inordinate propensity to fly off the handle. Also, the job wasn't really officially started and his delays haven't affected the construction time yet. Ivor promised himself to be more firm with his client in the future. After all, he owed it to the contractor.

* * *

So the job was finally ready to start on Monday, November 28th. Superior Builders started early, at 7 AM, cleaning vegetation and debris off the site in preparation for installation of the temporary fencing. On the evening of the first day, right after quitting time, it started raining. And it rained off and on for four days. The site was muddy and impassable for several days thereafter. The site was not dry enough to resume work until Wednesday, December 7th, 1994.

On the 8th Ivor received a fax from Superior Builders asking for a contract time extension for the rain and muddy site from November 29 through December 6, an eight day delay. Ivor agreed with the request and wrote a no cost change order extending the construction time 8 days. The original completion date would have been June 25th, 1995. Now it would be July 3rd. He mailed a copy of the extension to Hy and one to Karl, asking each to sign their acceptance of the eight day extension.

A couple of days later, he received a phone call from Lois. "Doctor Quagmire wants to talk with you. Here he is." Ivor waited for Karl to come onto the line. The wait was shorter than usual, only 2 or 3 minutes. "Ivor, what the hell are you doing? We've lost 8 days. That's 8 days more construction loan interest. Eight days more rent on our old offices. Eight days less rent from our future tenants. The job isn't even started and we've already lost 8 days. Are you trying to ruin me? Well, speak up."

"I'm sorry, Karl. My rulings have to be fair and honest. It rained and the site was in no fit condition for any kind of work. The contractor is fairly entitled to the 8 day extension."

"Well, next time, remember who's paying your fee."

"Karl, if you'd just talk this over with your lawyer, he'd tell you that during the construction period the architect's determinations must be even-handed, without favoritism shown to owner or contractor. It doesn't matter who's paying the fee."

"If I call my lawyer to ask questions about every aspect of the contract, his fees will get out of hand."

"You didn't send him the contract, bond, or insurance to review, did you, Karl?

"Um, well, uh, no."

* * *

Superior Builders resumed work on Wednesday, December 7th. They continued site clearing and started fencing the site, moving in chemical toilets, installing a temporary power pole, and moving in a prefabricated temporary office. A telephone and fax were hooked up and they were all ready to start laying out the building on Monday, the 12th. Superior's job superintendent, Brick Bennett, had the job and site firmly in his grasp.

The first week of December had been lost to rain followed by almost two productive weeks. But the Christmas holiday, starting on December 23rd, lasted until January 2nd. Superior's first payment request covered the period ending December 31st. It was delivered to Ivor's office on January 2nd. The amount requested, almost $49,000, included site mobilization expenses but very little for actual construction as there were only 13 working days on account of rain, weekends, and holidays.

Ivor visited the site, confirmed the validity of each line item of the request, approved it and sent it on to the owner. Upon its receipt, Doctor Quagmire immediately called Ivor. "Ivor, the job's hardly started and Hy wants me to pay $49,000. This is ridiculous. I visited the job on New Year's Day and all that's there are some trenches and piles of dirt. And you've approved this bill. What's the matter with you?"

"Karl, this is a fair bill, in accordance with the approved schedule of values, and is due and payable before the 10th. If you'll check the contract, Article 7.2 of A101, you'll see that 12% interest will be added to all overdue bills."

Doctor Quagmire paid the bill on the 10th.

* * *

The job proceeded smartly and efficiently until, in the 7th week of construction, the plumbing contractor ran into an unexpected snag. Princess Plumbing Company had subcontracted construction of the sewer line from 5 feet outside the building to the front property line and beyond to the main sewer connection in the street. When Bluebell Sewers applied for the sewer connection permit, they were told in the City Hall that the cost would be $2250. Bluebird, in their estimate had allowed only $50, the standard fee for a sewer connection permit and inspection. The $2200 difference was to discharge the property's share of a sewer district assessment. So Bluebell refused to proceed until the owner paid the $2200. This was all explained in a letter from Superior Builders to Ivor Judge. Ivor, in turn mailed a copy of the letter to Karl. Not having heard promptly from Karl, Ivor phoned him and was put through the usual rigmarole with Lois. Three days later, Karl finally returned the call.

When Karl came on the line he sounded displeased. "What's it this time, Ivor? Can't you contractors and architects build the building without constantly interrupting my busy practice?"

"Sure, Karl, we can build it. But, as the owner, you've got to do your part. Right now the plumbing connection is being held up because you neglected to pay off the sewer assessment on the property. Princess Plumbing's subcontractor, Bluebell Sewers, can't get the sewer permit until you pay the $2200."

"I thought Superior builders had to pay for all permits and fees for the construction. Why should I pay it?"

"The contractor has to pay for all building permits but the owner must pay for all charges and assessments against the property. This is your responsibility as the assessment is for providing sewer service to the property. This is clearly stated in the AIA General Conditions, Subparagraphs 2.2.3 and 3.7.1."

"Why are all your rulings against me? You're supposed to be my architect, Ivor. You always seem to side with the contractors."

"Karl, you know this is a fair ruling. Please send me your check for $2200 right now. Today. If you keep stalling, you'll only hold up the job and give the contractor a legitimate excuse for a time extension."

"Well, okay, Ivor. But you can't keep ruling against me every time the contractor makes a claim."

Ivor went to the City Hall with the sewer bond payment thereby clearing the way for Bluebell to get their connection permit. About three days later, Ivor received a fax from Hy Lowe requesting a 7 day construction time extension for the time lost in clearing up the sewer bond assessment. Ivor immediately dug out Superior's construction schedule and referred to the sewer work. He could see that the work was not on the critical path and that the time wasted by the owner didn't hold up overall completion. He telephoned Hy. "I just received your fax asking for 7 days on the sewer work. You know damn well that the critical path wasn't affected, so I can't allow any contract time extension."

"I know, Ivor. Bluebell asked Princess for it and Princess asked me. I told Princess that they'd never get it but that I'd give it a try. Just throw the request in the wastebasket. I'll understand. I'll explain it to Princess and they can lay it on Bluebell."

"No, Hy. I'm going to process it and turn it down. Doctor Quagmire has to see my ruling on this. He's convinced all my rulings are against him."

* * *

The work had been progressing smoothly and without further controversy when, in the 19th week of construction, Hy Lowe delivered his 4th payment request to Ivor's office. It was Monday, April 3rd. Hy phoned Ivor later to explain, "Ivor, because of the weekend, my April 1st billing is two days late, so I'd appreciate your asking Doctor Quagmire to pay this one promptly. I need to pay the subs on time or they'll give me trouble on scheduling. You can't blame them for going to the jobs where they're being paid promptly."

Ivor was sympathetic, "Sure, Hy, I'll do the best I can for you. You're doing an excellent job and you're right on schedule. I appreciate it. I'll be on the job at 8 tomorrow morning to verify the progress shown on your request."

After Ivor approved the request and, because of Hy's appeal for prompt payment, he personally delivered the payment certificate to Doctor Quagmire's office on his way home that evening. Lois was gone but Doctors Quagmire and Marsh were still there and getting ready to leave for the day. Ivor handed the envelope to Doctor Quagmire, saying, "Karl, Hy would appreciate your making this payment as soon as you can as he needs to keep his subs and suppliers happy and paid up to date."

Karl replied, "It's not due until the 10th, is it?"

"That's right. That's the due date according to the contract. That's all Hy is asking for. The last two payments were 3 and 4 days late. He could have billed for interest but he didn't."

"Dammit, Ivor, Hy owes me money and he isn't rushing to pay it. I'm not charging him interest. Why should I go out of my way to pay him on time? Right, Marsh?" She nodded agreement.

"How much money does he owe you, Karl."

"That's none of your business, Ivor. I can't reveal confidential client information. That'd be unprofessional and highly unethical."

The doctors left and Ivor continued on home wondering if he had done Hy's cause any good.

On the 11th of April, Hy called Ivor to ask about Doctor Quagmire's payment. Ivor agreed to call the owner to ask about it. When Ivor talked to Lois she said she knew nothing about it. Ivor asked her to remind Doctor Quagmire that the money was now a day past due. She said okay but didn't mention it to her employer.

The following day, Hy called again and asked about the payment. "Look, Ivor, I'm willing to go over to Quagmire's office and pick it up. I need the money right away. Will you please call him and tell him I'll be there to pick it up?"

"Knowing Doctor Quagmire as I do, I think that wouldn't be such a good idea. It would just infuriate him. He has an extremely short fuse." Ivor didn't feel comfortable making excuses for his intractable client.

"Well, what do you suggest, Ivor? If I don't pay the subs, they'll soon be walking off the job. They'll go to other jobs where they're getting paid."

"Look, Hy. I'll call him again and see what I can do. I'll call you back as soon as I know anything." Ivor placed the call and Lois said to call back at 2:50. Ivor went out for a quick lunch with Leo King.

Leo asked, "How're you getting along with Quagmire and Lowe and the Psi Creativity Center?"

"The building is looking good. Hy Lowe is doing an outstanding job with the construction and he's up to the minute on his time schedule. He's got a great superintendent and competent subs. But the fly in the ointment is Doctor Quagmire. He's too busy to take care of the owner's duties. He doesn't seem to think they're important. Right now I'm trying to get him to pay Payment Certificate #4. It was due two days ago. Quagmire seems oblivious to the contractor's cash flow problems."

Ivor interrupted his afternoon's assignments at precisely 2:50 to phone Karl. Lois reported that he was free and would come to the phone immediately. Ivor wondered how the word immediately could be construed so loosely when he was left to wait at least 5 minutes. Finally, "Hello, Ivor. What can I do for you? No more trouble, I hope."

"Karl, have you taken care of the contractor's payment yet? Hy Lowe called before noon and says he hasn't received it. He's deeply concerned. He has hungry subs and suppliers clamoring for their money. He can't afford to lose their cooperation"

"I don't know, Ivor. I'll have to check with Lois. She should have taken care of it by now. I'll check with her and make sure it's taken care of today."

"I'll call Hy and tell him he can pick it up at your office today. Okay, Karl? How about 5 o'clock?"

"No, Ivor. We're much too busy with clients. We're really bogged down. Make it 5 o'clock tomorrow."

Ivor called Hy and told him he could pick up the check at Quagmire's at 5 o'clock tomorrow. Hy was there at the stroke of 5. Lois was at her desk. "I have your check right here, Mister Lowe, but it hasn't been signed by Doctor Quagmire yet."

"When's he going to sign it?"

"He and Doctor Marsh left the office about 20 minutes ago for the Easter weekend and won't be back until 9 o'clock Monday morning." Hy couldn't believe that Quagmire had pulled such a shoddy stunt. But he didn't say anything to Lois as it wasn't her fault. He merely said, "I'll be here at 9 o'clock Monday morning."

Hy was waiting when Doctors Quagmire and Marsh arrived Monday morning. Lois had the check sitting on her desk. She put it into Doctor Quagmire's outstretched hand. He signed it and casually tossed it on Lois' desk without a word to Hy. Hy quietly picked up the check and delivered it directly to his bank, then returned to his office. He phoned Ivor and said, "Ivor, I collected Payment #4 this morning. It was 7 days late. I can't go through this every month. Something's got to be done about that miserable bastard, Quagmire. I've been studying the AIA general conditions and I want to exercise my right, in Subparagraph 2.2.1, to require the owner to furnish reasonable evidence to the contractor that financial arrangements have been made to fulfil the owner's obligations under the contract."

"Yes, Hy. You do have that right. Send me that request in writing and I'll take it up with Doctor Quagmire."

"And, Ivor, you might remind him, that if he fails to produce the evidence, I have the right to terminate the contract and recover for all work done, including all costs, overhead, profit, and damages. You'll find that in Clause 14.1.1.5 and Subparagraph 14.1.2."

"I'm acquainted with those provisions, Hy."

"Okay, Ivor, I'll deliver the request to you this afternoon. I'll also have a bill for him for the interest charges for this and all of the other late payments."

"Incidentally, Hy, Doctor Quagmire claims that you owe him some money. How much is it?"

"I don't owe him anything. My wife went to Doctor Marsh last year for two sessions of bereavement counseling when her brother was killed in a plane crash. The charge was $250 which I handed to Quagmire in cash. He put it in his pocket and said he'd send me a receipt. But he never did. Lois has been billing me every month ever since. I can't prove I paid it, but I'm not going to pay it again."

Ivor received Hy's two documents the next morning, the request for the owner's financial evidence, and the invoice for interest on past due amounts. He called Lois to make an appointment to see Doctor Quagmire. She started the usual routine of putting him off when he interrupted, "Lois, quit the stalling routine. I've got to see Doctor Quagmire today. When's his last client finished today?"

Lois was suddenly cooperative, "At 10 minutes to 6, Mister Judge."

"Tell him I'll be there at 10 minutes to six." Ivor hung up.

* * *

At 5:50, the last client of the day came out of Doctor Quagmire's office and Lois showed Ivor in. Ivor sat down, handed Hy's letter to Karl, and waited while he read it.

Karl read it quickly and then reread it more slowly, his face reddening. "Can that unspeakable pipsqueak really do this to me? I don't want him rummaging around in my financial affairs. Who the hell does he think he is?"

Ivor felt the best way to settle this whole issue was to lay it on Karl in the most direct way. So far, the kid glove approach hadn't worked. "Karl, you certainly must realize that it was your disregard for the contract procedures that finally triggered this request. I don't think Hy really doubts your ability to pay these bills. He just wants you to pay them on time."

"I didn't know a few measly dollars were so important to him. He must be very insecure. He could use some therapy. What's a few days, more or less?"

"I have another document for you to read. You'll notice that I've attached my certificate approving it for payment." He handed the invoice and architect's certificate to Karl. It took him only a minute to read and reread both sheets of paper.

"This is ridiculous, Ivor. He wants $1592.34 for interest on late payments. Those were just a few days over. It means nothing. He's nitpicking and gouging."

"Well, now you know the value of the late payments. He's got it coming, Karl. That's why I approved it. If you want to avoid any more interest charges, just pay the bills on time from now on."

"What about this request for financial information? What do I have to do?"

"Hy might be satisfied with your promise to make all future payments on time. Maybe I could talk him into it."

"Okay, tell him I'll pay on or before the tenth of the month from now on."

"That'll have to be in writing, Karl. I don't think he trusts you any more. After all, that's what was required in the contract that you signed in the first place."

"Okay, Ivor." Karl picked up his intercom and asked Lois to come in to take some notes. She appeared in a few seconds. "Lois, Ivor will dictate a memo for you to type up for me to sign." Lois looked expectantly to Ivor with pen poised over her pad.

Ivor slowly dictated, "Address it to Hy Lowe, Superior Builders. Dear Hy, This is to confirm that I will make all future certified payments on or before the due date. I am sorry if I have inconvenienced you by my late payments. Enclosed is my check in the sum of $1592.34 in payment of your invoice for interest on past due amounts on Payments #2, 3, and 4. To be signed by Doctor Quagmire. Type it up right now and write the check. I want to take them with me when I leave tonight." Karl wasn't smiling. Ivor was having difficulty suppressing elation over his near success.

In a few minutes, Lois reappeared with the memo and the check and handed them to Doctor Quagmire. He read them, audibly sighed, hastily signed both documents, and handed them to the architect.

Ivor kept going, "Now, Karl, what about this money that Hy owes you? He told me that it was only $250 and that he paid it to you. That he paid it in cash, that you put it in your pocket, and that you neglected to give him a receipt. Is that right?"

"Well, it could have been. I'm an extremely busy person. Lois probably neglected to send him a receipt and she keeps billing him. She still has his name on our list of delinquent clients. I'll talk to her about it."

Ivor, now on a roll, brought up another subject. "Karl, while you're at it, I'd appreciate your bringing my account up to date. You've been gradually falling behind ever since our last discussion about interest."

"Okay, Ivor, I'll have Lois send you a check tomorrow."

"No, Karl. I'll take it with me now."

* * *

Points of Law 17

PSI CREATIVITY CENTER II
Quagmire Employs his Patient as a Contractor

17.1 May an Owner Accept a Suspiciously Low Bid?

Let's suppose that you are taking bids to build your dream house. Your plans and specifications have been prepared by an architect and picked up by five prospective bidders. When the big day arrives for opening the bids, they are as follows:

Acme Construction	$575,000
Beta Construction	$492,125
Cray Construction	$489,000
Delta Construction	$472,895
Easy Construction	$325,592

Your architect has estimated that the job should cost $400,000.

When you review these bids, you say to yourself, "Hey, there must be some mistake." Easy Construction must have left something out. They couldn't possibly build the job for $325,000 and make a profit."

Your next thought, perhaps, is the evil notion that you could take advantage of the situation, and accept the low bid. Would the law let you get away with it?

Here, we deal with a segment of the law of contracts known as mistake. A contract is usually formed by a process of offer and acceptance. A contractor's bid is an offer that ripens into a contract when it is accepted by the owner. But what if the contractor makes a mistake in preparing its bid? Will the law allow the contractor to get out of its bid because it made a mistake? It all depends ...

First, it depends upon whether the mistake was a material mistake. Second, it depends on whether the mistake was a mistake in judgment or a clerical mistake. Third, it depends on whether the offeree should have suspected the mistake.

A mistake is material if it's a big mistake. Let's suppose that in compiling its bid for a job a contractor forgets to include the cost of installing a linen closet. The contractor's bid is $400,000 without the linen closet, but with the linen closet it would have been $401,072.15. The mistake is immaterial.

Now assume that in figuring the job the contractor leaves out the cost of the framing. The bid price without the framing is $400,000, but with the framing included it would have been $492,815. That mistake is material. The courts will relieve a contractor of a mistaken bid if the mistake is material, and if it is a clerical error rather than a mistake of judgment.

A clerical error occurs when a contractor, in compiling its bid, misplaces a decimal point or fails to get a number on the adding machine tape.

A mistake in judgment occurs when a contractor makes a reasoned but unwise decision to take a risk.

Let's suppose that an excavation contractor is figuring a job that requires imported fill. The contractor happens to know that a nearby job that is being performed by another contractor will produce dirt for export. The contractor assumes that because of this fortunate circumstance he can acquire the necessary import for next to nothing. After the contract is signed, though, he finds that the dirt has been committed elsewhere. The mistake adds $100,000 to the cost of performing a $300,000 contract. The mistake is material, all right, but it's not a clerical mistake. It is a mistake in judgment. The contractor takes the risk of mistakes in judgment.

Ivor, Quagmire's architect, has noticed that Hy's bid is about $200,000 low. As a sophisticated architect, he knows that Hy would have the legal right to withdraw the bid if it was submitted $200,000 low because of a clerical error. Ivor ascertains that Hy has carefully scrutinized his estimates, has not found any major errors, and is willing to stand by the bid. Therefore, he knows, the contract can be awarded to the low bidder and will be enforceable.

17.2 The Importance of the Bond

When Ivor notices that the contract package returned by Hy is missing the performance bond, he becomes skeptical about the viability of the contract. Superior Builders is a contractor unfamiliar to Ivor who was included on the bid list because its president's wife was a patient of Dr. Quagmire, the project owner. The bid of Superior Builders was $200,000 low, which made Ivor wonder whether Superior Builders was a competent bidder. At this point, with the bond missing from the contract package, he is wondering whether Superior Builders is even bondable.

Performance bonds are often confused with insurance because they are issued by insurance companies. However, a performance bond is not insurance: it is a guarantee. When it signs a performance bond, a surety company guarantees that the contractor will perform the contract according to its terms, in other words, finish the job as required by the contract per plans and specs, on budget, and on time. The surety is in the same position that you would be in if you guaranteed that the your brother-in-law would repay a loan to the bank: if your brother-in-law doesn't perform, you would have to perform on his behalf.

Naturally, bonding companies are unwilling to guarantee that contractors will perform their contracts until they have made a careful investigation to ascertain the financial strength of the contractor and the depth of experience of the contractor's directors, officers, and management. Inexperienced contractors, and those with inadequate capital, can seldom produce a performance bond from a legitimate bonding company.

A bonding company will not write a bond if it thinks there is any substantial chance that the owner will make a claim against the bond because of nonperformance by the contractor. Insurance companies are in business to take losses. Bonding companies are not. To protect themselves against loss, bonding companies require the owners and principal officers of corporate contractors to sign indemnity agreements under which they agree to hold the bonding company harmless from any loss.

The premium for a performance bond is usually about one percent of the contract price. The contractor naturally passes this cost on to the owner. Most owners find that the bond is well worth the price. Most contractors, however, do not have bonding capacity and therefore cannot provide performance bonds. Therefore, Ivor is pleasantly surprised when the performance bond and the insurance certificates are delivered.

17.3 The Importance of Certificates of Insurance

Ivor was also pleasantly surprised when the certificates of insurance arrived. Why are these so important? This is because construction is an inherently risky business, and a project owner is therefore exposed to risks that are not normally encountered in the conduct of its regular business. Here, the owner is a psychologist. The major risk in that profession is malpractice claims, so Dr. Quagmire probably has adequate malpractice insurance. But when a construction project is undertaken, Dr. Quagmire becomes vulnerable to all the risks of injury that are inherent in the construction process including jobsite injuries to workers, visitors, and bystanders, and the risk that construction activities will damage neighboring property because of such risks as overspray, cave-in, trenching, or fire. These risks can be covered by the contractor's liability insurance policy.

Under a liability policy, the insurance company promises to defend the contractor against claims for bodily injury or property damage, and to pay any judgment entered against the contractor because of such claims. The certificate of insurance provides that the owner is named on the contractor's liability insurance policy as an additional insured, and therefore the contractor's liability insurance policy protects the owner, as well as the contractor, from such claims.

When subcontracts are awarded, the subcontractors will likewise be required to supply certificates of insurance naming the contractor and the owner as additional insureds as to claims arising out of the activities of the subcontractors on the jobsite.

17.4 Performance Bond Forms

As a careful architect, Ivor will review the performance bond forms both for content and for authenticity.

The bond should be written on a form supplied by the AIA, which gives full protection to the owner. If the bond is written on a form supplied by the bonding company, Ivor will examine it carefully to make sure that it doesn't have loopholes written into it.

Ivor is aware that there is a substandard bond market in which certain surety companies will write bonds for contractors that would not normally be bondable because of lack of experience and financial resources. These substandard bonds can be purchased at an exorbitant price, and are written by bonding companies that are themselves substandard in their financial resources and in their willingness to respond to claims.

Ivor has also heard of cases in which performance bonds that appeared legitimate were actually forgeries. Although he is not a naturally suspicious person, Ivor makes it a rule to routinely authenticate the validity of a bond by calling the bonding company directly, and following up with a confirming letter. Careful architects also advise their clients, as Ivor did, that the insurance certificates and bonds should be reviewed by their insurance advisors and legal counsel.

17.5 The Nature of Builders Risk Insurance

"Builders risk insurance" is the nomenclature applied to insurance against the risk that a building may be destroyed or damaged by fire, earth movement, flood, wind, or other peril during the period when it is under construction. The owner and the contractor both have an insurable interest in the continued existence of a building while it is under construction.

Under the AIA contract documents, it is the owner's responsibility to supply builders risk insurance. In the event of a loss covered by the insurance (for example, a fire loss), the owner is required to process a claim with the builders risk carrier and to use the proceeds of the policy to pay the contractor and the subcontractors for the cost of rebuilding. If the proceeds of the insurance are insufficient to pay the cost of rebuilding, the owner takes the loss. If the peril is not covered by the builders risk insurance, the owner not only takes the loss but may also be subjected to a claim by the contractor and the subcontractors that the owner failed to provide the level of insurance required by the contract documents.

It is up to the owner, then, to decide whether to include the perils of earthquake and flood. These coverages can be very expensive.

An important feature of the AIA contract documents is that the builders risk insurance covers the interests of the owner, the contractor, and the subcontractors and that owner, contractor, and subcontractors waive rights of subrogation against each other. The question of subrogation arises after a builders risk insurer has paid a loss. Having paid the owner's loss, the insurance company steps into the shoes of the owner and can recoup the loss by "subrogating" against a person who is liable for causing the loss. Suppose, for example, that a careless welder starts a fire in a building under construction. After paying the loss, the builders risk carrier would subrogate against the welder. However, if the welder is a subcontractor and the construction was performed under the AIA documents, the right of subrogation has been waived and the welder is not liable to reimburse the insurance carrier for its loss.

Owners who dislike the responsibilities placed upon them by the AIA documents, require the contractor to carry builders risk insurance, with the premium added to the contract price. Thus, the owner avoids the responsibilities of acting as a trustee for the contractor and the subcontractors with respect to the collection and disbursement of insurance proceeds, and places upon the contractor the risk that the proceeds of insurance will be insufficient to pay for rebuilding.

17.6 Construction Loan must be Recorded Before Work Commences

Since project owners don't usually have money in hand to pay for an expensive construction project, most construction is financed by banks or savings and loan associations or insurance companies. Security for the loan is the very property under construction. The security is in the form of a mortgage or a deed of trust recorded in the county recorder's office. If the borrower (Quagmire) does not make the payments on the loan, the lender can foreclose its mortgage or deed of trust, sell the property, and use the proceeds of the sale to pay off the loan.

A construction lender does not hand over the full amount of the loan at the time when it is recorded, but advances "construction draws" periodically as the work goes on. Thus, theoretically, the value of the lot and the partially constructed building is always equal to or more than the amount of the loan proceeds that has been advanced. Dr. Quagmire is unhappy because of a delay in the start of construction which, Ivor explains, is Quagmire's own fault. The thing that is delaying the start of construction is the fact that Quagmire has not yet completed the arrangements for his construction loan, and therefore the mortgage or deed of trust securing the loan cannot be recorded. Why should this delay construction? It's a question of priority.

The contractor, subcontractors, and suppliers who will work on the project will have mechanics lien rights that take their priority from the commencement of construction. A mechanics lien is similar to a mortgage or a deed of trust because if the lien is not paid off, the holder of the lien can foreclose and have the property sold at public auction with the proceeds of the sale used to pay off the lien. When the same property is subject to multiple liens, the proceeds of the sale are used to pay off the liens in their order of priority. All too often, the value of the property is insufficient to pay off all the liens, so only the lien with first priority is paid off, while the other liens are "sold out" – in other words: worthless!

A deed of trust or a mortgage takes its priority from the date of recording, while mechanics liens take their priority from the date when construction work first begins. Therefore, before it will record its mortgage or deed of trust, the lender visits the jobsite (or has a title company do it) to determine that construction has not begun. If construction has started the work must be stopped for a month or more so that the loan can be recorded without losing its priority to mechanics lien claims.

17.7 Performing the Impossible

Right after quitting time, on the first day of construction, it started to rain. The site was so wet and muddy that it was "impossible" to work for eight days. Did this "impossibility" legally excuse Superior Builders from working?

Under the doctrine of impossibility the performance of a contract is excused if it is impossible of performance. For example, the broadcaster who signed a contract to announce the arrival of the Titanic in New York was excused from performance of his contract by the doctrine of impossibility. Since the Titanic never arrived, it was impossible to broadcast a description of its arrival.

Now let's look at Superior Builders' contract. Rain and mud made it impossible for Superior Builders to work for eight days. Nevertheless, this didn't make it impossible for Superior Builders to finish the job on time. They could have made up the eight days by accelerating the work.

As a matter of fairness, though, most construction contracts provide that the time for performance will be extended for delay caused by inclement weather, strikes, picketing, shortage of materials, fire, earthquake, or other event beyond the reasonable control of the contractor.

Now let's suppose that the building under construction is destroyed by earthquake. This would not make it impossible for Superior Builders to perform its contract. It would be a lot more expensive and might take more time, but it would be possible to clear off the jobsite and start all over again. For this reason, contractors sometimes seek to include a contract clause excusing performance in the event the project is substantially damaged or destroyed by fire, earthquake, or other casualty.

Now let's look at impossibility from a subcontractor's point of view. Suppose a building is destroyed by earthquake and not rebuilt. Such an event would make it truly impossible for the roofing subcontractor to install the roof, and would excuse the performance of that contract.

18

ABC Warehouse IX
Owner's Assignment of Subcontractor

Now that the ABC Warehouse building has been completed for a few years, and the business has developed into an unqualified success, one tends to forget some of the problems that had to be solved in the early days. Some are best forgotten, such as in this dismal story that occurred at the very onset.

It was early June 1989, and his fledgling warehouse business was taking root and beginning to thrive. Allen Brady, President of ABC Warehouse Company, was elated and almost floating on air. He'd just hired Judge & King, AIA, Architects, to design a wondrous state-of-the-art warehouse and offices on his recently acquired 10-acre site of prime industrial land in the San Fernando Valley. He was so excited he couldn't focus his mind on anything else.

Arriving home that evening, he greeted his wife with, "Get dressed, Alice, we're going out to dinner. Wherever you want to go. We're celebrating." He went on about the architects and chattered about the building project. He was full of it.

Alice quickly readied herself and they left for Shay New, a trendy new French restaurant in Encino. They were comfortably seated, had ordered cocktails, and were studying the menu, when Alice said, "It's fabulous about your new building and all, but what about my little brother? Can't you give some of the work to Sonny?"

Allen wasn't very diplomatic. "I wouldn't touch that lazy deadbeat with a 10 foot pole," he blurted.

Allen was totally fed up with his incorrigible brother-in-law. He had a so-called electrical contracting business called Speedy Electric. Sonny was in perpetual financial difficulty. Allen had tried to help him before by recommending him to friends and he always ended up regretting it. It was not only embarrassing, but once it cost him $1350 to finance completion of one of Sonny's jobs. It took him over a year to get his money back.

Alice persisted, "But, Allen, how can Sonny get ahead if he can't get work?"

"That's not my problem, Alice. He's a flake. An irresponsible incompetent. If he was any good he'd have no trouble finding work."

"How can you say that, Allen? You're cruel. He's my little brother. I love him."

"Well, I wouldn't hire him to wire a dog house."

"Why can't you give him a chance?"

"He's a first class jerk. And a royal screw-up. That's why." Allen was resolute.

Then Alice clammed up. It was not a pleasant meal. Thereafter the evening went rapidly downhill. Allen had descended from zenith to nadir in a matter of minutes. The celebration was definitely over. Neither spoke on the short, but seemingly endless, depressing ride home. The icy climate was no better at breakfast the next morning.

* * *

Four months later, one day in late September, Ivor Judge, FAIA was busily checking the conference room at Judge & King to make sure everything was in readiness for the meeting with Allen early that afternoon. They were going to discuss the contracting and bidding procedures to be used for the new 100,000 square foot warehouse and 5,000 square foot office wing for ABC Warehouse Company. The job was going out to bid on October 16, 1989.

In anticipation of the meeting with Allen, Ivor had sent him a copy of the latest edition of Recommended Guide for Competitive Bidding Procedures. This is a document developed jointly by the American Institute of Architects and Associated General Contractors, and issued simultaneously as AIA Document A501 and AGC Document 325(23). When Allen received it, although it was only eight pages, he was too busy to read it so he called his trusted warehouse manager, Carl Daly, into his office and handed him the slim booklet. "Here, Carl, read this before we go over to Ivor's office tomorrow. Let me know if there's anything in it I should know about. Okay?"

"Okay, Boss." Carl appreciated the honor of being the trusted key employee, but felt Allen took advantage of him.

Driving to Ivor's office the next day they discussed some of the items that Carl considered important. Allen was driving and Carl was thumbing through the booklet for the first time.

Carl said, "The bidding procedure looks pretty complicated and takes a lotta time. If we could cut out some of this useless red tape we could get the construction started sooner."

"Good thinking, Carl. Red tape always costs valuable time and money."

Promptly at 2 PM, Allen and Carl arrived at the Judge & King office. After the usual exchange of polite greetings, and the ritual pouring of coffee, Ivor started the meeting by getting out his copy of A501 and a pad to keep notes of topics discussed and decisions reached.

Ivor announced, "Our timetable is important. We expect to have the building permit in two weeks, on October 4, and we'll have the drawings corrected and all bidding documents completed by the 11th. So I propose to start the bidding period on Monday, October 16. The bids'll be due on November 7. That'll give the bidders three weeks to work on their estimates, gather their sub-bids, and prepare their proposals."

Allen thought that over and asked, "Isn't that wasting a lot of time? We wanta get the construction started. Why don't we cut it to a week?" Carl sagely nodded his concurrence.

Ivor explained, "For a job of this size a week is insufficient. The contractors need time to analyze the plans and specifications, solicit and correlate their subcontractors' bids, make their takeoffs, and prepare their proposals. They should also have enough time to carefully consider construction techniques and scheduling. If we don't give them enough time some of the bidders will drop out and those that do bid will probably be too high. It's in your interest to give the bidders enough time to properly analyze the job. Then the bids'll be more consistent and usually lower."

"Okay, Ivor, if you think that's what we ought to do." He glanced at Carl as though admonishing him for such a stupid suggestion.

Ivor continued, "The next thing we should discuss is the list of bidders. Do you have any general contractors in mind that you want to bid the job?"

"No, Ivor. I've a subcontractor in mind, but no general contractors. Who would you suggest?"

"I've drawn up a list of competent general contractors that we've had experience with or have received recommendations on. We've narrowed it down to five."

Carl interrupted, "Why limit it to five? We oughta get all the bids we can to make sure we get the lowest cost." Allen was thinking the same thing.

Ivor explained, "It costs money to bid a job and it isn't fair to contractors to waste their time and resources. If the bidders list is too long the incentive for any contractor to give it a proper analysis is diminished. Five competent bidders will be sufficient to insure competition even if one or two drop out. Here's my list of suggested bidders."

He handed Allen a list of five general contractors: Advanced Contractors, Certified Constructors, Fidelity Builders, Hyde Construction Company, and Industrial Erectors.

He continued, "If these contractors are okay with you, I'd suggest that you have your bank obtain an up-to-date credit report on each of them before we ask them to bid."

Allen replied, "Okay, we'll get that started later this afternoon." Allen handed the list to Carl.

Carl asked, "Do we have to go along with the low bidder? If we'd rather work with one of the other contractors, maybe we could get him to lower his price to the low bid?"

Allen added, "Or even lower?"

"That wouldn't be fair to the contractors who spend so much time, money, and effort on bidding. Besides, it makes no sense. If some firm is on the list that we wouldn't award the job to if they were low, then we should take them off the list. The list should include only contractors that are competent, experienced, financially capable, and acceptable. Then we'll be happy to go along with the low bidder, whoever it is. The understanding under which contractors are willing to incur the considerable expense of bidding is that the contract will be awarded to the lowest responsible bidder."

Carl asked, "What if we decide that the low bidder isn't responsible?"

Ivor explained, "That's why it's so important that we carefully qualify all the bidders before inviting them to bid. That way we won't have to eliminate the low bidder for some reason that we could have discovered before bidding."

Allen then asked, "How'll we know how long the construction will take?"

"Well, there are two approaches we could take. The first would be for us to estimate a proposed or desired completion date and impose it on the bidders. If it seems to be unrealistically short to the bidders, then costs will rise to make up for the expedited schedule. A better way would be to ask all of the bidders to state in their proposals how much time they will need. Then each bidder's proposed completion time will be an additional factor to take into account in evaluating the proposals."

Allen and Carl were carefully listening. Allen said, "It sounds okay to me. But how can we put some teeth into the required completion date. What if completion is late? What can we do about it? Shouldn't there be a penalty for late completion?"

"First, Allen, let me explain, we can't have a penalty. It wouldn't be enforceable. But, we can have a liquidated damages clause in the contract. This is a sum, chargeable to the contractor, which would approximate your actual damages for late completion, as nearly as we can estimate them. It would include such actual costs as additional rent on your old premises, interest on your invested capital in the new premises, and any other costs you expect to incur on account of delayed completion. It would be better for you, rather than me, to estimate the liquidated damages amount. It can be in the form of a lump sum for late completion plus an amount per day for continuing lack of completion."

"One more thing for me to do," said Allen. "Carl, make a note of that."

Ivor continued explaining the bidding process, "We'll issue five sets of the bidding documents to each of the bidders."

Carl asked, "How much does each set of plans and specifications cost?"

"About fifty or sixty dollars."

Carl, busily figuring on his note pad, asked, "Each of the bidders get 5 sets so that's 25 sets. If each set costs $60, that's a total of $1500. Why don't we cut it down to 1 or 2 sets? That'd save a lotta money."

Ivor explained, "The contractor who ends up with the job is going to need at least 25 sets of documents, so they can be used during the bidding period by the bidders. You'll get better bids if the bidders have enough drawings and specifications to work with. It would be false economy to try to save money on construction documents."

Ivor asked, "Well, are there any more questions?"

Allen decided this was the time to find out about getting Sonny to do the electrical work. "Ivor, my brother-in-law is an electrical contractor, Speedy Electric. I'd like him to do the electrical work."

"The best thing for him to do, Allen, is to submit his bid to each of the general contractor bidders, and if he's low, then he'll get the job," Ivor suggested.

"That's a good idea, Ivor. I'll give him the names." Allen was pleased with this simple solution to the problem with his wife and her flaky brother.

* * *

When Allen got home that night he told Alice about the meeting with Ivor and his suggestion about Sonny submitting his bid to each of the general contract bidders. "So, you can tell Sonny that's what he can do."

Alice wasn't impressed. "But, what if he's not low bidder?"

"Then he won't get the job."

Allen's home life was no better than it was. So, he'd have to talk to Ivor again. The next day he called Ivor and said, "Ivor, we've got to figure out a way to have Speedy Electric do the job. The low general contractor might not hire him."

"Well, Allen, if you really want Speedy on the job, then the best thing to do is to obtain their bid and assign it to the low bidder. This will be in the bidding instructions so the bidders will know all about it when they're preparing their bids. It's advantageous for your electrical contractor to be working under the supervision of the general contractor so that coordination with other trades and overall time scheduling will be looked after properly. The contractor's control and financial responsibility are important to the stability of the project. Of course, you won't know if your brother-in-law's price is competitive if electrical bids aren't taken."

Allen suggested, "Why not have the contractors take electrical subcontract bids as usual and then give the job to Speedy at the amount of the low bid?"

"No, Allen, we couldn't participate in that kind of practice. It would be unethical and unfair to the other electrical contractors. I doubt that any of the general contractors on our list would want to be involved in anything like that."

"Then, how can I be sure he's giving me his best price?"

"You won't know for sure. Usually, you negotiate the best agreeable price with someone you can trust."

"Well, I certainly don't trust Sonny, but I have no other choice."

So, Allen negotiated a deal with Sonny for $99,575. This wasn't too bad considering that Sonny's starting price was $125,000. He got the proposal in writing and gave it to Ivor to include in the bidding instructions. Alice was now satisfied. Allen was once again enjoying his home life.

* * *

The low bidder was Hyde Construction Company with a bid of £2,100,437 including Speedy Electric's subcontract. The general contract was signed on November 10, 1989. Allen and Ivor were very optimistic about the new building, as Hyde Construction had an excellent reputation and their bid was within the budgetary limitations. George Hyde assigned his best superintendent, Ezra Field, to run the job in the field. George and Ezra had devised a tight but realistic time schedule and notified all the major subs of their timing obligations. They were sure it wouldn't be difficult to meet the required May 18, 1990 completion deadline if all the subs and suppliers worked efficiently and cooperatively.

Judge & King had announced in the bidding instructions that there'd be a pre-construction meeting at the jobsite to discuss use of the site, answer questions about the documents, explain the design concept, discuss shop drawing and change order procedures, and for the participants to get generally acquainted with the project and each other. The meeting would take place at 9 AM on Monday, November 13, 1989. Hyde was asked to invite the major subcontractors to be present. The meeting was held as planned and it was considered helpful by all who attended. Unfortunately, Sonny wasn't able to make the meeting.

Ezra got the job mobilized on Monday, November 20, 1989. Speedy Electric was due to start the underslab conduit runs 10 days later. When they didn't show up, Ezra checked with his office to make sure that Speedy was properly notified as he had instructed. Finding that they had been duly notified, he called Speedy's number and got an answering machine. He left an urgent message but received no return call. The following day he called, and again got the answering machine. So, he then called the Hyde office and explained the situation to George. "What'll I do now, Chief?"

"I'll call Ivor and tell him we need a reliable electrician. We can't put up with this crap. I'll see if we can replace Speedy with someone we can work with."

"Good idea. If we don't have all the underground electrical in by Friday, December 8th, we'll be behind our schedule.

George called Ivor, "Ivor, where did you get that crappy electrical contractor? He hasn't shown up on the job and we can't find him. We're ready to can his friggin' ass and get someone we can count on. How can we get him outta our contract and off our damn backs?"

"Hold on, George, it's the owner's brother-in-law. I'll call Allen and see what I can do. I'll call you right back."

Ivor called Allen and brought him up to speed, "You better get Speedy out to the job fast. The job'll start suffering if his underslab work isn't in by a week from Friday. Where the hell is he?"

"How the hell should I know? I'll see if I can find him. I'll call you back as soon as I know." To say that Allen was displeased would be a gross understatement. He called Alice at home, "Alice, where's your flaky brother? He's supposed to be at the jobsite."

"Don't you remember? I told you he and his new girlfriend, Sapphire went skiing in Switzerland to celebrate his getting the ABC job. They got snowed in, in Zermatt, and won't be home until the day after tomorrow. Isn't that romantic? Snowed in. In Zermatt."

"Dammit, Alice. I knew it'd be like this. If he holds up the job, I'll make him pay plenty. The flaky bastard."

"Don't you want my brother to have a decent life? He's just trying to have a little fun, Allen. Don't be such a grouch. It isn't Sonny's fault they're snowed in. He'll be back. Don't worry, you'll see."

Allen relayed the message to Ivor who then called Hyde. George replied, "Okay, Ivor, he has 48 hours to show up. If he isn't here by 8 AM, Monday, December 4th, he's history. I'll get a decent electrician. Someone I can trust and rely on."

Sonny arrived home on Sunday night and he heard all the increasingly frantic messages on his answering machine. So he called his sister Alice first. She told him he better be on the ABC job first thing Monday morning or he could kiss the job goodbye. Then he called some of the people who worked for him from time to time. Several were busy but he was able to line up a crew of four to start the job. An electrician and 3 laborers.

Speedy's ragtag crew was waiting when Ezra showed up at 7 AM Monday to open the jobsite gate. Sonny showed up a few minutes later in a brand new, shiny, bright red, Dodge van with racing stripes and magnesium wheels. He had the electrical plans and specifications and a truck full of new tools. Soon his disparate band was organized and they started to make some headway. Ezra called George and reported that the crisis was over, for now.

* * *

A couple of weeks later, Sonny started making arrangements for building a subterranean concrete transformer vault, to be located near the main trucking entrance driveway. It had to be precisely dimensioned so the 30-foot wide driveway would fit between the corner of the warehouse and the transformer vault. He didn't want to get involved in excavating, forming, steel reinforcing, and concrete pouring, so he asked Ezra for a price to have Hyde do that portion of his work. A few days later, Ezra had an estimate prepared by Hyde's estimating department. They wanted $920 complete. Ezra said, "Sign here and we'll get right on it."

Sonny carefully examined the bid and said, "No way, Jose. This bid's ridiculous. I could've done it myself for half this much. You guys are too rich for my blood." Sonny took bids from a couple of other concrete contractors and found one, Zippy Concrete, that would do the job for $635. Sonny got them started right away and the vault was quickly built, forms stripped, and excavations backfilled.

A few days later Ezra asked Sonny if the transformer vault was properly measured, 35 feet from the corner of the warehouse. Sonny replied, "There's only one way to find out, Ezzy. Let's measure it." So Ezra got his 100 foot steel tape from the job office. Sonny held the end at the corner of the warehouse foundation and Ezra measured to the side of the vault. Ezra called out, "This is only 25 feet. This is wrong. Ten feet off. And it isn't square with the building. It'll have to be moved."

"I don't see how this could be wrong. I told Zippy to take it off the plans and measure it. It'll cost a fortune to move it. Maybe the driveway could be narrowed down a little."

Ezra told him what to do. "Take it out and rebuild it." Ezra, disgusted, strode off to his site office to make appropriate notes in his daily report.

That night, Sonny called Alice and told her about the problem. "Will you ask Allen if it'd be okay to narrow the driveway at that point? Then I won't have to move this perfectly good expensive vault."

"Sure, Sonny, I'll ask him." Then she explained the whole situation to Allen.

He immediately replied, "No! Hell, no!"

She pleaded. No. It'll cost Sonny a lot of money. No. You can't treat my brother like this. No. Be reasonable. No. Will you talk to him? No.

She called Sonny back. "He says no. He seems definite. He won't listen to reason."

"I didn't think he would. I'll just have to get Zippy to move it." But that was easier said than done.

Zippy said, simply, "Pay up. We put it exactly where you said to put it. You measured it and you drove the stake. That's exactly where we built it."

"Well, how much do you want to remove it and rebuild it?"

"We'll remove it and cart it away on a time and material basis. And then we'll build a new one for $950."

Sonny repeated, "$950 to rebuild it? You only charged $635 in the first place."

"Well, we made a mistake. We had a loss. Do you want us to do it or not?"

"Time and material for the removal? You must be crazy. Give me a price."

"Okay, $700, win or lose."

"Good enough. Go ahead with it. It's gotta be done." Sonny started off for his truck.

"Not so fast, Slick. Pay me for the first job before we go ahead with the demo. Then, you'll hafta pay me for the demo before we go ahead with the new vault."

Sonny knew he'd get no better prices elsewhere, so he gave Zippy a check for $635 and told him to go ahead. Of course the check wasn't covered but he was sure he'd be getting paid another draw in a couple of days.

***

When Ezra and Hyde's estimator, Sam Trent, were going over the percentages of construction progress for Hyde's December payment request, they got down to the electrical line item.

Sam asked, "What do you estimate the electrical to be at this point, Ezra?"

"The best I can tell, it's about 10%, give or take a tad."

Sam was shuffling through the current subcontractor bills and dug out Speedy Electric's bill. "Speedy's asking for 25%."

Ezra said, "No way can that be right. I could be a little low, maybe it's really 11 or 12%, but no way higher."

Sam explained, "We want to be fair to the sub, but if we go too high, the architect might reject our whole request. It'll just make us look bad. We'll put electrical down for 12% and hope it gets past Ivor."

A few days later when Sonny received his check from Hyde for his first payment, he was angered and disappointed when he saw that it was only $10,754 when he was expecting closer to $25,000. Also, he hadn't reckoned on the effect of the 10% retention. The checks he had already written were over $22,000. What could he do now? Better call Alice tonight after dinner.

"Hiya, Alice, how're things?"

"Just fine, Sonny. How's the ABC job coming?"

"Everything's coming along just great, Sis. Couldn't be better. We're making fantastic progress and I'm sure I'll make a decent profit."

"I'm glad to hear it, Sonny." She loved talking with her grown up little brother and hearing of his successes.

"Well, Sis, there is a slight problem you may be able to help me with. I've billed the job for 25% and they'll only pay 12%. It's a complicated technical thing. Basically, Hyde wants to build the job on my money. They do it to us subs all the time. Then, at the end, they'll want to hold 10% of my money until 35 days after the Notice of Completion is filed. It's terribly unfair. There really is no big problem with my finances, it's just a temporary cash flow glitch."

"I see. What can I do, Sonny?" She'd help him if she could.

"Well, maybe you could get Allen to speak to the architect and the general contractor and get them to release my money."

"I'll talk to Allen, Sonny. I'll call you right back."

She and Allen were just sitting down to dinner. "Allen, I was just talking to Sonny."

"That deadbeat? What's his problem, now? Has he moved the concrete vault yet?"

"I don't know about the vault. He has a simpler problem. It's a slight cash flow shortage. The architect and contractor are trying to cheat him out of money he has coming. He wants you to talk to them."

"No."

"Please, Allen?"

"No."

She turned red, left the table crying, ran to the bedroom, and locked the door. She dialed Sonny's number and waited. His voice answered, "Hello, this is Speedy Electric. I can't come to the phone right now. I'm on an important mission. Just leave your name, number, and time, and I'll call you back when I can." Alice started to leave her message when Sonny picked up the phone, "I'll always talk to you, Alice. What did Allen say?"

"No."

"No? Is that all?"

"You wouldn't want to hear the other things he says about you. He's a vicious brute. What'll you have to do now?"

"Well, Sis, I'll be okay if I can get my hands on $15,000. Maybe Allen will loan it to me. I'll pay it back as soon as I get my money."

"He doesn't sound like he'd loan you any money, Sonny. Whenever I mention your name he starts saying no."

"It's his fault I'm in this mess. After all, it's his building. He's got the advantage of my labor and materials and he hasn't paid for it. The lousy bastard."

After a few days of being locked out of the bedroom, Allen decided to loan Sonny the $15,000. He was embarrassed to go to the architect to pressure him to release the funds through the general contractor.

* * *

Speedy Electric was moving right along but not fast enough to meet Hyde's overall progress schedule. By the end of February, Speedy was two full weeks behind. George Hyde was fed up with all the time and effort that was going into supervising Speedy. Ezra was spending half his time riding herd on Speedy's inept crew. Sonny promised but didn't put on more help as suggested by Ezra. And materials were always slow in being delivered.

What finally burst Sonny's balloon was the impatience of his electrical material supplier, Consolidated Wholesale Electric Company, who filed a mechanics lien for $9,140.60. They were fed up with Sonny's broken promises and rubber checks and decided to put an end to their misery. Hyde's lawyers settled the delinquent bill to get the lien claim removed and notified Speedy that they were off the job. The subcontract was terminated and a lawsuit was threatened to collect the deficit from Speedy along with the costs of completion of the electrical contract in excess of the remaining contract funds.

Hyde's chief estimator and contract negotiator, Sam Trent, started discussions with Faithful Electrical Corporation about finishing the job. Hyde tried to get a lump sum proposal for finishing up the electrical work but Faithful declined, demanding instead a cost plus fee contract with no guaranteed maximum. That was the best they could do under the uncertain circumstances of a partially completed job. They also refused to guarantee any of Speedy's work and expressed reservations about guaranteeing their own interfacing work. Hyde trusted Faithful Electric and directed them to proceed with the work. Faithful poured on the labor and worked overtime until the job recovered the two weeks of schedule slippage.

George called Ivor and brought him up to date on all the details and ramifications of Speedy's departure from the job. He said, "We're glad to get rid of those people. Speedy was bad news from the start. Unfortunately, it's gonna cost Allen some money."

Ivor asked, "Do you have any idea how much?"

"No, not exactly. We won't know for sure until the end of the job. At that time we'll render an invoice."

Ivor reported this conversation to Allen. Allen listened to Ivor's report, and replied, "Good riddance. I hope that's the last I ever see of him. That damn brother-in-law of mine is a certified flake and a pain in the neck."

Hyde Construction, in spite of a few other problems, finished on schedule, thereby avoiding payment of any liquidated damages. The completed project was praised by owner and architect alike.

A few days after completion, Ivor received Hyde's invoice detailing the costs to conclude the electrical work:

Hyde Construction Company

Invoice for Final Electrical Costs
ABC Warehouse Company

June 10, 1990

Paid to Speedy Electric Company prior to termination of their contract $15,732.75

Zippy Concrete, to satisfy their lien claim $635.00

Remove and replace concrete transformer vault..$2,175.00

Consolidated Electrical Supply Corporation, to satisfy their lien claim............$9,140.60

Faithful Electric Corporation, to finish the electrical work $108,421.35

Subtotal $136,104.70

Hyde overhead and profit on additional electrical cost:

Final cost $136,104.70
Speedy contract..($99,575.00)
$36,529.70 @ 10% $3,652.97

Total electrical cost..$139,757.67

Additional bond premium on increased construction cost:

Final Cost $139,757.67
Original Electrical Cost ....($99,575.00)
Increased cost $40,182.67 @ 1% $401.83

Drummond & Eagle, legal costs to settle the lien claims
and terminate Speedy's contract $1,500.00

Subtotal $141,659.50

Less the Speedy Electric original bid...($99,575.00)

Net Increase in Hyde Construction Company contract $42,084.50

Ivor Judge checked the invoice thoroughly, signified his approval by preparing and certifying a change order in the sum of $42.084.50, and mailed it to ABC Warehouse Company.

* * *

Allen's secretary placed the bulky stack of morning's mail on his desk as usual. She hurried back to her own desk as she had quickly reviewed it all and didn't want to be nearby when he reached the bottom two items.

After quickly sailing through the top of the pile, he got down to the last two. The first of these was the electrical change order, which he expected, but he was astonished at the cumulative effect of the incremental increases. The last item in the pile was a letter from Sonny's lawyers, Darnel & Jimson, threatening legal action for improper termination, lost profits, termination costs, damage to reputation, and other unspecified damages, totaling $1 million. He later found out that similar letters went to Hyde Construction and Judge & King.

When Allen got home from work that night he told Alice about the bill for over $40,000 to finish up Sonny's electrical work and the threatened lawsuit from Sonny's lawyers. He also reminded her of the $15,000 Sonny borrowed. He knew he'd never collect from Sonny. "I hope I never see that flake again."

Alice sobbed, "Well, you've got your wish, Allen. Now, Sonny's disappeared and we don't know where he is. You've driven him away by your constant persecution. It's all your fault, you heartless beast."

* * *

Points of Law 18

ABC WAREHOUSE IX
Allen Helps his Brother-in-Law

18.1 Duty to Award Contract to Low Bidder

Allen is tempted to solicit competitive bids from five qualified contractors, open the bids, determine who the low bidder is, and then negotiate with the low bidder, other bidders, or even other contractors who did not participate in the bidding, for a still lower price. Ivor informs Allen that in the contracting business this is considered unethical. It is basically unfair to put five contractors to the trouble and expense of bidding a job, only to take advantage of the low bidder's price as a figure from which to negotiate a still lower price from another contractor. Is Allen under any legal obligation to award the job to the lowest bidder?

In law, a legal obligation is known as a "duty". Legal duties may be imposed by statute, by contract, or by the law of torts. A statute is a law enacted by Congress or a state or local legislative body. A contract is a legally enforceable agreement. Under the law of torts, duties are imposed by courts in order to advance social policy.

Most public agencies in the United States are required by statute to award public works contracts to the lowest responsible bidder. Competitive bidding protects the public interest by insuring that public works projects are accomplished by qualified contractors at the lowest possible price. Competitive bidding statutes exclude favoritism, graft, and corruption in the award of contracts for public works. Unfortunately, not all public projects are subject to competitive bidding, and therefore the opportunity for abuse persists.

Since Allen is not building a highway, a bridge, a dam, a post office, or a police station, but a private warehouse, competitive bidding statutes do not apply to him.

18.2 Law of Contracts

We look now to the law of contracts. Allen has no contractual relationship with the prospective bidders. A contract is defined as a promise to perform some act in exchange for a promise by another party to perform an act in return. For example, in a contract to build a warehouse, the contractor promises to build the warehouse in exchange for the owner's promise to pay the contract price.

Contracts are usually formed by a process of offer and acceptance. In the construction industry, a "bid" is an offer to perform construction work, but the bid does not ripen into a contract until it has been accepted.

18.3 Implied Contract

Even though there is no formal contractual relationship between Allen and the bidders, Allen does have the obligation to accept the low bid, under the doctrine of implied contract.

When a project owner invites bids from a select list of qualified contractors, to be opened in the presence of the bidders, the owner impliedly promises to award the contract to the low bidder. Otherwise the competitive bidding process would be a charade! Here, the owner's implied promise is an offer to enter into a contract with the low bidder and the bid is an acceptance of that offer. Normally, an offer can be withdrawn at any time before it has been accepted. Allen, however, does not have the legal power to withdraw the offer in this instance under a doctrine known in legal jargon as promissory estoppel. The reason is that Allen should anticipate that his implied promise to award the contract to the low bid would result in substantial and detrimental reliance by the bidders. The bidders rely on Allen's implied promise in a substantial and detrimental way when they incur the expense of preparing their bids. Therefore, Allen is required to award the contract to the low bidder.

If Allen wants the right to use the low bid as a benchmark from which to attempt to negotiate a lower price with another contractor, he could reserve that privilege in the Invitation to Bid. However, most qualified contractors would refuse to bid the job under such conditions.

18.4 Penalties and Liquidated Damages

In common with many owners, Allen wanted to put some teeth behind the clause of the contract that requires the contractor to finish the job on time. Allen wanted to include a penalty clause that would penalize the contractor for failure to finish on time. Ivor explained that a penalty clause wouldn't be enforceable, but a liquidated damages clause would be enforceable. This sounds like legal double talk. What's the difference between a penalty clause and a liquidated damages clause? Is it anything more than just a difference in terminology?

In order to understand this point, we must review some history of the law of contracts. As English common law developed in the sixteenth and seventeenth centuries, it was recognized that important commercial reasons support the notion that people should keep their promises. Take the situation of a merchant in Liverpool who wants to ship goods to Southampton. His willingness to entrust his merchandise to the ship captain depends on his confidence that the captain will keep his promise to deliver the goods. At the same time, the captain's willingness to deliver the goods depends on his confidence that the merchant will pay the freight. If contracts were not enforceable, commerce as we know it would be impossible. Courts recognized this.

18.5 Damages

The enforcement mechanism is called damages. If failure to perform a promise causes injury to a contracting party, the court will order the defaulting party to pay damages to compensate for the injury. The amount of damages is calculated by computing the amount of money it would take to put the injured party in the same economic position that it would have occupied if the contract had been performed. For example, if a merchant promises to pay £10 to deliver a load of coal to Southampton, the damages for failure to keep that promise would be £10, since an award of £10 would put the captain in the position that he would have occupied if the merchant had performed the promise.

Looking at the other side, it's not so easy to compute the damages the merchant would sustain if the captain fails to deliver the goods. Assuming that the captain fails to pick up the goods and they remain on the merchant's hands, the measure of damages would be the profit that the merchant would have made if the goods had been delivered as promised.

18.6 Penalties

As the law of contracts developed, the idea of penalties evolved. In order to insure timely performance of a contract, a penalty clause would require the defaulting party to pay a severe penalty, sometimes far exceeding the actual damages. For example, a shipping contract might require the shipper to pay £10 for shipping a load of coal to Southampton, plus a penalty of £100 for late payment.

Courts did not like penalty clauses because of their inherent injustice and because they may actually motivate contracting parties to try to induce a breach, since the penalty would be worth more than the performance. The rule evolved that the law abhors forfeitures and will not enforce penalties.

Now consider the following penalty clause:

> Contractor will pay the owner a penalty of $500 per day for every day after June 1, 2000, when the work is incomplete.

That is an unenforceable penalty.

Contrast the following:

> Contractor will pay owner as liquidated damages, and not as a penalty, the sum of $500 for every day beyond June 1, 2000, that the project is incomplete.

That is a valid liquidated damages clause.

There's more to it than just different terminology. Courts recognize legitimate occasions for the imposition of liquidated damages. Liquidated damages are simply damages that are ascertained in advance. Clauses imposing liquidated damages are enforceable in situations where actual damages would be difficult to compute.

Compare a six-month delay in the completion of an apartment building to a six-month delay in the completion of a bridge. It's easy to compute the owner's damages for delay in the completion of the apartment building: it is the rental value of the apartments during the period of the delay. Contrast the difficulty of computing the damages sustained by the public (organized as a Department of Transportation) from a six-month delay in the completion of a highway bridge. The actual damages would be difficult, almost impossible, to compute with any degree of accuracy. The law will therefore enforce a liquidated damages clause that the parties accept as a reasonable estimate of actual damages.

Therefore, Allen can include an enforceable liquidated damages clause in the contract provided it is a reasonable estimate of the damages he would actually sustain if the warehouse were not completed on time. Depending on circumstances, these damages might be very substantial. They could include the cost of renting temporary warehousing space and double-moving the entire contents of the warehouse!

18.7 Bid Shopping

Allen wants to give a job to his brother-in-law but doesn't want to pay more than a competitive price. How can he keep his brother-in-law honest? One good way would be to take bids from a half a dozen other electrical subs and then award the work to the brother-in-law at the low bid price. The reason this is unethical, if not just plain dishonest, is that it costs subcontractors a lot of time, effort and money to take the plans off and figure a job. They would never do it if they knew in advance that Allen was just using their bids in lieu of paying an estimating service to figure what the job is worth. Subcontractors are willing to figure a job only if they have a genuine chance to get the job!

Many states have enacted Subcontractors Listing Laws to require that prime contractors bidding public works list on their bid forms the names of the subcontractors they are going to use to perform the work. This prevents the contractor from using the low bid to shop for another, lower bid. Subcontractors Listing Laws, however, do not apply to private works. Nevertheless, it is a breach of construction industry ethics to shop subcontractor bids or to reject the low bid in order to award work to a favored subcontractor. Owners and subcontractors who get a reputation for bid shopping soon find that the best subcontractors refuse to bid their work.

18.8 Mechanics Liens

Allen caught a mechanics lien because Speedy Electric didn't pay Consolidated Wholesale for electrical equipment that Consolidated Wholesale supplied to the job. The fact that Speedy didn't pay its bill and therefore caused a mechanics lien to be recorded against the project justified Hyde in throwing Speedy off the job. This is because the prime contractor has an obligation to protect the owner against mechanics liens, and that obligation flows down from the prime contractor to subcontractors and their subcontractors.

A little basic knowledge of the mechanics lien remedy will help. Indulge yourself for a moment by imagining that you are in the market for a yacht. Money being no object, you have decided that the yacht should include a hot tub. The shipyard employs a hot tub company to furnish and install the hot tub. After you take delivery of the yacht you are informed that the shipyard didn't pay for the hot tub. You feel sorry for the tub company but you have no legal obligation to pay the shipyard's bill. If the hot tub company chose to deal with a deadbeat, that's their concern, and not yours. There is no legal relationship between you and the tub company at all.

The situation would be different, however, if you were building a house instead of buying a yacht. If the contractor you employ to build a house fails to pay for a hot tub, the tub company can record a mechanics lien against your property. If the lien is not paid off, the tub company could foreclose the lien and force a sheriff's sale of your house at public auction to the highest bidder, with the proceeds of the sale used to pay off the lien.

In the construction industry, vendors who make bad credit decisions and sell their merchandise to deadbeats can, in effect, collect their money from the unsuspecting property owner through the device of recording and foreclosing a mechanics lien claim. In effect, it becomes the owner's responsibility to make sure that the contractor, and all subcontractors and sub subcontractors, pay their bills on time.

19

ABC Warehouse X
Owner's Separate Contractor

As the new ABC Warehouse and Office building was nearing completion in the early months of 1990, Allen Brady was concerned about what the offices would look like and how they'd be furnished.

He'd seen the mess some of his friends in the warehousing business had made of their new premises. They bought perfectly good furniture and equipment but there was little, if any, attention given to efficient office layouts, and no coordination whatsoever in the color schemes and overall appearance. It seems that the overriding consideration was getting the lowest possible price, and nothing else mattered.

On March 6, Ivor Judge was making his weekly Tuesday morning site visitation. He was just finishing up with George Hyde and Ezra Field, when Allen drove up in his sporty new Mercedes convertible. He shouted to Ivor without leaving his car, "Meet me at Jake's Java and I'll buy ya a cuppa coffee. I wanta talk about something. Okay?"

"Sure, I'll see you there in 4 or 5 minutes." Ivor walked to his car while Allen sped off. Ivor threw his note pad and plastic hardhat into the trunk and changed out of his rugged tread-grip shoes into his clean shiny office shoes. He revved up and headed for Jake's, a new coffee shop about a half-mile from the jobsite.

When Ivor arrived at Jake's, he immediately went to the men's room to wash the construction grime off his hands and recomb his hair. Allen was already seated and had cups and a steaming pot of coffee on the table. They considered and decided against doughnuts and got down to the matter at hand.

Allen told Ivor of his concern about getting the offices furnished and fitted. "I want them to be tasteful and efficient but I don't wanta spend a fortune. What d'you think I oughta do?"

"Well, Allen, Judge & King offers that kind of service. We'll produce furniture and circulation layouts for your approval and design a color scheme. We'll select all the carpeting, window treatments, furniture, equipment, and fabrics. We'll do all the purchasing and arrange for installation of everything. At the end we'll present you with a complete list of everything purchased and a record of all the manufacturers, model numbers, guarantees, prices, and vendors. And, we'll coordinate the interior development with the work of the general contractor. All our services are described in detail in the AIA Standard Form of Agreement for Interior Design Services, Document B171, 1990 Edition, that we usually use. I'll send you a copy so you can see what's in it."

"Thanks. That sounds like a comprehensive service."

"Yes, we do it for many of our clients. I can make arrangements for you to look at some of our projects. We also have numerous photographs of our interiors work in the office."

"What do you charge for these services?"

"We charge on an hourly basis for our professional services and cost plus 10% for the purchasing service."

"Can you prepare a budget estimate for your services and the purchasing so I'll know approximately how much the whole thing will cost?"

"Sure, no problem. Most clients ask us to do that."

"Just as a rough guide, Ivor, what do people pay for office carpeting these days?"

"Well, Allen, there's a broad spectrum of prices available, but you can get a pretty good choice of moderate quality carpeting for around $15 a square yard.

"Thanks, Ivor. Let me think about it. I'll get back to you when I've decided what to do."

"Don't let it slide too long, Allen, time is getting away from us."

* * *

Upon arriving back at the ABC Warehouse in Pacoima, Allen found his warehouse manager sitting on a tall stool at his stand-up desk by the main loading door. He was eating his lunch and checking a pile of work orders for the afternoon's work.

Allen stopped at Carl's desk and recounted his conversation with Ivor at Jake's. "Whenever I talk to Ivor about anything, it always seems to get complicated and the price starts up. I've got to decide what to do about the interior decor and furnishings of the offices. It has to be decided pretty soon because it's got to be in place before we move in on May 29th."

Carl was referring to his pictorial wall calendar. "That's less than three months off. Is that enough time?"

"It's about 12 weeks, Carl. It shouldn't take any more time than that.

"Oh, hell no, Boss. You're right. That's more than enough time. Nothing to worry about. Are you going to let Judge & King do it?"

"Well, the way Ivor explains it, it sounds like it could cost a fortune. They do a lot of paperwork and record keeping and this is all charged for by the hour. I wish I knew some easier way to do it."

"You may be in luck, Boss. There's an interior decorator on my bowling team. I'll be seeing her this Friday night at the tournament. If you want, I can ask her to get in touch with you next week."

"That's a great idea, Carl. Nothing like a second opinion." Allen went on in to his office.

* * *

In mid-morning on the following Monday, Allen received a phone call. "Mister Brady, this is Steffi Thomas. I'm the interior decorator on Carl Daly's bowling team. I specialize in commercial interiors."

"Oh, yes. Thanks for calling. When can you come in to see me?"

"In a few minutes. I'm calling from my car. I'm about a block from your warehouse right now."

"Okay. C'mon over. I'll be waiting for you."

A few minutes later Steffi arrived. She came into the pathetic-looking office they had long ago grown out of. There were 4 old desks all pushed together in the middle of the room and a collection of mismatched filing cabinets around the edges of the office area. Cardboard transfer cases were stacked on top of the filing cabinets. The office was crammed into a corner of the corrugated iron warehouse, walled off with some old garage doors.

She looked around at the chaotic environment and thought to herself, "Whatever we come up with, it'll look a helluva lot better than this." She introduced herself, "I'm Steffi Thomas." She offered her hand and a business card to Allen. He offered her a decrepit, uncomfortable chair.

Allen explained what he needed. "Our new warehouse and offices are being built and will soon be finished. The offices must be ready for our moving in on May 29th."

"How much space must be furnished and decorated?"

"The office portion of the building is 5000 square feet. It all hasta be finished but about 50% won't need furniture yet, as it's to be set aside for future expansion. Meanwhile, we'll rent the space out to compatible office tenants. We must furnish our reception area, executive offices, secretarial, accounting and computer, sales offices, coffee room, and storage."

"Lovely. I can help you with your color scheme and the selection and purchasing of all of your furniture and carpeting."

"How much is your fee?"

"There's no charge for my services. It's all included in the cost of the furniture and carpeting."

"I'll have to think it over. I'll let you know what I decide."

After she left, Allen went out to the warehouse to talk to Carl. "Steffi just left."

Carl asked, "What'd you think of her?"

"Well, I'm favorably impressed. She isn't as complicated as Ivor. And her fee is lower. She doesn't charge anything for her services. The fee's included in the furniture cost."

"What if the furniture's too expensive?"

"Then I won't buy it."

"Oh. That seems pretty fair to me. Are you gonna hire her?"

"I think I will. I'll call her now."

* * *

Allen asked Steffi to come back for more conversation on Friday afternoon. She came struggling in lugging a huge cardboard box full of carpet and fabric samples and furniture catalogues. She dropped it on the floor next to her chair. She fished a book of plush carpet samples out of the box and fanned out the swatches in front of Allen.

"What do you think of these lovely carpet samples, Mister Brady? You'll note the fantastic range of colors, the smashing texture, and the luxurious rich pile."

"Yeah, it looks great. Oh, and please call me Allen."

"All right, Allen. You'll have to give me some guidance as to what quality or price range we should be striving for. Is this possible?"

"Sure, Steffi, our new project is costing over a million dollars. I'm a very high quality person. Nothing but the very best for me." Steffi smiled appreciatively.

Getting back to the plush carpet samples, Steffi asked, "And which color do you prefer, Allen?"

"That light brown looks pretty good to me, Steffi."

"That's Sahara Tan. An excellent choice, Allen. But I think you'll find that the Imperial Cerise will work much better in the contemporary business environment. Don't you agree, Allen?"

"I think you're right, Steffi. It's much better. How much does this type of carpeting cost?"

"It's $29.95 a yard including a high quality rubber padding, carpet stripping, and carpet laying.

"Isn't that a little steep? Don't you have anything for less money?"

She put the plush samples back and, rummaging around in the box, dredged up three more sample books. She laid them out on the desk and opened them so Allen could see all the colors. She explained, pointing, "This group is $21.95 a yard, this group is $14.95, and this group is $8.95."

Allen turned his attention to the $8.95 book. "This dark pink looks pretty good to me."

Steffi corrected him, "That's Rusty Cherry, Allen. It's a beautiful sophisticated color. A very wise selection."

She poked around in the box and came up with the lowest price range furniture catalogues. No point in wasting hours going through the expensive ones first. They spent the next hour looking through the furniture catalogues together and picked out desks, chairs, tables, and filing cabinets. They agreed on the colors from the color chips in each catalogue. She wrote up an order blank and listed the carpeting and all the furniture selected. The whole works added up to $29,750.00. Steffi said she would place the orders immediately but she'd need a check for $7500 as a deposit. She explained that most of the vendors would require deposits.

"There's one other thing, Steffi. The interior painting is in the general contract with Hyde Construction Company. But we hafta give them all the color selections."

"That's easy, Allen. Tell them to paint all the walls and doors Sioux White. It's a beautiful subtle off-white that will form a fantastic backdrop for the deep rich colors of the carpeting and furniture to the best advantage."

"Shouldn't we have a contract, Steffi?"

"No, Allen. That's not necessary. Just sign the order list and give me the check."

Allen signed the order, gave her the check, and carried her huge box of samples and catalogs out to her car.

* * *

After Steffi left, Allen went out to the warehouse to talk to Carl. "Well, I just made a deal with Steffi for all the carpeting and furnishings."

"Congratulations, Boss. That was quick. I knew she'd be okay."

"Yeah, her prices are much better than Ivor's. On the carpeting alone, I've saved a fortune. He wanted $15 a yard plus a 10% purchasing fee plus a professional fee for picking the color. Steffi's carpeting was only $8.95 including her fee. And not only that, she picked the paint colors for free."

* * *

On Ivor's regular Tuesday jobsite visit on April 3rd, he was talking with Ezra Field in the job office when Allen appeared. After the usual exchange of pleasantries, Ivor asked, "Say, Allen, what did you ever decide to do about the interior design and furniture purchasing?"

"Oh, I meant to tell you, Ivor. I've hired an interior decorator who specializes in commercial interiors. She's working on it."

"Good. Time is catching up with us. Move-in day is only 8 weeks downstream. Ezra reminded me this morning that the painter needs the office interior paint colors. We instructed Hyde Construction of the colors of everything else several weeks ago."

"That's simple, Ivor, Steffi said to paint all the walls and doors an off-white."

"There are dozens of off-whites. Which one does she want?"

"It was some Indian name. I'll think of it in a minute."

Ezra, being helpful, suggested, "Was it Navajo White? That's what we used on the Ajax job."

Allen said, "Yeah. That sounds like it. That must be it."

Ezra wrote Navajo White in his daily job log and Ivor put it in his notes for his weekly observation report.

Ezra said, "Thanks, Mister Brady, that'll keep us going."

Ivor asked, "Do you want me to take care of the coordination with your decorator, Allen?"

Allen thought about it briefly, visualized the complications, and answered, "No. I can take care of it."

* * *

At Ivor's jobsite visit on Tuesday, May 8, he was pleased to see that the office interior painting had been completed and it looked pretty good. He'd been looking forward to seeing the entire color scheme designed by Allen's decorator. He always found it refreshing to see an original approach by someone new.

That very day Steffi called Allen to inform him that the carpet would be laid the following week starting on Monday, May 14 and continuing on Tuesday and Wednesday. The draperies and blinds would be installed on Friday the 18th, and the furniture would arrive early the following week. Steffi seemed well organized. Plenty of time to meet the completion schedule.

She continued, "I'm going out to the job tomorrow to look at the paint job and to make sure everything is ready for the carpet layers."

"What time are you going, Steffi? I'll meet you there."

"About 10 o'clock. Is that all right for you?"

"Sure, Steffi, I'll see you there."

Allen was waiting in the new offices when Steffi came in. She took one look at the thousands of square feet of Navajo White walls and doors and came completely unraveled. She couldn't believe her eyes. The ordinarily composed and polite interior decorator flew completely out of control. Her face was getting red and her voice was loud and shrill.

"What the hell is this, Allen? This isn't Sioux White! What's wrong with those damn idiotic painters? This is an absolute atrocity! It's appalling! It's outrageous! It's got to be repainted."

"What's wrong with it, Steffi? It looks fantastic. It's beautiful."

"This is the wrong damn color, Allen. When the carpeting, window coverings, and furniture arrive, they'll look like hell. It'll be nauseating. This color has a strong yellow-green tinge to it. It'll clash. Sioux White has a subtle rosy-pink cast that will blend and harmonize with everything else we picked."

Allen suggested, "Well, why can't you just change the carpet and furniture colors."

She looked at Allen like he was an idiot. "Too late for that, Allen. Everything's ordered. It's on the way. We're stuck with it. The paint's got to be changed. It's dead wrong. Where the hell is the moronic painter? Where the hell is Hyde's dim-witted superintendent? We've got to talk to someone." She was totally unhinged.

By chance, Ezra came in to the office wing to see Allen about signing up for the electrical service. Steffi attacked him before he had a chance to say hello.

"This idiotic paint job has got to go. It's the wrong damn color. It has to be redone. I've never before seen such incompetence!"

Ezra, defending the work, "What are you talking about? This is Navajo White, exactly what you specified. I saw the paint cans before the work was started. I checked. Lady, you don't know what the hell you're talking about."

"What do you mean, Navajo White? It was supposed to be Sioux White." Then, turning to Allen, who was already turning an unusual shade of pale green, "What did you tell them, Allen?"

"I may have made a mistake."

Steffi persisted, "Well, it'll have to be repainted. It can't stay like this! And it'll have to be done before Monday when the carpeting goes in. That's all there is to it."

Ezra, always practical, suggested, "The painting foreman is working with his crew at the loading dock. I'll go get him." And he left to return a few minutes later with Glen Hughes, Painting Foreman for Spectrum Painters.

Allen asked, "Can you repaint the offices with Sioux White before next Monday?"

"Why? What's wrong with it the way it is? This is good work." He was genuinely puzzled and concerned that his work was being criticized.

Steffi said, "Your work is wonderful. It's just the wrong color. It should have been Sioux and it's Navajo."

"That's what we were told to use."

"I know. But it's wrong. It's not your fault." She glanced meaningfully at Allen. Allen wished he were somewhere else.

Ezra said, "Don't worry about it, Glen, you'll be paid to repaint it, but when can you do it?"

"We can't get to it right away because of all our other commitments. I'll have to phone my boss to find out when we can work it in. I'll call right now." He left to go to the mobile telephone in his truck and was back in less than 10 minutes.

He reported, "We can work this weekend, Saturday and Sunday, and with a second crew, we can work 16 hours a day. Of course, you'll have to pay time and a half for both days and double time for the second crew. Otherwise, we can't do the work until next week. It would take all week with one crew and no premium time."

Allen was getting weak. He could see expenses mounting by the minute and the schedule going rapidly down the tube. He finally recovered enough strength to say, "There's no choice. It's gotta be done. Do it this weekend so the carpeting can go in on Monday."

The painting foreman was filling in an extra work authorization form and asked Ezra and Allen both to sign it.

Allen asked, "What kind of money are we talking about?"

Glen replied, "I don't know exactly, but it won't be far off 5 or 6 grand. You'll know when you get the bill."

* * *

Early Monday morning, the 14th, Snappy Carpet Mechanics arrived with a crew of four and several huge rolls of carpeting and padding on their truck. They parked as near the office entrance as possible and started unloading the heavy, bulky rolls of 12-foot wide material. It was a nice, sunny, dry day so they would use the asphaltic concrete parking area for unrolling and cutting off the padding and carpeting. The foreman, Brendan, went in the open front door of the office wing and scouted around to determine the best starting point.

Brendan got his crew organized and they started to make progress right away. They spent the first couple of hours cutting and installing the tackless stripping in all the areas to be carpeted.

Ezra arrived around mid-morning and asked one of the carpet crew who was in charge. The man pointed at Brendan, who said, "I'm Brendan, carpeting foreman. Who are you?"

"I'm Ezra Field, Job Superintendent for Hyde Construction. I'm sorry but you're gonna hafta move your truck and all those rolls of padding and carpeting. We're gonna stripe the lot today and you're in our way."

Brendan complained, "No chance. We've been scheduled for weeks to install this carpeting today and the next two days. Then we're on to other jobs. We can't do the work if we hafta move our truck and materials."

"I can see that, but I've gotta build this project and meet the deadline. Parking lot striping has been on our schedule for months. Today is the day for painting the stripes. The men, materials, and equipment are here and ready to work. You're in the way and you'll hafta leave. Pronto."

"Well, we're not moving. Go talk to Steffi Thomas. We're working for her. Now, if you'll just get the hell out of our way, we'll get back to work."

So, Ezra left and went back to his jobsite office. He called George Hyde, brought him up to speed, and asked him to follow through with a call to Ivor. Maybe the two of them could unravel the stalemate. George called Ivor and Ivor called Allen. Ivor described the situation to Allen.

"Ivor, how can we move in on time if Hyde's people are interfering with our interior work schedule? They're supposed to be helping, not hindering."

"Hyde's people are asking the same question about you, Allen. Your carpet people are obstructing Hyde's work progress. The general contractor is in charge of the building and site and the overall schedule. You should have cleared your interior work schedule with Hyde's schedule. This is covered in the contract."

"Where? I didn't see anything like that."

"Look in the AIA General Conditions of the Contract for Construction. Subparagraph 6.1.1 gives the owner the right to hire separate contractors."

"I didn't hire any separate contractors. Who says I did?"

"Steffi Thomas, your interior decorator is a separate contractor, Allen. But your present problem is covered in Subparagraph 6.1.3, which requires the owner to provide for coordination of the owner's separate contractors with the work of the contractor. That's Hyde. You've had Hyde's schedule for months and you've received all the updates. You should have checked with Ezra before sending anyone to do work on the job."

"Shouldn't you be doing that?"

"No, Allen. You didn't hire us to coordinate the work of your separate contractors. You're doing that yourself."

"So, what should I do now?"

"Check in with Ezra and coordinate Steffi's schedule with Hyde's. Then, everything should be all right."

So, Allen called Ezra to see what could be done. Ezra made it clear. "Your carpet layers hafta go. We're hanging electrical fixtures in the offices on Tuesday and Wednesday and my carpenters are reinstalling the door hardware on Friday, the 18th, our last day on the job. Then we're outta here. Your carpet people can have the space on Saturday, the 19th, and any time next week."

"Okay, Ezra. I'll call Steffi and see what I can do. You can tell the carpet people they'll have to leave but that we're working on the rescheduling."

Allen called Steffi and told her what had happened. She said she'd call Snappy Carpet Mechanics to get them rescheduled. Snappy agreed to put the schedule off one week to Monday, Tuesday, and Wednesday, May 21, 22, and 23, but he demanded extra compensation for the false start and rescheduling of other jobs to accommodate. He insisted on $750 and was adamant. She saw there was no point in arguing about it. She knew she was over a barrel. When she reported back to Allen, she told him about the extra but was more concerned about the rest of the schedule. "The drapery and blind installers will be there on Friday the 18th. I hope nothing stops them from working. You better check on it."

"Okay, Steffi, I'll check with Ezra."

She continued, "And then I'll have to reschedule the delivery and installation of the furniture. They were supposed to come in on Monday and Tuesday but it will have to be put off to Thursday and Friday."

Allen's mind was becoming cluttered with scheduling details. He hoped Steffi had it straight in her mind because he didn't. Maybe Ivor was right about keeping a comprehensive and coherent written record. He was also losing track of the costs. The penalties he was paying were rapidly soaking up any savings he had visualized.

The drapery and blind installers started on Friday morning but finally gave up. Although the job superintendent said they could work, there was too much competition for the space, as carpenters, electricians, and clean-up people were seemingly everywhere. They left and phoned Steffi for a new date. She said, "Now you'll have to wait until after the carpeting is in. You can finish your work on Thursday or Friday."

The drapery installer explained to her, "We're working at the new Supreme Hotel on those days. Our next available time is June 1st."

"Well, that's out. That's the date of the grand opening party."

"We haven't any other time. We're up to or elbows in work right now."

"How about Saturday, the 26th?"

"That's the first day of the three-day Memorial Day weekend. I'd have to ask our installation people to see if anyone wants to work. I'll call you back."

Later that day he called Steffi back to report that two of his people were willing to work on Saturday, the 26th but they demanded double time. Steffi agreed to it and told him to go ahead. She then called Allen to report the additional cost, which she estimated to be approximately $300.

* * *

Hyde Construction finished their work on May 18th as scheduled. The carpet laying proceeded on Monday and was completed on Wednesday. Allen and Steffi were pleased with the result. Allen had to admit that the new paint color was just right with the carpeting.

Steffi brought Allen up to date. "The desks, chairs, and tables will be arriving tomorrow morning. I have an installation crew lined up to put the desks together and put everything in place in the right rooms. And on Friday the filing cabinets will be arriving. Then I'm through. You can then move in."

The next morning, Thursday, the 24th, Steffi arrived at 11 o'clock to see how things were progressing. Her two-man installation crew, Tom and Jerry, was sitting in the reception room on the new carpeting playing Gin Rummy. She asked, "Where's the furniture?"

Tom answered, "We don't know. We thought you'd know."

She left and went to her car phone. She called the furniture supplier and found that they were going to deliver a day later, on Friday, the 25th, as they had some kind of emergency that required a change of schedule. Steffi complained, "You could have called me. I'm paying two installers to sit here and play cards. You better be here with the furniture tomorrow morning at 8 o'clock." She added, "Or else," and hung up. She didn't exactly know what the "or else" would be. She told the installers to come back the next morning at 8 o'clock.

Steffi was there on Friday morning at 8 to find that the furniture truck was there as well as her two installers. Under her direction, they got the whole truck unloaded and the truck left, after requiring her signature on the bill of lading. She checked the list and couldn't find any chairs. The driver said, "There's nothing else on the truck and everything we unloaded tallies with the list. Just sign here and we can leave."

"But where are my chairs?" Steffi was starting to get a piercing headache.

"How would I know? I just deliver what they give me. Call our office." They left.

Steffi was directing Tom and Jerry on the location of each of the desks. "Tom, that's supposed to be a left hand desk. Move that one to the opposite side of the room. Bring a left hand desk in for there."

"Okay, Steffi, will do."

He went out to find the left-hand desk and couldn't find any. Steffi found that they had sent the proper number of left and right handed desks but the returns were all right handed. There should have been 6 left-handed returns.

"What'll we do now?" asked Jerry.

"All we can do is install the desks without the returns where we haven't got the correct ones. Put the wrong handed returns in one of those empty offices."

In the afternoon, the filing cabinets arrived. They were a nice quality cabinet but the color was off. It was a different manufacturer from the desks but the color name was the same. Dusty Rose. The color clash was hideous and disconcerting. The headache got worse.

After unwrapping all the desks, tables, and filing cabinets, the pile of wrappings just about filled one small office. Steffi told Tom and Jerry to move it all out and pile it on top of Hyde's trash pile next to the loading dock.

Just as they were finishing up, Allen arrived to look around. He was impressed with the beautiful furniture and didn't notice at first that there were no chairs. He noticed the missing left handed desk returns first. He asked, "Steffi, those desks look incomplete. Isn't there supposed to be something on the left side where the drawers are missing?"

"Yes, Allen. The shipper made a mistake and we have to send back 6 right hand returns."

"When will that take place?" Allen was sitting on the end of a desk.

"It's too late today to phone them so we'll have to wait until next Tuesday. Everyone's off on their Memorial Day holiday already."

"There are a couple of other things, Allen. The chairs aren't here yet. I phoned earlier today and found that they were out of stock and the new stock won't be available for 30 days. So, you'll have to rent some chairs temporarily."

Steffi showed him the filing cabinets and pointed out the color problem. It didn't really need pointing out, though, as it looked horrible. She explained, "I'll contact the vendor on Tuesday and find out what they're going to do about it."

Allen asked, "How about the draperies?"

"They're lined up for installation at 7 o'clock tomorrow morning. Someone has to be here with the key to let them in. I'd do it myself but I have another appointment. Could you do it, Allen?"

"Uh...Sure...I'll be here."

"You're a dear, Allen."

* * *

On Tuesday, May 29th, the ABC offices were moved to the new premises. Steffi dropped by to talk with Allen and to resolve the loose ends.
"How are you getting settled, Allen?"

"When we got here yesterday, we had to clean up after your drapery installers. They'd left screws and bits of drapery cord all over the floors, but, what was worse was little piles of plaster dust in every room where they had drilled into the plaster. We had to vacuum it all up.

Allen continued, "And this morning I got a call from Ezra complaining about all the trash your furniture installers left in Hyde's trash pile. He's sending me a bill for $60."

Steffi didn't say anything as Allen continued his grousing, "Some of my people have complained that the office layouts aren't convenient. The desks seem to be on the wrong walls and it's not handy. The tables and filing cabinets are in the wrong places. The telephone and electrical outlets are all in the wrong places."

"Well, Allen, the furniture is all portable. You can put it any place you want." Steffi was trying not to lose her optimism and her smile.

"Allen, I've talked to the filing cabinet manufacturer and they've admitted that the color is off. They're willing to repaint them with the proper color, at no additional charge. You'll have to empty the cabinets so they can be picked up next Monday, June 4th. They'll be returned repainted a week later on the following Monday."

Allen wasn't too enthusiastic about having stuff in transfer cases for a week in the new office. But, what could he do?

"Allen, I've got a bill for you. I'll need a check today because I have to pay the vendors."

"But, Steffi, the job isn't done. All the loose ends hafta be tied up first."

"I need the check today, Allen. The vendors are expecting it."

"No, Steffi. Not until the chairs and desk returns are replaced and the filing cabinets are back. Also, I expect credit for the extra labor for the carpet layers and the drapery installers and the Hyde trash pile."

Steffi objected, "Why should I have to pay for the extra labor, Allen? That was caused by you and Hyde Construction."

"Another thing, Steffi, your furniture installers dinged the newly painted walls and doors and Hyde is going to charge us for the repairs. We're going to deduct that from your bill, too. You may as well stop arguing, Steffi. I'm not paying anything today."

She got up and left. She went straight to the Law Offices of Darnel & Jimson, which she'd redecorated last year.

Allen called Ivor when the lawsuit arrived. "Ivor, I thought we had arbitration clauses in all these contracts. Why has Steffi filed a lawsuit?"

Ivor asked, "Did you have an arbitration clause in your contract with Steffi?

* * *

Points of Law 19

ABC WAREHOUSE X
Allen Hires a Decorator

19.1 The Owner Must Not Interfere with the Contractor's Work

Allen has decided to employ a member of Carl's bowling team as his interior decorator. Ivor Judge, FAIA, offered to coordinate Steffi's interior decorating activities with the rest of the work, but Allen decided to take care of it himself. As we shall see, this apparently simple decision is fraught with legal danger. Coordination of a construction project is a complicated job that should usually be left to experts.

In the lingo of the construction trades, Steffi, Allen's interior decorator, is known as a separate contractor, and it is Allen's obligation to coordinate the work of the separate contractor in such a way as to avoid any interference with the contractor's work.

Many construction contracts provide that if a contractor's progress is delayed by inclement weather, picketing, casualty, or acts of the owner, the contractor will be entitled to an extension of time but no damages for delay. Such no damages for delay clauses are very strictly construed by the courts, because they tend to put the contractor at the mercy of the owner. Most courts hold that such a clause will not exonerate an owner from liability for damages if the owner actively interferes with the contractor's performance.

19.2 Measure of Damages

The damages sustained by a contractor because of interference by the owner must be computed according to the circumstances of the case.

In some cases, when crews or equipment are actually prevented from working, the cost of the downtime can be computed from equipment rental records and payroll records.

In other cases, where interference by the owner affects the efficiency or productivity of crews and equipment, the damages must be estimated according to expert testimony by multiplying the percentage of reduction of productivity against the hourly cost of performing the work. For example, supposing that interference by the owner reduces crew productivity by 30 percent, and the hourly cost of the crew is \$1,000, and the interference persisted for ten hours, the damages would be $.3 \times \$1,000 \times 10 = \$3,000$.

In some cases, the contractor's damages cannot be measured by the cost of standby time or the cost of reduced productivity because the jobsite records are simply inadequate to the task. Suppose, for example, that working drawings supplied by an owner are incomplete and full of errors so that the contractor makes false starts on different elements of the work, removes and replaces work that is improperly designed, and has to move crews around the jobsite while waiting for corrected drawings from the architect. Crews must disrupt their work, move from one part of the jobsite to another, assemble the tools, materials, and equipment required for the different work, and become oriented to new locations before they can resume operating at full efficiency. Combine this with disagreements as to the reasonable cost of extra work ordered by the owner and disputes as to the scope of the work required by the contract documents, and it becomes literally impossible to accurately establish the amount of damages sustained by the contractor because of each separate instance of interference, lack of information, and extra work. In such cases, the contractor may prove the total cost of performing the work and subtract the contractor's estimated cost (using the estimate prepared by the contractor at the time when it originally figured the job) and submit the difference between the estimate and the actual total cost as the measure of damages. (This is known as the total cost method of computing damages.)

19.3 Owner Must Coordinate

One of the important functions performed by a general contractor is to schedule the work. The object of scheduling is to insure that the various trades follow each other in an orderly manner and avoid interfering with each other. Superintendent Ezra scheduled the 14th of the month for striping the parking lot. Unanticipated by Ezra, the floor covering contractor employed by Steffi showed up on the morning of the 14th and laid out its carpet on the parking lot for cutting and trimming. This made it impossible for the striping contractor to proceed. Although Allen perhaps had the right to assume that Steffi would check with Ezra before moving her floor covering contractor on to the jobsite, the legal fault for the interference falls upon Allen because the AIA documents require the owner to coordinate separate contractors with the work of the contractor.

When Snappy, the floor covering contractor, had to pull off the job and reschedule, he ended up having to pay a crew for showing up at the jobsite and starting work only to have to stop work, roll up the carpet, load it back on to the trucks, and take it back to the warehouse. This cost Snappy $750, and Steffi had no choice but to pay it or lose her floor covering contractor.

19.4 Acceleration of the Work

Because of Allen's scheduling problems, the drapery and blind installers had to perform two days' worth of work on one Saturday. This brings us to a discussion of acceleration and constructive acceleration. When an owner requires a contractor to advance the completion date of the project, it is known as acceleration. It is recognized that acceleration often increases a contractor's costs because it may require work to be performed after hours or on weekends at time-and-a-half or double-time, or it may require the contractor to increase crew sizes by employing inefficient workers or inefficient crew sizes. A contractor has the contractual right to consume all of the allotted time and therefore if the owner wishes to accelerate performance it is a change to the contract and the contractor need not agree to the change unless the owner pays additional compensation.

Constructive acceleration occurs when an owner delays a contractor's performance but refuses to grant to the contractor an extension of time. A contractor is entitled to damages to compensate it for the costs incurred as a result of a constructive acceleration. It was because of constructive acceleration that Steffi had to agree to pay an additional $300 to have the drapery and blind installers work on Saturday.

19.5 Backcharges

Allen has told Steffi that he will charge her for trash removal caused by her drapery installers and for damage to walls and doors caused by her furniture installers. In construction lingo, such charges are known as backcharges. The term backcharge applies not only to charges imposed by an owner against a contractor, but also to charges by a contractor against a subcontractor, or by a subcontractor against a sub subcontractor. In legal terminology, backcharges are called offsets. In law, an offset is any claim that a defendant has against a plaintiff to offset the plaintiff's claim against the defendant.

Steffi has performed work for Allen and therefore has a claim against Allen for money due under the contract. At the same time, Allen has a claim against Steffi for trash removal and the cost of repairing damage to his walls and doors. Steffi's claim and Allen's offset both arise out of the same transaction, which is Steffi's interior decorating work.

A legal offset can also arise out of two separate transactions. Therefore a contractor with a legitimate offset against a subcontractor on one job can offset it against payments due to the subcontractor on a different job.

19.6 Oral Contracts

Samuel Goldwyn said oral contracts are not worth the paper they're written on. This is not an accurate statement of law. A contract, in its legal definition, is an objective manifestation of an intention to be bound. A legally enforceable contract can arise without any words at all. For example, suppose a customer points to an item in a deli case which the server prepares and hands over. By their conduct, these parties entered into a contract under which it was agreed that the server would prepare a bowl of matzo ball soup and the customer would pay for it. With a few exceptions, such as real estate contracts and a contract to make a will, oral contracts are just as enforceable as written contracts. Therefore, if Steffi fulfilled her side of the deal by supplying the furniture and floor covering listed on the order, she would be entitled to be paid the agreed price of $29,750.

19.7 Insurance for Property Damage

Steffi's furniture installers dinged the newly painted walls and doors. Allen has said he is going to backcharge Steffi for the cost of repairing the dings. At this point, both Allen and Steffi should be thinking about their insurance coverage.

Two types of insurance might apply to this particular loss: property insurance and liability insurance. Both types of insurance apply to property damage, and Allen's property has been damaged by Steffi's installers.

Many types of property insurance are available, including named peril and all risk. A named peril policy insures the property owner against damage caused by a named peril, such as fire.

An all risk policy covers the owner against all risks of physical loss to the insured property except loss caused by excluded perils. (The term all risk could be considered somewhat deceptive, since the typical all risk policy contains a long list of exclusions including such things as earthquake, soil settlement, and faulty workmanship.) Most property owners carry all risk insurance and therefore

Allen can probably seek reimbursement from his carrier for the cost of repairing the dings caused by Steffi's furniture installers.

Steffi should also be thinking about insurance. She may possibly be aware of Allen's property insurance, but her interest would focus mainly on her own liability insurance. Steffi, like most businesses, probably carries a commercial general liability insurance policy that insures her against claims for bodily injury or property damage. Allen has made a claim against Steffi for property damage and therefore Steffi should turn the claim over to her carrier for adjustment.

19.8 Subrogation

The issue is subrogation. When one person satisfies a debt that is properly the responsibility of another, that person steps into the shoes of the creditor by a process legally called subrogation. Thus, after reimbursing Allen for the cost of repairing the dings, Allen's insurance carrier can subrogate against Steffi, who, in her turn, will tender the claim to her carrier.

19.9 Arbitration

When Allen refused to pay Steffi for the furniture, Steffi sued. Allen protested the lawsuit, since he thought that any disputes on the job would be resolved by arbitration. Architect Ivor had to inform Allen that arbitration is a matter of contract. Does Allen have an arbitration clause in his contract with Steffi?

Well, there's obviously no arbitration clause because there's no written contract. But how about an oral contract? Couldn't Allen and Steffi orally agree to arbitrate disputes? The answer is yes, they could, but that oral agreement would not be enforceable. In order to be enforceable, an agreement to arbitrate must be in writing.

Why might Allen prefer arbitration to litigation? Mainly because of speed and expense. For disputes such as those between Allen and Steffi, arbitration would be much faster than a lawsuit, and the legal expense would be less. In fact, if they wanted to, Allen and Steffi could authorize an arbitrator to decide their dispute without any lawyers being involved at all. The arbitrator would convene a hearing (probably at the warehouse), Allen would tell his story and Steffi would tell hers, after which the arbitrator would make a written award. If the losing party didn't comply with the award, the winning party could have it confirmed by a court and thereby turn the award into an enforceable court judgment.

20

Pacific Horizon I
Value Engineering: The Contract Negotiation Phase

Leo King, FAIA, was gravely concerned with the disappointing bidding results that afternoon. Still sitting in Judge & King's conference room, he was again reviewing the summary of general contractor bids for the shell construction of the Pacific Horizon Professional Office Building to be built at Malibu, California.

Super Constructors	$8,098,505
Construction Engineers	$7,607,900
Excel Builders Incorporated	$7,510,835
Peerless Contracting	$7,502,945
Delta-Apex Construction	$7,119,000

The top bidder was obviously in outer space, at almost $500,000 over the second highest. But the next three bidders were very close together, within a range of less than a 1.5 % spread. However, the low bidder was far too low, almost $400,000 below the nearest competitor. But what was worse, the low bidder was still way over the owner's budget. Actually, it was $439,000 over the architect's estimate of $6,680,000.

Leo knew there would be serious repercussions for Judge & King. The client would be extremely unhappy and would want to hold the architects responsible for the high bids. The only way the project price could be lowered at this point would be to cut back the program or lower the quality of construction, or both.

Representatives of the general contractor bidders had all just left and the architect was nervously waiting for the client to show up. He dreaded breaking the disappointing news.

Leo's secretary disturbed his misery to announce the arrival of Alexander Brink, President of Horizon Development Company. He was accompanied by his Financial Vice President, Cyrus Flint. Leo gestured to his secretary to show them in and rose to greet them.

The architect tried, but failed, to sound casual. "Well, Alex, the bidding results are disappointing, to say the least." He handed a copy of the bid summary to each of them. Alex went immediately to the bottom line and saw that they were over $400,000 off the mark. Cy was assimilating and digesting the information and had already calculated the rate of overrun and the percentage differences between the bids. Cy dated his sheet, March 17, 1992.

Alex broke the silence. "Okay, Leo, what do we do now? We sure as hell can't go ahead at this price, can we, Cy?" Alex was clearly displeased.

"No, Alex," Cy agreed, "Our financial projections are based on a building cost of $6,680,000. That's all we can lay out for the shell and site development. We're boxed in by the rents we can reasonably expect in this market."

"So, Leo, what'll we do? Re-bid the job?" Alex again asked.

"No," Leo explained, "There's no point in re-bidding unless we change the drawings and specifications. Then we'd lose even more time on a new bidding procedure. Also, some of the bidders could lose interest and drop out. Our best course would be to meet with the low bidder and see what we can negotiate. They may have some ideas on how we can reduce costs."

"That sounds like a good idea. What do you think, Cy?"

"Well, we've got to get $439,000 out, one way or another. Is this possible?"

Leo answered, "The only way to know for sure is to talk with the low bidder."

Alex concluded, "Okay, Leo, let's do it." Cy nodded his approval.

Leo picked up the telephone and got the offices of Delta-Apex Construction in Phoenix, Arizona. He was told that Mike Van Buren, Vice President, Sales, would call him back shortly. A few minutes later Van Buren called Leo and made an appointment to come to the Judge & King office for a meeting with Alex and Cy the very next day at 2 o'clock. Leo said, "We'd like to talk to you about your bid."

Mike Van Buren was staying at a Santa Monica motel while trying to get a Los Angeles office established for Delta-Apex. If they could land a decent contract in the Southern California area it would justify their starting a Los Angeles operation. He was working on establishing contacts and finding jobs to bid. The bidding and proposal compilation was done mostly in the Phoenix office but Mike was talking to Los Angeles area subcontractors and suppliers. He was also studying for his California contractor's license so he could serve as Responsible Managing Officer of Delta-Apex's Los Angeles operation.

Mike phoned Kyle Lincoln, President of Delta-Apex in Phoenix. He proudly announced, "Kyle, we're low bidder on the Pacific Horizon job and I have an appointment to meet the owner in the architect's office tomorrow."

"Great. Maybe we'll have our L.A. Office sooner than expected. Good work, Mike." After hanging up, Kyle immediately walked to the office of his Chief Estimator, Elmer Taft. "Well, we're low bidder on the Pacific Horizon job, Elmer. Will you please double check our bid proposal figures to make sure we're not out on a limb? And let me know right away where we stand."

* * *

The meeting at Judge & King got underway promptly after introductions. Leo directed a preliminary statement to Mike, "Delta-Apex is low bidder, but the bid is considerably higher than our estimate. We've got to cut almost a half million out. Do you think it can be done?"

"That's quite a bit but I'm sure we can do it. We have a process we call value engineering. We do it all the time."

Cy said, "I've never heard of value engineering. What is it?"

Leo explained, "The definition in The AIA Architect's Handbook of Professional Practice is 'The process of analyzing the elements of a project design in terms of their cost-effectiveness, including the proposed substitution of less expensive materials or systems for those initially suggested.'"

Mike continued, "We'll talk to our subs and suppliers and our estimating department to see where the money is. We may have to cut out some things. We may have to lower quality. We may have to find simpler systems. We'll just start looking."

Alex said, "We don't want to cheapen the building, Mike. We just want to get the cost down."

Mike hastened to assure him, "We won't cut anything out without your approval, Alex."

Alex emphasized, "But we've got to get the cost down or we can't go ahead with the project."

Cy laid it on the line. "At the present cost figure, we'd be better off to sell the property and forget the whole damn thing."

Leo suggested, "As soon as you get the information together, Mike, just let me know and we can have another meeting."

Alex was getting concerned with timing. "How long will this take? We've got to get moving. We didn't expect to have this delay."

Mike promised, "I'll be back in a week."

* * *

Pacific Horizon was the third office building project of Horizon Development Company. Behind them were the supersuccessful Beverly Hills Horizon and the impressive Santa Monica Horizon.

This latest project, was a three story speculative professional office building of 56,800 square feet gross with a net rentable area of 48,280 square feet. In the center of the building was a large atrium, 40 by 100 feet, covered at the third story roof level with a huge skylight. A restaurant on the first floor would use part of the atrium as a landscaped outdoor dining area. Adjacent to the office building was a two story pre-cast concrete parking garage to accommodate 200 cars and a surface parking lot for another 50 cars. The balance of the prestigiously located site was to be luxuriantly landscaped.

* * *

A week later, the principals of Horizon Development and the contractor were back at Judge & King's conference table. The architect asked, "Well, Mike, what've you got for us?"

"We and our subs are still working on the problem and I've a few ideas to discuss. But we're nowhere near our goal yet. We'll continue working on it and no doubt we'll find other significant savings."

Alex asked, "What've you found so far, Mike?"

"We've talked to our air conditioning sub, Air Unlimited, and they tell us that the A/C system is way over-designed."

Leo's knee-jerk reaction was predictable. "Whatta ya mean, over-designed? In what way?"

"Alan Condon, of Air Unlimited, says he can cut out $83,000 without hurting the job."

Leo, now sounding defensive, asked, "How, may I ask, would he accomplish this?"

"I don't have all the details, but he mentioned a few items like fewer zones, simpler equipment, and less temperature differential."

Leo said, "In this climate, a good air conditioning system is crucial to the success of a rental office building. This is no place to cut."

Mike added, "Alan also recommended installing a passive solar energy water heating system so we can eliminate the gas water heating equipment and thereby lessen operating costs. The initial costs are nearly the same."

Horizon's Financial Vice President asked, "The A/C reduction accounts for nearly 20% of our savings objective. What else have you found, Mike?"

"Our painter, Manny Hughes, of United Painters, has news of a recently developed exterior paint coating that is long lasting and economical, and can be applied in one coat. No primer is needed and the savings in labor and material translates to a net reduction in the painting bid of $21,675."

Alex and Cy were each keeping a tally of the proposed savings. Cy had computed that this was another 4.9% of the total savings required.

Leo asked, "How long has this paint been on the market, Mike?"

"I don't know for sure."

"How do they know it's long lasting?"

"I don't really know. I'll ask Manny. He'll know, I'm sure."

Alex asked, "What's next, Mike?" The developer was warming up to this fellow.

"The skylight over the atrium is costing a small fortune. Our original cost breakdown was based on a separate bid directly from the specified manufacturer. Our glazing contractor says he can cut a bundle off by changing from the proprietary system to a built-up system that he can make from standard parts from a glazing catalogue. The difference would be $16,800."

Cy added it to his list and computed an additional 3.8%. The three items, so far, added up to 27.7% of the required reduction. He was impressed. Now we're headed in the right direction.

The architect, defending the original specification, explained, "The specified skylight is a carefully engineered and tested assembly that has been proven by experience. How do we know that your glazier can build it properly?" Leo felt that he was being placed in the position of a killjoy. He didn't want to sound negative but he saw potential problems in many of these suggestions. All of these possible savings would require a lot of research to establish their validity.

Alex asked, "What's next on the list, Mike?"

"Our estimating department thinks that the two elevators could easily be reduced to one. We've seen many buildings similar to this and they get along just fine with only one elevator. After all, a third of the building's population is on the first floor and won't even use the elevator. This would cut $31,250 plus the related structural and electrical costs."

"Great. That's another 7%." Cy was getting enthusiastic. They were clearly on a roll.

Leo was skeptical. "That could be a serious mistake, Alex. If your only elevator is out of order, you have none left. I wouldn't like to do this."

Alex warned, "Look, Leo, we can't be too idealistic. If the building is perfect, we can't build it. We have to make sacrifices so we can build our investment." Cy was vigorously nodding agreement.

"What else have you got?" prompted Cy.

"I have nothing else at the moment. Our estimating department is continuing to work on it and I'll be back with some more proposals for your consideration. How about a week from today, here, at 2 o'clock?"

The architect asked, "Mike, will you please give me all your proposals in writing so we can study them and also pass them on to our consultants?"

* * *

In the Phoenix office, Elmer Taft and two estimators were busily sifting through the plans and specifications and talking to subs. They were finding many more items to consider, some fairly minimal, but others quite substantial. In the process, Elmer discovered a serious mistake in their bid. In the heat of bidding, someone, probably at the last moment, had transposed a figure when transferring subcontract bids onto the master tally sheet. Samson Engineering's structural steel bid was $969,669 but was entered as $699,669, a difference of $270,000. As soon as Elmer found it, he went directly to Kyle Lincoln's office.

"Kyle, I found a hole in our bid. We made a $270,000 mistake." He showed Kyle the Samson bid and the Delta-Apex master tally sheet on which the proposal was based.

"Do you think we ought to try to get out of the bid, Elmer? We might be able to get out of it on the basis of that clerical error." Kyle was sick. He really wanted this job to get the L.A. office launched, but he didn't want to lose $270,000.

"Well, Kyle, I've been thinking about that, and I think it could be a terrible mistake to bail out if we're really serious about getting a Los Angeles office established. If we drop out, we might not get any more work to bid. It would ruin our reputation."

"But, Elmer, we can't take a $270,000 hit like this. We're not that big. And we can't ask Samson Engineering to help on it. It wasn't their fault."

"We might be able to find enough savings in the value engineering process to cover our error," Elmer thought aloud.

"Do you really think we can?" Kyle was dubious.

"It's possible. We should at least give it a try. We'll just have to put a little more effort into it."

"Okay, Elmer, this is your project. Do whatever you have to. See if you can pull it out of the fire."

It didn't take Elmer long to figure out the parameters of the problem. The owner required a price reduction of $439,000 and Delta-Apex needed an additional reduction of $270,000. So, he needed to find savings of $709,000 to put the deal together. If he gave the owner about two thirds of all savings, the other one third would cover Delta-Apex's bidding error. Thus, all he had to do was simply find $709,000 in cost reductions. He figured that shouldn't be too difficult.

* * *

Elmer Taft contacted Delilah Samson, President and General Manager of Samson Engineering. They spent a good two hours going over the structural steel drawings and specifications. Samson, being a civil engineer, had some practical ideas and they worked up a price that would save $228,000 on the structural steel. The savings were all predicated, of course, on the architect's and owner's acceptances of the proposed changes.

Elmer could see the advantage of in-depth extensive discussions with the subs for ferreting out the fat in the subcontracts. He talked to the window wall, electrical, plumbing, roofing, waterproofing, and asphaltic concrete paving contractors. He talked to the suppliers of hardware, concrete, lumber, and millwork. Practically all of them had something to offer in the way of money-saving ideas.
Elmer briefed Mike Van Buren on the fine points of value engineering. Mike was ready for the next meeting.

* * *

On Wednesday, April 1 at 2 o'clock, the four determined men reassembled in the Judge & King conference room. All eyes were on Mike Van Buren, as only he held the solution to their mutual problem. If he and his estimators and subcontractors could squeeze enough dollars out of the cost breakdown, there would be a Pacific Horizon Professional Office Building. If not, they could all kiss the project good-bye.

Mike didn't disappoint them. He had a smile on his face. He explained to them how thoroughly the Delta-Apex estimators had examined the plans and specifications and analyzed the construction processes. "Our estimating department, headed by Elmer Taft, has been working extremely hard on this project. Their criterion is to ascertain that the value of each item is commensurate with the cost. This, after all, is the objective of value engineering. We don't have the figures perfected yet, but we're well on our way."

Alex couldn't wait. "What've you got to tell us about today?"

"We've been talking to the steel fabricator. She has rechecked the engineering and feels that the structural frame is somewhat over-designed. She can use slightly smaller members and simplify the connections and cut $141,360 out of the steel bid."

Cy was really impressed. This item alone was a whopping 32.2% of the needed savings. Cy regarded these savings more as changes in numbers than in the building. He shoved his calculations over to Alex so he could see them. Alex looked at the figures, winked at Cy, and asked, "Anything else, Mike?"

"Yes, the reduction in the heights of the steel floor and roof beams would reduce the building height nine inches, thereby lessening some of the costs of masonry, painting, and window wall. We haven't completely figured it out yet but we think it's about $4500."

"Wow, you guys really know how to cut costs." Cy thought this was marvelous.

Leo didn't appear to be as pleased as the others who were reveling in the savings. "Now, wait a minute, Mike. We'll have to have our structural engineering consultants look over any proposals to change the structural system. The structural integrity of the building is top priority. Saving a few nickels and dimes isn't worth the risk."

Mike quickly reassured him, "Naturally, we'd want you and your engineers to check it out, Leo. That goes without saying."

Cy said, "Lighten up, Leo. Mike's not talking about a few nickels and dimes. This is significant money. We should check it out very carefully."

Alex asked, "What else, Mike?"

Mike mentioned the ongoing discussions with the subs and suppliers and promised to be back in a week. Wednesday, April 8, same time, same place.

Alex reminded him, "Let's not kill too much more time, Mike. I know what you're doing is tedious and difficult but we have to get this construction started. We don't want to lose our financing."

* * *

The day after the meeting, Leo asked Myles Nolan, his structural engineering consultant, of Owens & Nolan Engineering, to drop by for some discussion. Leo outlined the Samson suggestions and asked Myles what he thought.

Myles asked, "Do you have anything specific? Any facts and figures?"

"Not yet. They'll be submitting them in the next few days. I just want to know. Is it possible to reduce the cost of the structural frame without endangering the structural integrity?"

"Well, yes. You can skin anything down to the bare minimum. In a steel frame, you could lessen the depth of the beams without collapsing the building. But you'd pay for it in springy floors and excessive seismic movement. It's not unsafe, but it's aesthetically displeasing and generally unacceptable in a good quality office building. In an earthquake you'd have more damage. It might be acceptable in a more utilitarian setting."

"When the specifics come in, Myles, I'd like to have you look them over."

* * *

At the meeting of April 8, Mike had an impressive list of further cost reductions. By changing the subterranean waterproofing to damp proofing they'd save a good chunk. Cheaper electrical switchboards and distribution panels saved a bundle. Changing the waterproofing system in the garage structure saved another significant amount. Downgrading the window wall system contributed a major saving. By the time they got to the bottom of the list they had also cut the quality of the doors, electrical fixtures, plumbing fixtures, toilet partitions, carpeting, roofing, and sheet metal. The grand total of value engineering savings now amounted to just over $525,000. Mike had all these cost reduction items listed with the technical backup details so the architects and engineers could say yes or no to each item. Alex and Cy were overjoyed. The project was nearing reality.

As they were leaving, Alex shook Mike's hand and congratulated him, "You've saved our bacon, Mike. We owe you a lot. Our investment was literally down the tube. Now you've saved it for us." Cy heartily agreed.

Leo reminded the others, "We'll have to have a little time to evaluate these suggestions and determine which ones are worth taking. How about next Wednesday, same time, here?"

"Can't you make it any quicker, Leo? We're wasting time. Let's not lose our momentum," pleaded Alex.

The architect said, "This checking is important. Some of these savings can be realized, of course, but others could be unwise and foolhardy. We should check them out."

* * *

When they reassembled in Judge & King's conference room for their meeting at 2 o'clock on April 15, Alex, Cy, and Mike were brimming with optimism. They were in high hopes of winding up the deal and proceeding into the construction phase.

Leo explained what he and the firm's engineering consultants had been doing for the past week. They had carefully checked out all the money saving opportunities presented by Delta-Apex, item by item, accepting some, rejecting others. The architect summarized the week's work by explaining, "Some of the proposals had to be rejected out of hand, since the effect on the building was too damaging for further consideration. Others were generally acceptable. The rest are acceptable only with the owner's express understanding and acceptance that the quality of the building will be seriously reduced. Those suggestions that were in violation of the building code had to be eliminated. We also rejected any that compromised structural integrity, fire safety, or security."

He passed out summary sheets that detailed the rejections and acceptances. The savings rejected added up to $165,000. The net effect on the total savings, in round figures, was a reduction from $525,000 to $360,000. Alex and Mike were sitting there, as though stunned. Cy was in a near-catatonic state.

Alex spoke first, "Dammit, Leo. We practically had a deal. Now, you've screwed it up. Your obsessive idealism is costing us our investment. We can't build a monument to your ego." He was disappointed and upset with the architect.

Leo looked crestfallen, as he and his engineering consultants had tried to be as understanding as possible of the owner's economic interests and, at the same time, be responsible.

Mike, recognizing that he would have to be on friendly terms with the architect if the building went ahead, tried to smooth out the situation. "I'm sure, Alex, that Leo had good reasons for rejecting some of our savings. We knew ourselves that some of them were a little questionable. All we have to do now is go back and study the plans and specs a little more and find an additional $80,000 in acceptable savings. Now that we know the types of things that Leo won't accept, we can avoid them this time."

Cy asked, "Do you really think you can find $80,000 more?"

Mike replied, "I'm not sure, but we'll try."

Leo made the usual announcement, "Same time, same place, next Wednesday, April 22."

Alex asked, "Isn't there any way we can get back sooner? That'll be our sixth meeting and we've killed over a month since we started." He and Cy were getting impatient.

Mike said, "I'll need the week. Elmer and our estimators will have to cut pretty close to the bone to get another $80,000 out. This takes a lot of time and plenty of skill and hard work."

* * *

On the 22nd they were back to the conference table. Mike was broadly smiling, a dead giveaway that he'd probably achieved what they were all hoping for.

Alex asked, "Well, Mike, are we going to build this building or not?"

"It's all up to your architect. We have a little over $88,000 in additional savings. If they're okay with you and Leo, we've got a project." He passed out his new list of proposed savings.

Leo quickly skimmed the sheet and commented, "We're really scraping the bottom of the barrel now. Look at this $37 item, and here's one for $89. I'm sure I can get through this list fairly quickly. I'd suggest we meet again tomorrow, here at 2 o'clock. Okay?"

So the contractor and owners left in renewed good spirits. They'd probably sew up the deal tomorrow.

After the others left, Leo went into Ivor Judge's office. He was frank with his partner. "I'm getting sick of this value engineering exercise on the Pacific Horizon project, Ivor. All we're doing is squeezing out all the quality. There's no spare capacity left in anything. Every electrical circuit and panel is loaded to the limit and the A/C is skinned to the bone. Much of the expected savings will be dissipated in increased operating and maintenance costs in perpetuity. And some of the savings will be used up in re-engineering and changing the construction documents."

Ivor was concerned and sympathetic, as he'd been through the same procedure himself so many times. "Have you mentioned all this to the owner?"

"Yes, but Alex and Cy are under so much pressure to keep initial costs down that they don't want to hear about the downside."

"It's always very difficult to get an owner to accept that some money-saving measures will prove uneconomical in the long run."

* * *

So, on Thursday, April 23 they met for what would be their last meeting in the value engineering phase. Leo produced copies of his report and passed them out. Of the $88,000 in possible savings, the architects had accepted only $79,000.

Cy, after reviewing the list, said, "That's all we need. Now, we're on budget. We have savings of $439,000 and a contract price of $6,680,000."

Alex was effusive in his praise, "Mike, you've done a stupendous job. We don't know how to thank you and Delta-Apex. You've saved our investment."

Leo reminded his clients that the drawings and specifications would now have to be changed to reflect all the money-saving adjustments and resubmitted to the building department for approval. "This will take at least two weeks for our work and another week for the building department."

In the next few days Leo prepared the construction contract and had it delivered simultaneously to Delta-Apex and to Horizon Development Company with instructions to have it reviewed by their legal and insurance counsel. In a subsequent meeting they resolved the remaining minor differences and signed the contract. The building permit was issued on May 15 and construction got under way on Monday, May 18, 1992.

* * *

Points of Law 20

PACIFIC HORIZON I
Selling the Job

20.1 The Doctrine of Illegality

Mike is studying to take the test for a California contractors license. If he carefully studies, he will learn that he is breaking the law. Even though his company may be licensed in Arizona, that does not authorize it to bid work in California.

The purpose of contractors licensing laws is to protect the public from incompetence and dishonesty in contracting. It is not just the individual project owner who has an interest in the competence of the contractor. The community has an interest in sound, safe construction of roads, bridges, housing, and commercial and industrial buildings. Some would say that this interest is adequately served by the inspectors who enforce the building codes and that it would be sufficient to rely on the marketplace to control incompetent and dishonest contractors. But all state legislatures have gone along with the organized construction industry and require contractors, just as physicians, lawyers, real estate brokers, and dogs, to be licensed.

If Mike drives his pickup truck (or keeps his dog) without a license, he commits a misdemeanor – a violation of criminal law – and is subject to fine or imprisonment. If he pursues the business of contracting in California without a license he would also violate the criminal law and be subject to fine or imprisonment. There is, however, another penalty that is more severe and also much more likely to be imposed: a penalty that is imposed under the law of contracts. The elements of an enforceable contract are 1) competent parties, 2) mutual assent, 3) consideration, and 4) a legal object. A contract with an unlicensed contractor does not have a legal object because it calls for the performance of construction services by an unlicensed contractor. An unlicensed person is prohibited by law from performing construction services, and therefore the "legal object" requirement for enforceability of a contract is not fulfilled. From this it follows that a construction contract entered into by Delta-Apex Construction will be unenforceable. This is called the doctrine of illegality.

The doctrine of illegality has ancient roots in English common law. It originated in smuggling cases. A smuggler sued to recover the price of smuggled goods sold and delivered. Judges refused to involve themselves in the enforcement of such contracts. The law leaves parties to an illegal contract where it finds them.

20.2 The Exception

Cases soon arose in which the doctrine had to be modified for the protection of innocent parties. Suppose that B signs a purchase order to buy twenty cases of Sinatra CD's from S, and makes a down payment of $50,000. S violates the purchase order and fails to deliver the CD's. Unknown to B, the CD's are bootlegs manufactured by S. S refuses to return the $50,000 and pleads in defense that he manufactured the CD's in violation of the copyright laws.

Courts will not allow S to set up his own criminal activity as a defense to a legitimate claim. But the defense would be good if S could prove that B knew (or should have known) that he was dealing in pirated goods. This exemplifies a legal doctrine called in pari delicto: the defense of illegality applies against one who is a party to the crime, and not against an innocent party. Thus, if B is innocent he can get his $50,000 back, but not if he knew that he was participating in illegal commerce.

Now let's go back to Delta-Apex Construction. Suppose that Horizon Development Co. signs a construction contract knowing all along that Delta-Apex is an Arizona contractor unlicensed in California. Now suppose that Delta-Apex breaches the contract by failing to install shearwalls as they are designated on the drawings. This weakens the structure so it deflects in a windstorm. Horizon sues for the cost of repair and Delta-Apex defends on the ground that it was an unlicensed contractor and Horizon knew it all along. Under the doctrine of illegality, the contract would therefore be unenforceable because Horizon was in pari delicto.

20.3 The Exception to the Exception

Now we confront an example of legal thinking that bedevils the studies of every student of the law: the exception to the exception. The rule is that the doctrine of illegality bars the enforcement of illegal contracts. The exception is that under the doctrine of in pari delicto the rule does not apply against an innocent party. The exception to the exception is that the doctrine of in pari delicto is inapplicable against a party for whose protection a statute was enacted. Here, the contractors license law was enacted for the protection of property owners and they are protected by the contractors license law even if they deal knowingly with an unlicensed contractor.

This gets us back to Delta-Apex Construction and Horizon Development. Delta-Apex is bidding a job without a contractors license. This is a violation of the contractors license law. If Delta-Apex signs a contract and starts performing it without a license, the consequences will be serious because Delta-Apex will be an outlaw and the doors of the courthouse will be closed. This means that if Horizon fails to pay for Delta-Apex's work, whether for a legitimate reason or not, Delta-Apex can't enforce the contract.

What if the contract has an arbitration clause? Could Delta-Apex collect the debt by going to arbitration rather than by going to court? The answer is no. An arbitrator has no more power than a court to enforce illegal contracts.

20.4 Horizon Also Needs a License

Let's take an even closer look at the contractors license law and analyze Horizon's position. Now wait a minute! Didn't we just say that Horizon was the innocent party, the property owner for whose protection the contractors license law was enacted? True, but consider, dear reader, that Horizon is building a speculative building. The building is for sale and therefore Horizon is a developer. The contractors license requirement applies to developers: owner-builders are exempt from the requirements of the contractors license law, but only if they build for their own use. When an owner builds for sale, it becomes a merchant builder, a developer that is required to be licensed.

20.5 The Exception to the Exception Does Not Apply

So maybe the doctrine of in pari delicto will apply after all. Since Horizon is a merchant builder, a developer, the contractors license law was not enacted for its protection, and it is not an innocent party. Such being the case, neither party can enforce the contract against the other. If Horizon fails to make a progress payment Delta-Apex can't sue to collect it. And if Delta-Apex fails to install a shearwall per plans and specs, Horizon can't sue for cost of repair.

20.6 Fraud and Misrepresentation

In order to sell the job, Mike has to squeeze $439,000 out of the contract price. This Mike proposes to do by the process of value engineering. The value engineering concept is presented to the owner as a way of reducing the cost of a project without reducing its value. Realistically, it should be defined as a process of reducing the quality in such a way that cost reductions are greater than the decrease in the project's ultimate value. For example, if Mike can squeeze $439,000 out of the cost while reducing the project value by only $100,000, the owner may be willing to accept the reduction in quality because it makes economic sense.

Now Elmer has discovered a $270,000 bust in the estimate, which means Mike has to squeeze another $270,000 out of the cost in order to build the project without taking a loss. We don't know if he intends to inform Horizon of the bust. Horizon's attitude toward value engineering might be different if it knew that 25% of the "savings" was going to go to compensate Delta-Apex for its own bust. This could be cheapening the job for the benefit of the contractor.

Here we must deal with fraud and misrepresentation. Fraud occurs when a person makes a false statement of fact to a party who relies on the false statement to its detriment. For example, a jeweler who passes off cut glass as a diamond commits fraud. A statement of opinion, even if exaggerated, is not fraud. If a salesman tells a potential customer "this car is worth $5,000" it is not a statement of fact but a statement of opinion.

20.7 The Underestimate

Architects are not often held legally liable for underestimating the cost of a construction project. Courts say that an estimate is not a guarantee. Architects seldom can control the ultimate cost of a project because they don't control the award of the contract (whether by competitive bid or by negotiation), they don't control the construction process, and they don't control the size or quality of the project. Underestimation may not cause any actual damage to the project owner. If a project is well designed and it costs $500,000 to build, then it is presumed to be worth $500,000 even if the architect estimated that it would only cost $400,000.

In our case, Horizon is a speculative builder and the sales price of the completed project will depend upon the rental income that it can produce. Therefore even if the completed building is worth more (because it costs more) that economic value cannot be realized if the rents can't be raised and therefore if the architect's estimate is accurate Horizon might make the decision to abandon the project. In such a case, Horizon would be entitled to reimbursement of architect's fees. Horizon might seek to recover the cost of acquiring, holding, and selling the property but the court would probably not award such damages because those expenses would have been incurred whether the architect underestimated the cost of the project or not.

20.8 Withholding Information

We have said that fraud is defined as a false statement of fact that is relied upon by the plaintiff to its detriment. Here, we are not sure it can be proved that Mike specifically represented that all the savings from the value engineering process would be passed on to the owner. Perhaps the equivalent of that representation could be inferred from Mike's conversations and conduct during the whole value engineering process. Indeed, the very phrase "value engineering" seems to imply a process that by definition would assign all savings to the owner whose property is cheapened by the process rather than to increase the profits of the contractor who is in charge of the cheapening process.

Fraud may consist of withholding information as well as providing it. So far as we know, Mike has withheld the information about the bust in the estimate and the necessity that Delta-Apex reserve some of the savings generated by value engineering to itself rather than pass them on to the owner. Mike knows that Horizon has no way of learning about this. He also knows that the bust, and the determination of Delta-Apex to cheapen the job in order to protect its own profits, would be crucially relevant information to Horizon's decision whether to contract with Delta-Apex or to abandon the project. Therefore Mike has a duty to disclose the facts, and the failure to do so is fraud.

20.9 Plenty of Blame

The parties here, architect, owner, and contractor, are working themselves into a situation fraught with business and legal danger. They all bear some responsibility. The architect underestimated the job by about $439,000. The developer seems to be ignoring signals that the job is being cheapened almost so much that life safety is at risk. The contractor, in its anxiety to get a job in California, is violating the contractors license law. The developer, which is, or should be, a licensed contractor itself, doesn't seem to know or care.

Yet each party can point to some extenuating circumstances. Delta-Apex is not dealing with a neophyte, but a sophisticated developer that should be able to look after its own interests. Judge & King, because of the greed of the contractor and the developer, seems to be losing control of its own design. Horizon is entitled to look to Judge & King to protect its interests and yet the Judge & King, in order to avoid potential liability for the underestimate, might be tempted to cooperate with Delta-Apex to save the job even if that means sacrificing the interests of Horizon in attaining a quality project.

21

Pacific Horizon II
Value Engineering: The Construction Period

The two partners at Judge & King, AIA, Architects, personally performed most of the construction administration duties on their jobs. They generally visited each job at least once a week all during the construction period. It was usually established as a uniform time and day each week so the subcontractors and suppliers could be there if they needed the architect's input. Owners were always invited to attend. Some did.

The general contractor and the job superintendent were expected to be there for every visit to explain progress and future scheduling, to clarify misunderstandings, and to receive instructions. The architect would write a Job Observation Report for each visit to keep the owner informed up to the minute on job conditions. A copy of the report would always be sent to the owner and the contractor.

Recently, one of their young project architects, Inigo Johnson, desiring to expand his professional experience into the area of contract administration, had been pressuring the partners to let him have the Horizon job at Malibu. Inigo was the project architect for design and preparation of construction documentation and thus had a sound grasp of the contract requirements. He pointed out that it would be handy from his home in Venice.

Leo King had planned to administer this job himself, but agreed with his partner, Ivor Judge, that this project would be ideally suitable for Inigo to pursue his on-the-job training as an architect. The weekly site observation visits had been set up for 8 o'clock every Tuesday morning. Leo would attend some of the site meetings and keep a watchful eye on Inigo's performance.

The job was mobilized on Monday, May 18, 1992, and had been progressing right on schedule. Delta-Apex's superintendent, Fred Buchanan, experienced and capable, had the job well in hand. He knew how to deal with the subcontractors effectively to get the most cooperation and production from them.

Mike Van Buren, Vice President, Sales, in charge of Delta-Apex Construction's Los Angeles office always attended the weekly meetings. Sometimes Alexander Brink, President of Horizon Development Company, the owner, also attended. On some occasions he brought Cyrus Flint, Horizon's Financial Vice President, along with him.

It was in the 11th week of construction, July 28, when Inigo arrived on his regular Tuesday morning observation visit. Mike Van Buren was in the site office talking to Fred Buchanan and Alan Condon, owner of Air Unlimited, the air conditioning subcontractor.

Inigo had his clipboard for taking notes for his report. Alan Condon asked to be heard first so he could get on with his day's business.

Mike said, "Sure, Alan, spill it. What's on your mind?"

"Well, do you remember during the bidding period I suggested installing a passive solar energy water heating system?"

Mike nodded, confirming his recollection of the suggestion.

"Well, do you want to do it or not?" Alan stood expectantly, looking first at Mike and then at Inigo, then back to Mike.

Mike asked, "How much would it cost?"

"I haven't figured it out yet. I didn't want to waste any time on it unless you wanted to do it."

Inigo asked, "What all does it involve?"

Alan described the system. "We'd install two arrays of solar panels on the roof, one to supplement the hot water boiler in the space heating system and the other, a smaller one, would replace the gas fired domestic water heating system for the toilet room lavatories, janitor sinks, coffee sinks, and the restaurant. We'd need to install two circulating pumps, some valves, some piping, and tie it into the controls for the two systems."

Mike replied, "My recollection was that you said it wouldn't add very much to the building cost. What you've just described sounds complicated, Alan. And complicated usually means costly."

Alan clarified, "Well, there'll be a deduction first. The water heater in the penthouse and all its gas piping can be left out. So, you'll have to get a credit from the plumber for it. The biggest reason for doing this is that solar energy's free and therefore less fuel gas has to be paid for."

Inigo explained, "We can't decide this just standing here. It has to be taken up with the owner. And we'd have to run it past our mechanical and electrical engineers. So, I'd appreciate it if you'd write out a detailed description of what you propose to do and work up the cost figures. Submit it through Mike's office. He'll send it on to me."

Mike added, "And do it fast. The construction's moving swiftly and we don't want to hold up the schedule."

Inigo continued, "And Mike, will you please get a quote from the plumber for leaving out the water heater and gas piping? We'll have to consider the whole ball of wax at one time."

Fred Buchanan was listening to the interchange of ideas and added his contribution. "Alan, how would you plan to attach the solar panels to the roof?"

"We'll have to think about that."

"Would you install them before or after the roofing installation?"

"That's another thing to think about."

"While you're at it, think about how the piping will be routed and how it'll be flashed at roofing penetrations."

"Okay. I'll think about that, too."

Fred asked, "Will we need to add any roofing pads for maintenance traffic paths?"

"I'll let you know when I find out, Fred."

Inigo wrapped up the subject by requesting Mike to follow through on all aspects of the proposal so Leo King can take it up with the owner and consulting engineers. He also included it in his observation report.

* * *

About three weeks later, the solar heating proposal arrived in the mail at Judge & King's office. Leo passed it on to Inigo for review. Delta-Apex's proposal totaled $5786.80 and was backed up with the paperwork from Air Unlimited, Lilac Plumbing, and Tops Roofing.

> # Delta-Apex Construction
> ## Change Order Proposal # 7
>
> Pacific Horizon Professional Office Building, Malibu, California
>
> August 17, 1992
>
> For adding solar energy heating systems to supplement hot water space heating system and domestic water heating system. Omit specified water heater, vent, and gas piping.
>
> Air Unlimited Solar panels, mounting frames, circulating
> pumps, piping, controls, and labor $4780.00
>   Overhead & Profit, 15% $717.00
>   **$5497.00**
>
> Tops Roofing Pitch pockets and traffic pads $320.00
>   Overhead & Profit, 15% $48.00
>   **$368.00**
>
>   Subtotal..........................**$5865.00**
>
> Delta-Apex Construction Overhead & Profit, 10%.................................$586.50
>
> Lilac Plumbing Credit for omission of water heater and gas piping.................. ($722.00)
>
>   Subtotal..........................**$5729.50**
>
> Bond Premium.. $57.30
>
>   **Total Change Order....$5786.80**
>
>   Contract time extension ....10 calendar days

Inigo reviewed the proposal, brought it back to Leo's office, and asked, "What should I do with this?"

"Check it over for completeness, feasibility, and reasonableness, and then send it on to Horizon so Alex and Cy can decide if they want to go ahead with it."

"What do you think of this price, Leo?"

"Well, it may be reasonable but it's more than I thought it would be. When Alan Condon brought up the subject during our value engineering discussions about four months ago it was my impression that it would be a somewhat incidental cost and that the plumbing credit would pretty well cover it."

* * *

About a month later, Inigo came to Leo's office and asked, "Have you heard any more from Alex or Cy about Delta-Apex's solar panel proposal? Fred Buchanan's been asking about it."

"No, Inigo, I'd forgotten all about it. Maybe I better call Alex and ask him what they want to do about it."

"That'd be a good idea. I've had it in my pending folder since it came in on August 19th."

Alex wasn't in when Leo called but returned the call the next day. "What's up, Leo?"

"I was just checking to see what you want to do about that solar energy proposal."

"I'll have to check with Cy. He was looking into it. I'll call you right back."

About ten minutes later, Alex was back on the phone. "Cy tells me that he's been checking into it and the gas company has estimated that savings at the gas meter would average about $1450 per year. If this is true, we could pay for the $5800 extra cost in less than four years. So, we'd like to go ahead with it."

Leo replied, "We'll prepare a change order for your and Delta-Apex's signatures and get it going in the field. This'll be Change Order # 7."

* * *

The following Tuesday, September 22, Fred Buchanan told Inigo that there was a slight problem with the solar panel change order. The proposal had been pending for so long that Lilac Plumbing had already installed the gas piping to the penthouse water heater location and the water heater had already been purchased. "Lilac doesn't want to give the $722 credit. Do you think the owner would still want to go ahead with the change order?"

"I don't know. I'll have to check. I'll let you know as soon as I find out." Inigo put this in his notes for the observation report. That afternoon he asked Leo to inform the owner that the cost had gone up $722, although there would still be some credit for returning the water heater after restocking and shipping charges.

Leo called Alex and explained the situation. Alex was somewhat displeased that the cost had risen so rapidly but acknowledged that conditions change quickly on a building under construction and conceded that they hadn't reacted promptly enough to the paperwork. He said, "Well, we'll go ahead with it anyway, as it will extend the payback period only 5 or 6 months."

A few days later, Delta-Apex's invoice arrived detailing the change in price for adding the gas piping back into the contract.

> **Delta-Apex Construction**
> Change Order Proposal # 8
>
> Pacific Horizon Professional Office Building, Malibu, California
>
> September 25, 1992
>
> For furnishing and installing gas piping for water heater in Penthouse.
>
> Lilac Plumbing Gas piping and valves for water heater..........................$600.00
>   Overhead & Profit, 15%... $90.00
> $690.00
>
> Delta-Apex Construction Overhead & Profit, 10%............................... $69.00
>
> **Subtotal**............$759.00
>
> Bond Premium.. $7.59
>
> **Total Change Order**......$766.59
>
> Contract time extension ......0 calendar days

Leo sent Change Order Proposal # 8 on to Horizon for approval.

The work of Change Order #7 had been in production on the job and eventually the circulating pumps had to be hooked up to the electrical system. Alan Condon approached Fred Buchanan at his site office. "Say, Fred, we're ready for the electricians to hook up the circulating pumps for the solar energy systems. We'd appreciate their getting right to it as we'd like to test the systems as soon as we can."

Fred replied, "Okay. I'll go talk to the electrical foreman."

Fred went into the building and found Graham Buckley, Power Electric Corporation's foreman on the job. "Graham, when are you going to hook up those circulating pumps for Air Unlimited?"

"I don't know what you're talking about. I didn't see any circulating pumps on the drawings. Where are they?"

"The circulating pumps are located in the penthouse. They're on the solar energy change order. Change Order #7."

"We didn't get any Change Order #7. The last one we got was #6." Graham went back to marking circuits in the second floor main distribution panel.

Fred returned to the site office and Alan Condon was gone. He phoned Mike Van Buren and brought him up to speed on the missed electrical connection of the circulating pumps. Mike said, "Dammit. We forgot to get the electrical quote when we put the change order together. Well, we'll just have to ask Power Electric to give us a price. And then we'll have to ask the architect to get the owner's approval."

Fred volunteered, "I'll get Graham Buckley to prepare a quote and send it to you through his office. It shouldn't be very much. It's only two pumps. A couple extra circuit breakers. No big deal."

"Thanks, Fred."

When Fred went back into the building to talk to the electrical foreman, he was astonished at Graham's answer. "It's going to cost quite a bit. The penthouse power panel's up to the absolute limit. There's no spare capacity in it. So we'll have to enlarge it or add a panel. The feeder cables will have to go to the next larger size and we'll have to increase the conduit size. The basement power panel will also have to be enlarged as well as the main feeders. All of this work is already installed so we'll have to take it all out and reinstall it. It'll be fairly costly."

"Well, give us the quote as soon as possible. Keep it as low as you can. Send it to Delta-Apex's office to Mike Van Buren's attention."

The Power Electric quote arrived the next day via fax. The price, including overhead and profit, was $9027.50. They also wanted 5 calendar days extension to their schedule.

Mike called the architect and tried to break the news gently, "Leo, we have a slight problem here. We forgot to include the electrical hookup to the pumps in the solar energy change order. We just got a price from Power Electric for making the necessary electrical changes and hooking up the pumps."

"Oh? How much is it?"

Mike mumbled, "A little over nine thousand."

"What? Someone here coughed and I didn't quite hear the figure."

"$9027.50." Mike held his breath.

Leo woke up. "That's ridiculous! Why would it cost that kind of money? There must be some mistake."

"Well, the circuits and panels are full and quite a number of conduits, cables, and panels that are already installed will have to be changed."

"Isn't there enough spare capacity in the system to add a couple of small circulating pumps?"

"No. All the spare capacity was taken out during the value engineering process. Don't you remember? It was taken out and the owner got credit for it. And remember, you approved it."

"Well, send it on to me. I'll have to convince the owner that they should pay another $9027.50."

"It's a little more than that now, Leo. Delta-Apex will have to add its overhead and profit and the bond premium.

"It gets worse and worse."

"And, Leo, we can't wait very long for this approval. We gotta keep going or we'll fall behind on our schedule. I'll fax you a new proposal right away."

Delta-Apex Construction
Change Order Proposal # 9

Pacific Horizon Professional Office Building, Malibu, California

October 9, 1992

For electrical hooking up of two circulating pumps for the solar energy system and revising the existing panels, wiring, and conduits.

Power Electrical Corporation
Installing larger circuit breaker panel in penthouse, revising existing conduits and feeder cables, revising main panel in basement,
and hooking up two circulating pumps in penthouse......$7,850.00
    Overhead & Profit, 15%......$1,177.50
                              $9,027.50

Delta-Apex Construction Overhead & Profit, 10%......$902.75

                           Subtotal......$9,930.25

Bond Premium...... $99.30

                    Total Change Order.$10,029.55

                  Contract time extension ......5 calendar days

---

Leo faxed a memo to Alex along with the new proposal. He thought this would be easier than trying to explain it on the phone. In it he explained the electrical hookup problem and the consequent costs.

A few minutes later Alex called Leo. "What in hell is going on, Leo? I've just received this ridiculous change order proposal for another $10,000 on the solar energy systems. We approved the original $5800 because it would pay back in fuel gas savings in four years and we paid extra for the gas piping, but now it's over $16,000. That would take 11 years to pay off. Let's just cancel the whole damn thing."

"We can't cancel it, Alex. All the work of Change Orders #7 and #8 is done. All that's left to do is the electrical hookup."

"Well, we never would have approved the original change if we knew it would cost 16 grand. Just cancel it."

"That can't be done, Alex. You don't have to go ahead with the electrical hookup if you don't want to, but you'll have to pay for the work already done under the original approved change orders."

"Well, we won't pay it. Just tell Delta-Apex to stuff it." Alex slammed the phone.

Leo called Mike Van Buren and recounted the conversation with Alex. Mike said, "Well, I'll tell Air Unlimited and Power Electric to forget the electrical hookup. But you tell Alex Change Orders #7 and #8 must be paid."

"There's another problem, Mike. If the solar energy system isn't going to be completed, the water heater will have to be put back and hooked up."

"I'll get Lilac Plumbing to go ahead with the work and to give me a quote. At least the gas piping is in"

"We already have a quote. It was $722 including the piping."

"Yes, but that no longer applies. The water heater was returned for credit, so there will be shipping and restocking charges. Also, Delta-Apex will add on its overhead and profit and bond premium."

A few days later a fax arrived at Judge & King's office detailing the costs for adding the water heater back into the contract.

Delta-Apex Construction
Change Order Proposal # 10

Pacific Horizon Professional Office Building, Malibu, California

October 25, 1992

For installing the originally specified water heater in the penthouse and installing the vent.

Lilac Plumbing Furnish and install water heater and vent in penthouse.	$272.00
Overhead & Profit, 15%	$40.80
	$312.80
Tops Roofing Repairing roof at new water heater vent.	$80.00
Overhead & Profit, 15%	$12.00
	$92.00

Subtotal............$404.80

Delta-Apex Construction Overhead & Profit, 10% $40.48

Subtotal............$445.28

Bond Premium $4.45

Total Change Order.......$449.73

Contract time extension........0 calendar days

Leo sent it on to Horizon. He knew it was heaping more fuel on the fire.

On the tenth of November, when the next contract payment came due, Delta-Pacific included the original solar energy Change Order #7 and the supplementary Change Orders #8, #9 and #10 and Leo approved it for payment.

Alex instructed Horizon's Financial Vice President, Cyrus Flint, to deduct the amounts of the four change orders from the payment. When Delta-Apex received the short payment, they did not pay Air Unlimited since the sum attributable to their work had not been collected. Alan Condon's immediate reaction was to accost Mike Van Buren at the jobsite. "Mike, why didn't we get paid for our work on the solar energy change order?"

"Simple. Because the owner didn't pay us."

"Well, we gotta be paid. And damn soon."

"We'll try to collect it but the owner appears to be stubborn on this. You'll probably have to eat it."

"Like hell we will." He went to his truck and picked up his mobile phone.

Two days later Air Unlimited filed a mechanics lien claim.

Horizon's lender promptly contacted Cyrus Flint and told him that the lien would have to be cleared immediately or they'd deduct one and a half times the amount from the next loan draw. Alex called Leo and ordered him to get Delta-Apex to clean up their act. Leo called Mike and relayed the message. Mike called the president of Delta-Apex, Kyle Lincoln, in Phoenix, and explained the whole complicated mess.

Mike asked, "So, what do we do now, Kyle?"

"Why don't you call our Los Angeles lawyers and see what they have to say?" Kyle hoped Mike and the lawyers could straighten out this little squabble and keep him out of it.

Mike then called Pat Jimson of Darnel & Jimson and made an appointment to see him and talk over the situation. The lawyer suggested that Mike bring all the relevant documentation along with him.

Darnel & Jimson was highly regarded in the Southern California construction industry, but especially among general contractors, as a practical and effective law firm. Patrick Jimson had been working with Mike Van Buren on various legal matters that had arisen since Delta-Apex Construction had opened their Los Angeles office.

Comfortably seated in Pat's office, Mike explained the whole situation surrounding the solar energy change order and the subsequent change orders for the electrical hookup and water heater installation. "So, to cut a long story, Horizon doesn't want to pay us the $17,032.67. What can we do about it?"

"You've already tried to collect the money in the usual way, but to no avail. Therefore, the best recourse available to you is to demand arbitration. The AIA General Conditions sets out the procedures for making claims, obtaining the architect's ruling, and filing the arbitration claim. We must follow those procedures. They're part of the contract."

"Okay, Pat, so what do we do next?"

"Let's start getting all the documentation organized and I'll get the ball rolling. But first, we have to make sure Delta-Apex's contractor's license is in order."

"Delta-Apex has been licensed in Arizona since 1960."

"That won't do you any good, Mike. What about the California license?"

"We got our California license last June. Since I've been in L.A. I've been studying nights and took the license exam last February. I didn't pass it until my third try in June. But now we've got it. I'm the Responsible Managing Officer."

Pat was hastily sifting through the pile of documents, looking for the contract. "What's the date on the contract, Mike?"

"May 15, 1992."

Pat couldn't believe what he was hearing. "That means Delta-Pacific didn't have a license for the first month of the contract. We can't file a lawsuit without alleging and proving that Delta-Apex was licensed when the contract was signed and continuously all during the contract.

"Strictly speaking, the license may not be required in an arbitration, however that depends on the arbitrator.

"But I think it would be too risky to go into the arbitration with this hanging over us. If Horizon ever finds out you weren't properly licensed, they may decide not to pay you any more for anything. And there isn't much you can do about it."

"Is that right? Are you sure? What'll we do now?"

"I'd suggest that you pay off the subcontractor liens, forget about the $17,000 owing, and try to finish up the job without any more disputes.

* * *

Points of Law 21

PACIFIC HORIZON II
The Construction Period

21.1 The Lawyers and their Clients

The following are edited transcripts of notes taken by the lawyers who were consulted by the characters in our story. Notice that lawyers identify with their clients and often refer to them as "we" and "us."

21.2 Lilac Plumbing $1,002.80

> We signed a standard form subcontract to install a plumbing system for a rental office building. The system included a water heater with gas piping and a vent in the penthouse. The architect asked us to give a price to delete these. We offered a credit of $722. No change order was issued. We went ahead and installed the piping. Then a change order deleted the water heater only. The architect changed his mind again and put the water heater back in which cost another $312.80. The owner got disgusted, refused to pay for anything, and the prime contractor shorted us $1,002.80!

COMMENT: The contract price of the water heater piping and vent was $722 and now the plumbing contractor is charging $1,002.80 for the same scope of work. This is legally justified because at the time the subcontractor figured the job, it did not anticipate the cost of removing, restocking, and then replacing the water heater. The plumbing subcontractor has a legitimate, legally enforceable extra.

The lawyer will probably record a mechanics lien and send a letter demanding payment but not pursue foreclosure because the attorneys fees for filing foreclosure suit would exceed potential recovery.

21.3 Air Unlimited $5,497

> We entered into a subcontract to install HVAC system in an office building. We made a proposal to substitute solar water heating for the conventional system that was specified and the owner accepted the proposal. The architect issued a change order to the prime contractor, Delta-Apex, and Delta-Apex issued a change order to us. We installed the solar system and it passed inspection. The owner changed his mind, went back to the conventional gas-fired water heater, and refused to pay for the extra work.
>
> Delta-Apex refuses to pay us because it hasn't been paid by the owner and relies on a pay-if-paid clause in the contract. The pay-if-paid clause is void in California and I am to record a claim of mechanics lien and make a demand for payment on Pacific Horizon and Delta-Apex. If no pay, we sue.

COMMENT: The claim of lien is legitimate and justified. The pay-if-paid clause is void under a California Supreme Court ruling which rules that the enforcement of such a clause would improperly restrict the subcontractor's mechanics lien remedy, and thus violate the California Constitution.

Under California law, the lien would expire unless a foreclosure suit is filed within 90 days. Since the cost of filing a foreclosure suit (using form pleadings) is only about $1,000, the investment in attorneys fees would be justified.

21.4 Pacific Horizon

We employed Judge & King, Architects, to design an office building and administer the construction contract. The HVAC subcontractor, Air Unlimited, proposed that we substitute solar heating for the gas-fired water heater. We accepted the proposal, based on a form of change order prepared by Judge & King, which would have cost $5,786.80. We signed the change order based on an analysis that the solar heating would save fuel expenses of $1,450 per year and therefore would pay for itself in less than four years.

After the solar heating is already installed the architect informs us that he left out the electrical to hook up the circulating pumps and this will cost another $10,029.55. We refused to sign that change order.

Then we get hit with another change order from the plumbing subcontractor to restock and replace the water heater ($449.73).

We caught a mechanics lien of $5,497 from Air Unlimited. The construction lender is withholding 1-1/2 times that amount from the proceeds of the construction loan.

The client investigated the cost of a release bond. The bonding company won't issue the bond unless we collateralize the full amount, so we would have to give them a certificate of deposit for $8,246 as security against any loss on the bond.

I advised the client that the lien is probably valid but we should wait 90 days to see if Air Unlimited files a foreclosure suit. Meanwhile, we should demand restitution from Judge & King. It was their responsibility to advise us of the full cost of the change, especially since they knew that we were analyzing the cost benefit. We had to know the full cost to determine whether the benefit would justify the cost.

I am to write letters to Delta-Apex and Judge & King demanding that they remove the lien and do whatever is necessary to hold us harmless from any expense beyond the original contract price.

TO DO:

- Write demand letters.
- Check contractors license status of Delta-Apex and Air Unlimited.

21.5 Delta-Apex

Client installed solar heating system pursuant to a change order that was signed by the owner, owner reneges. Owner says the change order proposal was misleading because it left out the cost of electrical hookup. Had it been informed of this cost, owner claims it would never have approved the change order.

Since the owner hadn't paid us, we refused to pay the HVAC subcontractor $5,497. Air Unlimited has liened the job. I advised client that the pay-if-paid clause in the subcontract is void.

Client screwed up on contractors license. I advised that we can't sue because we can't allege and prove that we were properly licensed when the contract was signed and at all times during the performance of the contract. Therefore, our strategy must be to settle even if we have to absorb the whole cost of the change.

The omission of the electric work is mostly our fault because it was up to us to include all necessary costs in the change order. Even so, Judge & King should pick up some of the loss because it was also up to them to assemble all the costs before presenting the change order to the owner. Nevertheless, we should have made sure that the electrical hookup was included in the original change order.

The change orders contained a total of $1,002.80 for the plumber. They have no fault. Advised client to pay plumber in full and take them off the table. We will then hold a meeting with Judge & King and Air Unlimited and try to negotiate the remaining $5,497. Maybe we can get them to go 1/3, 1/3, and 1/3.

21.6 Judge & King

We have a bust on a change order. We designed an office building and are administering the contract. The HVAC sub proposed to change from a gas-fired water heating to a solar system. We approved a change order for $5,786.80 and the owner signed the change order based on a cost benefit analysis. (The savings in gas would pay for the change order in less than four years.)

After the system was installed it was discovered that the prime contractor left out the electrical hookup. We prepared a change order for the electrical hookup, $10,029.55, and owner refused to sign it because that would destroy the cost benefit. HVAC sub has liened the job and the prime contractor is demanding payment.

We have potential liability for not making the owner aware of the total cost before processing the first change order. The owner is withholding our payments until the lien is removed and the change order claims cleared up.

I evaluated our potential liability to the owner as 80/20. But the potential liability of Delta-Apex is also 80/20. They should have included the missing electric before we even paid the change order. In effect, we just missed their mistake. If the lien isn't taken care of promptly the attorneys fees and costs will go up. We should promptly negotiate a settlement. Maybe Air Unlimited will cut its bill. Delta-Apex should throw in their overhead, profit, and bond premium claim and pick up about 60% of the lien.

21.7 Mediation

Pacific Horizon, Delta-Apex, Air Unlimited, and Judge & King submitted the dispute to a neutral mediator. The notes of the mediator follow:

> Bust in a change order. Air Unlimited submitted a proposal to substitute a solar water heating system for the gas-fired design. Delta-Apex forwarded the change order proposal and Judge & King approved it, but everybody forgot to include the cost of the electrical hookup. Mediation commenced at 9:00 a.m. and finished at 11:30 a.m. when a settlement agreement, written in longhand, was signed by all parties.

21.8 Settlement Agreement

> This document memorializes a settlement reached at mediation on December 19, 1998. The parties are Pacific Horizon, Inc., owner, Judge & King, AIA, Architects, Delta-Apex Construction, contractor, and Air Unlimited, subcontractor.
>
> 1. Owner shall make no extra payment in respect of any change order related to solar heating, but will pay the full contract price with no adds or deducts.
>
> 2. Delta-Apex will pay the roofing subcontractor Tops Roofing and the plumbing subcontractor Lilac Plumbing the full amounts of their change orders: $368 and $1,002.80.
>
> 3. Delta-Apex Construction will pay Air Unlimited the sum of $3,780.
>
> 4. Judge & King does not admit liability, but will pay Air Unlimited the sum of $1,000.
>
> 5. Upon receipt of those sums, Air Unlimited will waive its overhead and profit of 15% ($717) and will release of record its claim of mechanics lien.
>
> Dated: December 19, 1998
>
> _____
> Pacific Horizon
>
> _____
> Judge & King, AIA
>
> _____
> Delta-Apex Construction
>
> _____
> Air Unlimited

22

Pacific Horizon III
Value Engineering: After Completion of Construction

Pamela Lyons was beginning to detest her job as building manager for the sparkling new Pacific Horizon, an upmarket rental office building in Malibu. At first, the constant excitement and activity of accommodating the needs of the original tenants and getting them moved in made the job intensely interesting. She liked meeting new people and dealing with the architects and builders. They treated her with respect and she was welcome in every office in the building. She enjoyed the upbeat banter with the carpenters, electricians, and painters finishing up the tenant improvements.

Now that the building was close to a year old, and the initial flurry had subsided, some unpleasant facts were becoming increasingly obvious. The stream of new tenancy had dried up and stagnated at about 60% occupancy. They hadn't signed a new lease in 5 months. When the construction was first completed they'd had plenty of interested lookers and inquiries. But now the real estate brokers and rental agents completely ignored the building. It was common knowledge among the office leasing fraternity that there was something seriously wrong with the air conditioning system at Pacific Horizon, not to mention various other rumors.

There was a constant flow of tenant complaints about the environmental conditions in their suites, environmental conditions being a euphemistic expression for too hot or too cold. Some of the tenants were even talking to their lawyers about withholding rents, claiming damages, or rescinding leases.

Alexander Brink, President of Horizon Investment Company tried to stay on top of everything, telephoning Pamela daily and visiting the building once or twice a week. He had to hold it all together, as this was an important investment of his company. He tried to avoid the tenants, and Pamela kept him up to date on the status of complaints. Alex couldn't help noticing that his personable, normally enthusiastic building manager was growing more somber by the day. She was beginning to stay home sick fairly often, 4 or 5 times a month. He intended to speak to her about it if she didn't improve markedly in the very near future.

Alex was not used to such severe vacancy problems, as Horizon's two other buildings had leased up quickly and had experienced very little tenant turnover. In these trying days at Pacific Horizon he spent a lot of his time talking with real estate agents as well as Leo King, the architect, and Mike Van Buren of Delta-Apex Construction. He also spent considerable time with Alan Condon of Air Unlimited and some of the other subcontractors. But this job fell mostly on Pam's shoulders. The building had an almost constant presence of air conditioning workers servicing the equipment in the penthouse and the basement. They examined the heating and refrigeration equipment and adjusted the ductwork, the supply diffusers, the return air grilles, and the thermostats. Whenever a complaint came in from a tenant, Pam would call Air Unlimited and most of the time they would send someone to look around and make some adjustments. But, whatever they did, it didn't help much, if any. The complaints kept coming in.

***

The air conditioning situation first came to Pamela's attention when Karen Brent got fired. Karen had been Malcom Wilson's secretary for 15 years. They were the first tenants to move into the building. Malcom is an independent consultant economist to the scrap metal industry. They had a fairly small second floor suite consisting only of two fairly large rooms. His office was 16 feet by 22 feet and faced the southerly oriented windows. Her office, which also served as the reception room, was 16 feet by 18 feet and had windows facing a cantilevered walkway and the building's three-story skylighted atrium. A thermostat in Karen's office determined the air conditioning temperature in the Wilson suite.

Karen arrived in the office each morning promptly at 8 o'clock. She would check the thermostat and, usually feeling cold, would set it at a warmish 75 degrees. About an hour later, Malcom would arrive and, with a cheery, "Good Morning, Karen," would walk directly to the thermostat. He'd generally say something like, "I don't know how you can stand it when it's so hot in here. It's like an oven." He'd then reset the thermostat to a cooler 65 degrees and continue on in to his office. When he was out of sight and had his office door closed, she would turn the thermostat back up to a comfortable 75.

After about 20 minutes, his door would burst open and he'd come out to examine the thermostat. He'd turn it back down to 65 degrees, glance sternly over to Karen's desk, and return to his office. In a few minutes, she would reset it to her choice. Shortly, he would be out again to look at the thermostat. "Karen, if you're so damn cold, why don't you get yourself a sweater?" He'd reset it and go back to work, slamming the door behind him. A few resentful minutes later she'd again reset it. It would go on like this every day until, in the second week of their occupancy, Malcom came in one morning and, as usual, he reset the thermostat without speaking to Karen and went on into his office. He was back in less than a minute with a roll of scotch tape and wound it around the thermostat completely taping it into position. In a stern voice he instructed, "Now, Karen, you leave this thermostat alone." He returned to his office and closed the door. She immediately rose, went to the thermostat, unwound the scotch tape, and reset it to 75 degrees. She rewound the scotch tape and went back to the report she was working on.

Malcom came out of his office in about a half hour. His necktie was off and his sleeves were rolled up. Standing at the doorway, he said, "There must be something wrong with this air conditioning system. Will you please call the building office and ask them to have someone take a look at this damn thermostat?"

Karen knew that it would be pointless to turn in a complaint since she had changed the thermostat, so she let it slide. That afternoon, when Malcom returned from lunch, he commented to no one in particular, "This place is like a damn blast furnace. It's cooler in a steel mill. That thermostat's gotta be fixed." He then unwound the scotch tape to reset it. When he exposed it and saw that the thermostat had been reset to 75, he moved it back to 65 and angrily turned to Karen. A monumental disagreement ensued and she quit her job about a minute before she would have been fired. On her way out of the building, Karen stopped off at the Office of the Building and told Pamela what an ungrateful, uncaring, selfish brute her ex-boss was.

"All he ever cares for is himself, the lousy bastard. I hope he freezes to death in that refrigerator he calls an office."

Pamela assumed that Karen and Malcom's air conditioning complaints were simply a disagreement over the temperature of the office. With Karen gone, everything would be all right.

During the next couple of weeks, while he was interviewing prospective secretaries, Malcom had no problem with the temperature now that he had sole control of the thermostat.

***

Omicron Media, one of the most important tenants in the building, occupied over half of the third floor, 7680 square feet. Their office manager, Sally Blunt, was very forthright in transmitting any dissatisfactions they had with their leased premises. She was gradually driving Pamela up the wall. She telephoned Pamela at least once a day, and often more than that, to report imperfections. The standards of the janitorial maintenance service were never up to her exacting criteria. If a wastebasket was unemptied, Pamela was the first to know. Sally invariably wrote memos to confirm all conversations, whether telephonic, or in person. Almost daily she transmitted complaints about the unsatisfactory room temperatures in various parts of their suite. It was always either too hot or too cold. Sometimes both at the same time. The window offices would be too cold and the inner spaces too hot, or possibly the reverse. Sally always used colorful language and expressed herself in extremes as she felt it would be more effective. It was always "hotter than hell" or "cold as an iceberg." She blamed all personnel illnesses on the building's heating and cooling systems.

Pamela usually had lunch at Pacific Patio, the ground floor restaurant. She always sat at the same table at the far end of the atrium patio where she could see what was going on. Sally often dropped by her table to drop off memos or to bend her ear. "Hi, Pammy, here's a memo of this morning's phone call. Will you drop by our office after lunch so I can show you something? I know you will, Pammy."

"Sure, Sally. I'll drop by." Pamela abhorred being called Pammy. But what could she do? She lost interest in the rest of her salad. She didn't finish it or the coffee. She went back to her office on the Second Floor. Twenty minutes later the phone rang. It was Sally. "Hi, Pammy. Aren't you coming up? You said you would."

"Yes, I'll be there in a few minutes." A headache was coming on. She might have to leave early.

When she arrived in Omicron Media's waiting room, a concerned Sally was waiting for her. "Hi, Pammy. We've got a problem. A gale is blowing in Mister Klinggmann's office. He's our Creative Director, you know."

Pamela figured it was just another of Sally's overstatements. She followed Sally into Mister Klinggmann's office. Sally steamed forward like the QE Two, clearing the way. "Pardon us, Mister Klinggmann, Pammy and I've come to rescue you." Pamela cringed and smiled bravely. The headache worsened.

When they got close to the window wall, Pamela noticed that the drapery was billowing out and rustling. She saw an open slot of bright blue sky at the top of the grey tinted glass. The glass seemingly had dropped in the frame leaving it completely out of the vinyl gasket at the top. There was an open slot about 1/2 inch high the full width of the window. Indeed, a blast of cold air was rushing through and noticeably lowering the room temperature.

Pamela had become practically immune to Sally's constant stream of complaints. But this one was different. It looked like a serious problem. Pamela asked, "When was this first noticed?"

Mister Klinggmann answered, "About a week ago, I heard a whistling noise at the top of the window and I noticed that the glass had slipped down and out of that rubbery stuff that holds the glass. Each day the crack seems to get a little bigger. Today the wind came up and now it's cold and windy in here. If the crack gets any larger, I expect birds'll be flying in." Pamela hated tenant complaints clothed in sarcastic terms. But she smiled appreciatively.

As they were leaving, Sally said, "I knew you'd want to know about this, Pammy. I'll send you a memo later this afternoon."

Before leaving the Omicron office, Pamela and Sally examined all of the window offices, and found three other windows that had a similar slight open crack at the top. They were still in the preliminary whistling stage.

Pamela returned to her office and called Alex. She briefed him on the new problem, and asked, "What do you want me to do about this?"

"Call Mike at Delta-Apex. Tell him to get hot on this. And let me know what transpires."

She asked one of the air conditioning mechanics to bandage the tops of the windows with duct tape as a temporary measure to keep the wind out and the air conditioning in.

* * *

Pamela assumed that Omicron Media's environmental temperature complaints were really caused by the window glass slippage problem. So, it was somewhat of a disappointment when the third tenant moved into the building. Computer Systems Incorporated was in their new office only two days when they first reported their dissatisfaction with the thermostatic controls. Although the office consisted of over 4,000 square feet and 16 offices and spaces, there were only 3 thermostats. Of the 23 people in the office, the only ones that were happy were the outside sales personnel who phoned in every day. When Pamela visited the office after the first complaint, she silently had to agree that the situation was pretty bad. She told Mister Grimsby, CSI's Office Manager, that she'd get someone to look right into it.

The next four tenants had much the same problems and none were the least bit reticent to report their dissatisfactions to Pamela. Her headaches were growing in intensity and coming on more frequently.

***

On her way through the lobby one morning, she stopped off in Pacific Patio to get a cup of coffee to bring up to her desk. There, Sally Blunt from Omicron Media neatly collared her. "Hiya, Pammy. What's happening? I'm getting complaints from practically everyone in the office. The windows are still slipping and the place is always sweltering or freezing. And what about all the other problems I've sent you memos on?"

"We're working on them." Pamela was getting sick of being the scapegoat for others, she didn't know whom. She had to present a reasonable front for this miserable building but hadn't the stomach for it much longer. She had totally lost the zest for this job. Upon returning to her office, she called her boss.

"Look, Alex, I can't stand this any more. The tenants are sick of my stupid alibis. They don't take me seriously anymore. They want action. Something positive has to be done about this damn place. There's been nothing but buck passing between the architect, the contractor, and the subcontractors. Everyone agrees that the air conditioning system isn't worth a damn but no one does anything about it. The glass is slipping in the windows and all we've got so far is duct tape. What're we gonna do?"

"I'll do what I can, Pamela, and I'll get back to you."

Pam thought, "I'll bet. I won't hold my breath." She went home early.

***

Alex walked into the adjoining office to talk it over with his associate, Cyrus Flint, Financial Vice President. After a lengthy discussion of all that had thus far transpired, they decided it was time to call their legal adviser. "Maybe Virgil will have some useful ideas."

"Yeah," agreed Cyrus, "maybe we'll have to get tough with the architects and contractors."

"I'll call him now, Cy." Alex went back to his office.

Virgil Byron Jennings was a sole practitioner, representing a number of real estate developers, lenders, and investors. He was intimately acquainted with the problems peculiar to real estate investment. He had represented Horizon Development Company since its inception 6 years previously. Although Virgil had a satisfactory relationship with Alex and Cy, he knew that they only called him after a problem had developed. They seldom asked him to review anything in advance. He knew they had entered into the architectural and construction contracts for Pacific Horizon without his review of the contracts. They hadn't run any of the leases past him either. As an experienced lawyer, he realized that nobody likes to see legal fees mounting up, but Alex and Cy didn't seem to consider that the relatively small fees up front could often prevent avoidable problems and larger fees downstream.

"Virgil, I need to talk to you."

"I haven't heard from you in a long time, Alex. What's up?" Virgil had heard about the physical difficulties and tenant problems at Pacific Horizon. Who hadn't?

"Virgil, we have some serious difficulties at our new building at Malibu." He described the physical problems, the deteriorating attitudes of the tenants, and the seeming inability of the architect, contractor, and subcontractors to solve any of the problems.

Virgil referred to his calendar. "I can see you tomorrow afternoon at 2 o'clock, here in my office. Okay?"

"Good, Virgil. We'll be there."

* * *

Alex and Cy unloaded the whole tale of woe on Virgil, who listened attentively and took copious notes. After nearly an hour of hearing about one sad situation after another, Virgil suggested, "There's not a whole lot we can do before we know what exactly is wrong with the building and who's likely to be responsible for it. The best thing we can do is to call Frank Grimm."

Alex and Cy looked at each other. Alex reacted first, "Who's Frank Grimm? Have you heard of him, Cy?"

"No. The name's new to me."

Virgil explained, "Frank Grimm is a forensic architect. He'll examine the construction documents, inspect the building, study the applicable standards, and report back to us."

"That'll cost a lot of money, won't it?" asked Cy.

"Well, it has to be paid for. No question about that. The important thing is, we need the information. Without it we won't know which direction to go."

Alex asked, "Well, why can't we get Leo King or Mike Van Buren to do it? They already know the ins and outs of the building. They're acquainted with everything."

Cy added, "We'll have to pay Frank Grimm for the time he'll need to get up to speed."

Virgil explained, "It would be best to get an objective person who's not trying to cover up or minimize any of his own mistakes. Frank is a real expert and it won't take him very long to get acquainted with the whole situation."

"Well, okay, Virgil, if you think that's the best thing to do."

"All I need now is your approval to hire him."

"Okay, you've got it."

* * *

Frank Grimm, FAIA, was making his first visit to Pacific Horizon. He was somewhat out of breath from the hike up the flight of stairs to the second floor Office of the Building. Pamela told him that the building's only elevator was out of service temporarily for maintenance but should be back in action in a couple of hours. Frank knew from studying the construction documents that one elevator had been eliminated in the value engineering phase before the contract was signed.

Pamela gave him a comprehensive tour of the building showing him all of the problem areas and letting him into the various mechanical and electrical equipment rooms with her master key. He stayed on a couple of hours longer taking closer and more intensive looks at items of particular relevance. He took numerous photographs and measurements and made several sketches. When he was through with his physical examination he returned to Pamela's office and questioned her about all that had transpired in connection with tenants' complaints. She gave him the names and addresses of all the subcontractors and maintenance companies.

Frank Grimm spent the next couple of days interviewing Leo King, Mike Van Buren, Alan Condon, and the key people in all of the principal subcontracting firms involved in Pacific Horizon. Then he called the Law Office of Virgil Byron Jennings.

"Virgil, this is Frank. Can I come in to see you?"

"Of course. Have you written your report yet?"

"Not yet. I thought it would be advisable for us to have an informal discussion before I run the bill up any higher.

Virgil asked the architect, "Do you mind if I invite the owner, architect, and contractor to attend? It'll cut down on a lot of explanations later. Also, they can ask questions if they wish."

"That's a good idea. It's okay with me."

"Then how about tomorrow afternoon at 4 o'clock?"

* * *

Although Frank arrived on time, the others were all seated and waiting for him in Virgil's conference room. When Frank was seated, he opened his battered leather briefcase and extracted a thick sheaf of notes, sketches, and photographs.

"I thought it would be a good idea if we had an informal discussion before I write my report. A complete report takes a lot of time and can be fairly expensive." Alex and Cy exchanged apprehensive glances, but both were relieved that the expensive report might be avoided.

Virgil agreed. "That's a good idea. The report won't even be needed if no lawsuits are filed."

The others all heartily agreed that there was no need for anyone to file any lawsuits.

Frank started, "I've learned a great deal about Pacific Horizon in the last few days. The most obvious problem seems to be the air conditioning. The main difficulty appears to be insufficient temperature control of individual spaces. Some areas of dissimilar heat load are improperly grouped together. There's also insufficient central system capacity in the heating system, although the refrigeration capacity seems adequate."

Mike asked, "Are you a mechanical engineer?"

"No, Mike. I've only made rough calculations. I'll confer with mechanical, electrical, and structural engineering consultants before finalizing my report."

Alex directed a question to the building's architect. "Leo, how did this happen? I thought your engineers knew how to design environmental systems."

"Our original design would have been okay, but you let Air Unlimited reduce the quality of the system to save money. When our engineers and we objected and pointed out the deficiencies, you and Cy overruled us. You said we were being too idealistic."

Alex turned to Mike, "I thought Air Unlimited would give us a good system They said the system was overdesigned."

Mike replied, "Well, you got the $83,000 credit. Surely, you didn't think that the system would be the same if we took that much money out?"

Alex looked at Cy who was studying his note pad.

Alex asked Frank, "What were you saying about the heating system being inadequate?"

"Well, in the process of comparing the design drawings and specifications with the installed system, and talking with the Air Unlimited people, I discovered that they reduced the boiler size expecting the solar energy panels to be installed. The smaller boiler was already in place before the solar panel change order was cancelled."

As each new bombshell was dropped, the note taking around the table increased in ferocity.

Alex asked, "Can the air conditioning system be rectified?

"Yes, it can be altered and improved, although it would be hard to bring it up to the standard of the original design. For one thing, some of the zoning flexibility will have to be restored by the use of electrical reheat coils. This is inefficient, not only because of having to reheat air that has already been cooled, but the electrical system is already up to its complete capacity. Nothing electrical can be added without radically changing the system. New panels, conduits, and main feeders. This will be expensive."

Virgil, with a sidelong glance at his wristwatch, suggested, "Let's move on to the next item, gentlemen. We can defer lengthy discussion on individual items to another meeting. Okay, Frank?"

"The window wall problem is next. The glass is slipping down in the frame. It starts on the first floor and, over time, works its way up towards the top floor. The horizontal aluminum member that supports the glass and separates the glazed panels from the opaque glass spandrel panels sometimes twists and allows the panel above it to drop slightly. The open crack on the top floor keeps getting wider up to a maximum of about 3/4 inch."

Leo asked, "How does this happen?"

"It is an inherent problem in some generic window wall parts if the ends of the snap-in horizontal members aren't securely screwed to the verticals."

Alex asked, "Leo, why did you specify that type of material?"

"We didn't. This was chosen by the glazing contractor."

Alex looked at Mike who shrugged his shoulders. Alex said, "I know, we saved a lot of money on it."

Frank added, "We may as well take the skylight over the atrium next."

Alex asked, "What's wrong with the skylight? I didn't know anything's wrong with the skylight, did you, Cy?"

"No, I haven't heard anything about it," Cy answered quizically.

Frank continued, "I noticed that the skylight was leaking. You wouldn't particularly notice it because of the patio landscaping. But it'll get worse. Also, the openable sections specified on the drawings have been omitted. This will cause future problems and will have to be rectified. Also the double glazing has been reduced to single glazing. This will add to the heat loss problems in the heating season and increase heat gain in the cooling season."

Alex observed, "This building's getting shoddier by the minute." Addressing Mike, he asked, "Why did you let the glazier make these inferior substitutions?"

Mike coolly replied, "To save money. You approved it all. Don't you remember? You said the building was too idealistic. You'd rather have a building you could afford than one that was perfect but couldn't be built."

Virgil, hurrying it along, "We can discuss that later. Next item, Frank."

"I couldn't help noticing the exterior paint job. In several places on the office building as well as on the parking garage, the paint is cracking and bulging. Large pieces are bridging over the substrate. It's only held in place because it is a continuous membrane. As it gets older, it will soon fall off in large chunks. It's a fairly thick coating of something I've never seen before."

Mike said, "Well, there's no problem there. We'll just call Manny Hughes, our painting sub. He'll take care of it. He's good for it."

Frank said, "I've been trying to locate him. He seems to have gone out of business. No forwarding address."

Mike suggested, "Then we'll contact the manufacturer. The paint's guaranteed to be long-lasting. That's what Manny said."

Frank added, "I tracked down One World Paint Manufacturing Company, the maker of the paint, and they too are gone. Bankrupt."

Cy, in a feeble attempt at optimism, guessed, "It shouldn't be too big a deal to repaint the building, should it?" He looked around the table expectantly at the construction industry people.

Frank explained, "The present coating will all have to be removed first. That might be a tough and lengthy job. Then the surfaces will have to be properly prepared and repainted, a primer and two finish coats. It'll cost a lot more than the original job."

Alex asked, "How much did we save on that?"

Frank concluded, "There are a few miscellaneous items, like the asphaltic concrete paving subsiding in the driveway and parking lot and the garage floor waterproofing allowing rainwater to seep through and drip onto the cars parked on the lower levels, and a few more minor items we can discuss later."

Virgil announced, "Well, it's getting late and we've been talking a long time. We'll plan another meeting."

As they were all leaving, Virgil asked Alex and Cy to remain for a moment. The three were left seated at the circular table.

Virgil tried to summarize the realistic position for his clients.

"Contractors are obligated to carry out the construction depicted in the contract documents. We can hold Delta-Apex to that standard. If they will not voluntarily assume that obligation, we can make a legally enforceable claim through arbitration.

"However, we can't ask them to comply with any higher standards in any situation where you approved the downgraded substitutions. The architect warned you about the negative effects of each and every change that was made from the original documents. You overruled the architect and took a monetary credit.. But the contractors were then obligated to follow the substitute designs, use appropriate materials, and produce results suitable for their intended purposes.

"I'll ask Frank Grimm to clarify which defects are the contractor's responsibility and to identify which should be voluntarily upgraded at your own expense."

Cy looked at Alex and wryly observed, "Maybe we got carried away with the value engineering."

* * *

Points of Law 22

PACIFIC HORIZON III
Living with the Building

22.1 Defective Material

Instead of supplying a regular paint job the painting subcontractor sold Pacific Horizon on a thick coat of exotic material. The defective material will have to be removed and replaced.

Parties potentially responsible are the architect, the prime contractor, the painting subcontractor, and the paint manufacturer.

Architect Judge & King is not responsible because it didn't specify the defective material and probably even warned Pacific Horizon against using it. Pacific Horizon, however, accepted the change in order to reduce cost. An architect is not responsible for construction defects that are caused by deviation from the architect's plans and specifications.

How about prime contractor Delta-Apex? Its obligation to Pacific Horizon was to provide the paint job that was specified by the contract documents. They did so and they are therefore not responsible for the unsatisfactory result.

It's true that as a licensed contractor Delta-Apex is responsible for complying with the requirements of the building codes. Building codes, however, are written for public safety, and since public safety is not involved, it's unlikely that use of the unfamiliar paint material violated any building code requirement.

Delta-Apex, as prime contractor, is responsible for the activities of its subcontractor. Is it possible for Pacific Horizon to make a valid claim against the subcontractor?

22.2 Misrepresentation, Fraud, and Breach of Warranty

We may infer that when the subcontractor sold Pacific Horizon on the idea of using the exotic material it made some representation as to the performance of that material. It may, for example, have represented that the material would "last a lifetime" or that the paint job would "last for ten years." Such a misrepresentation of fact would give Pacific Horizon a cause of action for fraud. The painting subcontractor has taken bankruptcy, but bankruptcy does not discharge a fraud claim.

Pacific Horizon might also contemplate making a claim for fraud or breach of express warranty against the manufacturer of the paint. A claim of fraud or breach of warranty can be founded on misstatements made in advertising brochures. A statement in the manufacturer's brochure, if false, will support a claim for breach of warranty and, if intentionally false, supports a claim of fraud.

22.3 A Worthless Claim

A claim that is legally sound may be worthless if it is to be asserted against an irresponsible party. Here, the painting subcontractor has taken bankruptcy and the paint manufacturer has gone out of business. Next time around, Pacific Horizon will probably get a credit report and check references before entrusting work to a subcontractor whose main selling point is that he can do the job cheaper than anyone else could.

23

ABC Warehouse XI
Substitutions in Lieu of Specified Materials

"It seems that whatever we specify, the contractors always want to use something else!"

Ivor Judge, FAIA, was sounding off in the drafting room of Judge & King for the benefit of any who would listen. He was painstakingly researching the suitability of several suggested substitutions on the ABC Warehouse project. These substitution requests, although originating primarily with the subcontractors, were submitted to the architect's office by the general contractors who were bidding the job.

His partner, Leo King observed, "It only seems that way, Ivor. You can't blame the contractors for wanting to use familiar materials or those of manufacturers they're accustomed to working with."

Leo always looked at things rationally and tried to be sympathetic to the contractor's viewpoint. Both architects recognized that subcontractors could get better prices from suppliers that they most often dealt with. They were aware that some manufacturers and suppliers paid bonuses or rebates to subcontractors who met certain purchasing targets over a year's time. Naturally this creates a strong incentive for them to limit purchases to such companies whenever possible.

The ABC Warehouse job was a 100,000 square foot concrete tilt-up warehouse building and 5000 square foot office wing on a 10-acre site in a suburban master planned industrial park near Los Angeles. The job was now out to bid and the general contract proposals were due in a little over a week. Ivor was working on an information bulletin that had to be sent out to all bidders by the end of the day. Judge & King always observed the formality of collecting the bidders' questions and then answering them all at one time in a bulletin issued simultaneously to all the bidders. This provided a level playing field with no advantage to any of them.

The substitutions approved so far included two additional paint manufacturers, a T-bar ceiling system, lighting fixtures, fire doors, and roofing materials. Several requested substitutions had been rejected as not being equivalent to the specified items. This third and final bulletin also clarified all other pending questions raised by the subcontractors, suppliers, and general contractors. Bids would be due on November 2, 1989.

* * *

When the bids were opened, Hyde Construction Company, at $2,100,437, was the low bidder. The general contract was duly signed on November 10, 1989, in a meeting at the offices of Judge & King. After the contract was signed, mutual congratulations exchanged, and hands sincerely shaken all around, Ivor Judge announced that the pre-construction jobsite conference promised in the supplementary general conditions would be held on the following Monday morning at 9 o'clock. He asked George Hyde to have his superintendent and representatives of all the key subcontractors present. Hyde had attended many other such jobsite conferences and knew pretty much what to expect.

* * *

The pre-construction jobsite conference got underway promptly. Ezra Field introduced all the subcontractor representatives to the architect and to each other. Business cards, in ritual fashion, were exchanged all around. The architect, using his prepared agenda, chaired the meeting. He was also taking notes to aid in preparation of the ensuing Jobsite Observation Report that would be sent to the owner and all in attendance. George Hyde, his superintendent Ezra Fields, and representatives from most of the major subcontractors were present. All were enthusiastic about starting a new job and the buoyant air of optimism was palpable. Ivor started down the list, ticking off the agenda items as they were discussed.

"The next subject is Substitutions. We'll entertain requests for substitutions only for the next couple of weeks. After that, you'll play hell getting a substitution approved. The only legitimate reason for a substitution at this point is unavailability of a specified item. You had your chance during the bidding period for substitutions on other grounds such as our having specified outdated products or omitting to specify all the acceptable brand names."

Ivor was laying out the ground rules for substitutions on this job. He explained the importance of getting all shop drawings underway immediately and all materials and equipment ordered.

"Order your long-lead materials immediately so we'll know as soon as possible if any specified product is unavailable. In this way, there'll be no legitimate excuses for not being able to obtain the specified materials and equipment on time and the job won't be held up. If you find that the specified materials are not available for any reason, we'll review proposed substitutions. If you come to us weeks downstream looking for substitutions on account of non-availability, we'll know you didn't place the material orders. Does everybody understand?"

All present nodded assent. Several subs had reasonable questions. Kyran Zeno, the representative of Splendid Air, the A/C contractor, asked, "What'll have to be submitted for approval of a substitution?"

Ivor explained, "We'll need complete manufacturer's literature including all sizes, capacities, model numbers, and engineering characteristics. Everything we need to know to enable us to compare it with the original specification. The proposed substitution must be substantially similar to what was specified. Read the definition of 'or equal' in the Supplementary General Conditions."

Jasper Kass, job foreman for Aardvark Plumbing, asked, "You specified copper piping for the hot and cold water lines. Is it okay to use ABS plastic?"

"No. Any more questions?"

The conference agenda moved on down to other topics of interest to the contractor and subcontractors in getting the job into efficient operation. Construction was started a week later on November 20, 1989.

* * *

In the following week, Judge & King started receiving submittals of shop drawings, product data, and samples from Hyde Construction Company. These had been submitted to Hyde by the subcontractors and suppliers as required by the trade sections of the specifications. The procedure for submittals was specified in the general conditions and the supplementary general conditions of the contract.

The AIA General Conditions, Document A201, Subparagraph 3.12.5, requires the contractor to review and approve shop drawings, product data, and samples before they are sent on to the architect for approval. This would theoretically cull out inappropriate requests or those that would not fit in with field conditions as construction progresses.

Among the product data submittals were Aardvark Plumbing's manufacturer's catalogue cuts and specifications for all the plumbing fixtures and a list of all the materials to be used. Everything appeared to be in good order, strictly in accordance with the architect's drawings and specifications, so Ivor Judge duly approved the submittal, dated it, assigned it the next shop drawing approval number, and returned two copies to Hyde Construction.

* * *

Ivor visited the project every Tuesday morning at 8 o'clock. Subcontractors with problems would plan to be there to confer with the architect and to get their pending questions answered. In the ninth week of construction, on Ivor's regular Tuesday Morning job-walk with Hyde's superintendent, he noticed a two-inch steel gas line running from the gas meter vault to the corner of the building. It was lying on the bottom of a two-foot deep trench that was being backfilled. He asked the superintendent, "Ezra, who's the manufacturer of that piping?"

"I believe the specifications require Universal Steel, Scott Industries, or Midwest Piping."

"That's right. That's what's specified. But what is this?"

"I don't know, Ivor. Maybe we can see a trade mark somewhere." He was down in the trench looking for some markings. The asphalt and paper wrapping prevented easy examination.

Just then the plumbing foreman, Jasper Kass, showed up. He looked extremely concerned, seeing that the architect and job superintendent were examining his work. He smelled trouble and warily asked, "Can I help you with something?"

Ezra replied, "Yeah, Jasper, what kind of piping is this, anyway?"

"It's the best gas piping money can buy. We use it on all our jobs."

Ezra persisted, "Yeah, but what kind is it?"

"It's called Superior piping."

"Who makes it?"

"It's made by East Asian Superior Steel."

Ivor asked, "I've never heard of that company. Where's it located?"

The plumber replied, "I believe their main plant's in Northern Cambodia. It's the finest steel piping of its type made anywhere in the world."

"Well, it may be, but it's not an approved substitution, so it'll have to come out and be replaced with one of the specified products." Ivor was jotting notes for his Jobsite Observation Report.

The plumber was looking uncomfortable. Jasper knew that they shouldn't have tried to get away with it. That's what he had told his boss. But he felt an obligation to Aardvark Plumbing to put up an argument anyway. Jasper replied, "Other architects always approve this piping. It's ridiculous to remove it. This is equal to or better than any of that piping that was specified. We use this piping all the time."

Ivor explained, "According to the AIA General Conditions, Paragraph 3.5.1, any unauthorized substitutions may be considered defective. You might as well stop backfilling and get that piping out of there." He turned to the superintendent and firmly warned, "And, Ezra, make sure that it's removed from the jobsite and replaced with one of the specified brands."

"Okay, Ivor. I'm really sorry about this. It escaped my attention. I'll keep a better watch on these things from now on." He scowled at Jasper who was just quietly standing there thinking about what his boss would say.

When Ivor left, Ezra turned to Jasper and read him off. "You idiot, what're you and Emmet trying to do? I get blamed because you clowns are trying to cheat the job. Now the architect thinks I don't know what I'm doing or that Hyde Construction is in on your shoddy scam. How much'll you save after you've redone the job properly?"

Jasper was standing there uncomfortably, trying to appear credible and contrite. Just then, Emmet Pismire, owner of Aardvark Plumbing drove onto the jobsite and, seeing his foreman engaged in earnest conversation with the job superintendent, walked over to join in. Ezra Fields greeted him with, "I've had just about enough of your shabby crap, Emmet. From now on you do your job right or I'll can your sorry ass off the job." He left, walking angrily towards his job office shack.

Emmet asked Jasper, "What was that all about? I've never seen Ezra so steamed."

"The architect saw the Cambodian gas piping. He says it all has to come out. We hafta use one of the specified brands of piping."

"The specifications say 'or equal' so that's what we used. Why didn't you tell him that?"

"Look, Emmet, you don't hafta convince me. You gotta convince the architect. And he's not buying it."

Emmet blamed Jasper for the situation. "It's all your fault, Jasper. I told you to backfill that piping last Friday."

So Emmet reluctantly ordered removal of the Cambodian gas service piping and replacement with one of the specified brands. Emmet considered himself lucky. Although it cost well over $1200, it could've been a helluva lot worse. He'd just have to make it up on something else.

* * *

Ivor was interrupted by his secretary on the intercom. "Allen Brady is on the line and he sounds agitated."

Ivor thought, "What is it this time? He's always agitated."

He punched the lighted button and cheerily inquired, "What's new, Allen?"

"I'll tell you what's new, Ivor. I just received my copy of your latest Jobsite Observation Report and I see where you condemned 200 lineal feet of steel gas piping because of some stupid technicality."

"That's right, Allen. That piping wasn't in conformance with our specifications. It had to come out."

"But, why? You could've negotiated a coupla hundred bucks credit from Aardvark Plumbing and leave the piping in."

"But Allen, we don't know anything about that piping. It could fail in a few years and you'd have to pay for replacing it. Leaking gas could even cause a fire or an explosion. There's a lot of substandard piping around with thin spots and other serious defects and irregularities not found in the specified piping."

"Well, Ivor, it was just an idea."

"Thanks for calling, Allen."

At least Ivor knew that someone was reading his Jobsite Observation Reports.

* * *

A couple of weeks later Ivor received a request from Aardvark Plumbing for the approval of a Superhot Water Heater, Model # 86 on the basis that the specified water heaters on the previously submitted material list were now unavailable. The request had been forwarded by Hyde Construction Company. Ivor immediately called Sam Trent at Hyde Construction and asked, "Sam, how come Aardvark Plumbing wants to use a substitute water heater?"

"They say they've called all over town, Ivor, and can't find any of the specified heaters. Emmet said they need a quick approval so the job won't be held up."

"Did you check with any of the plumbing wholesalers to see if the specified heaters are in stock?"

"No, Ivor, I'm up to my elbows in estimating a new office building job right now. Sub-bids are still coming in and our bid is due in less than an hour."

"Did you check to see if the Superhot Water Heater is equal to what's specified?"

"Look, Ivor, I haven't got time to check everything. The phones are all ringing and the fax is running steadily. I gotta get back to this bid, Ivor. Sorry."

Ivor then dialed an old friend at Western States Wholesale Plumbing Supply and asked him about the availability of the specified water heaters, United Thermal, Commercial Union, or Dynamic Pacific.

"We're up to our ears in all models of all three of those brands. We can ship as many as you want immediately. We'd love to get rid of them."

"Thanks, Rob, that's what I suspected. One more question. Why would any plumber want to use Superhot Model # 86? Frankly, I've never heard of them."

"They don't make them any more, Ivor. Superhot went out of business about three years ago. Too many claims of defective burners and controls. They went bankrupt. Your plumber must have found some of Superhot's old stock at a good price or was stuck with one they couldn't sell."

* * *

Jasper Kass was in Aardvark Plumbing's warehouse loading materials at the time Emmet Pismire received Ivor's disapproval of the Superhot water heater from Hyde. Emmet shouted to him, "You might as well take that water heater off the truck, Jasper. The architect didn't approve the substitution."

"Do you blame him? This water heater's no damn good. Why don't you junk it? We'll never sell it." Jasper and his helper were manhandling it off the truck.

"Just put it over there against the wall. We'll sell it to someone eventually."

"What'll we do for a water heater now?"

"On your way to the job, swing by Western States Plumbing Supply and pick up one of the specified water heaters, whichever is cheapest."

Emmet stood in the doorway of his small warehouse watching his ancient truck go down the street. He was wondering what additional economies he could put into effect to improve profitability and keep the job in the black. That gas piping fiasco had cost him dearly and the water heater switch had now fallen through. Fortunately Jasper had gotten the main water piping buried before anyone saw it.

About a half-hour later, Jasper phoned Emmet from the jobsite to tell him that Western States Wholesale Plumbing wouldn't let them have the water heater on credit. The Aardvark Plumbing account was frozen until it was brought current.

* * *

When Ivor told his partner about the Cambodian gas piping and the Superhot water heater, Leo advised, "You better keep a wary eye on everything Aardvark does from now on. They're liable to try anything to pick up a buck."

The following Tuesday morning, ABC's president, Allen Brady, was present when Ivor walked the job with Ezra Field and George Hyde. Ivor asked Ezra, "Did you see the water main piping from the meter before the trench was backfilled?"

"No, it was suddenly backfilled before I had a chance to examine it." Ezra felt he was under criticism again. He fervently promised himself to watch Aardvark's operation more closely from now on.

Ivor suggested, "Well, I think we ought to have a test hole dug and see what this piping looks like."

George said, "I saw Emmet here a little while ago. I'll get him over here. We can talk about it."

In a few minutes he returned with Emmet and Jasper.

Ivor asked them, "What kind of piping did you use for the main water line from the meter?"

Emmet promptly answered, "The specified piping, of course. Isn't that right, Jasper?"

"Yeah, sure thing, Emmet. That's what we used all right. Absolutely. I'm sure of it."

"Well," said Ivor, "I didn't see it. So, we better dig it up and have a good look."

Emmet, defensively, protested, "Now, wait a damn minute. Somebody's gonna hafta pay for digging that hole. We're not doing it for nothing. And it's gonna hold up the job. We haven't got time to be digging holes when we should be doing plumbing work. We've got a schedule to meet. This is ridiculous." He was righteously indignant.

The architect removed the Project Manual from his briefcase, turned to the AIA General Conditions, Document A201, Subparagraph 12.1.2, and read aloud, "If a portion of the work has been covered which the Architect has not specifically requested to observe prior to its being covered, the Architect may request to see such work and it shall be uncovered by the Contractor. If such work is in accordance with the Contract Documents, costs of uncovering and replacement shall, by appropriate Change Order, be charged to the Owner."

Allen, now suddenly interested, interrupted, "Charged to the owner? Charged to the owner? Why should I pay? And what if it holds up the job? I'm getting screwed again." He was plainly upset with the system. He didn't like the way it was heading.

Ivor continued reading from A201, "If such Work is not in accordance with the Contract Documents, the Contractor shall pay such costs unless the condition was caused by the Owner or a separate contractor in which event the Owner shall be responsible for payment of such costs."

Allen, still incensed, said, "Somebody else pays, because I sure as hell won't."

Ivor was persistent. "Well, let's start digging. George, who's going to dig the hole?"

George Hyde replied by directing Ezra to go get a shovel. But as soon as Ezra left for the shovel, Emmet and Jasper started walking away. George followed and asked Emmet, "Where are you going, Emmet?"

"There's no point in waiting to see you guys dig up the Cambodian piping. If we hafta replace that piping, we're broke. Just pay us what you owe us and we're gone. You'll hafta get another plumber to finish the job." Emmet and Jasper continued edging their way toward the truck.

George Hyde carefully explained the situation to the plumbers, "If you leave, we're paying you nothing. We'll hire a replacement plumbing contractor and charge the completion costs to your contract. If there isn't enough left in the contract we'll charge the rest to you."

* * *

Hyde Construction quickly found a reliable replacement plumbing contractor, W.C. Plumbing Company. But, as Hyde expected, they insisted on a cost plus contract with no guaranteed maximum and would give no guarantee on any work done by Aardvark.

George Hyde asked Sam Trent, his chief estimator, "Where the hell did we find Aardvark Plumbing?"

Sam replied, "Well, they submitted their bid at the last minute and we didn't have time to check them out."

* * *

Points of Law 23

ABC WAREHOUSE XI
Substitutions

23.1 Deviation from the Specifications

As this story illustrates, there are often differences between what an architect specifies and what a subcontractor would like to supply. A subcontractor can increase profit by introducing less expensive materials and methods of construction. Although there are occasions when this economic benefit can be attained without reducing the quality of the end product, or even while increasing quality, it more often happens that the subcontractor's savings cheapen the job in a qualitative, as well as in a monetary sense.

23.2 Or Equal

Aardvark Plumbing argues that the less expensive pipe is an "or equal." This ungainly term comes to us from the law of public contracts. In many jurisdictions the law wisely requires that specifications for public projects that use brand names must specify at least two brands followed by the words "or equal." The object of such laws is to prevent favoritism and corruption. Such laws apply only to public projects and not private ones such as the ABC Warehouse job. A private owner has the right to specify any brand or product for any reason. Therefore even if Aardvark were able to prove that the Cambodian pipe was equal, the installation of an unspecified brand would be a breach of contract.

23.3 Substantial Performance

It is the normal rule that A cannot require B to perform its obligations under a contract unless A has performed its own obligations. A party cannot ignore its own obligations under a contract while at the same time demanding punctilious performance from the other party! But there are occasions when rigid insistence upon the punctilious performance of a construction contract would cause injustice or economic waste. In such circumstances, under rather narrowly circumscribed conditions, courts will treat a contract that has been substantially performed as if it had been fully performed. If a contractor's breach is unintentional and does not reduce the economic value of the job, or if the nature of the breach is such that it can be fully remedied by an award of monetary damages, and the cost of rendering full performance would be economically wasteful, courts may accept substantial performance.

Let us suppose for example that a masonry contractor employs a mortar mix that is marginally different from that specified by the contract documents, but just as good. The mortar can't be replaced without tearing down the building. Since the cost of repair would be grossly uneconomic and the quality of the building has not been affected, the doctrine of substantial performance would apply and the contractor would be entitled to be paid for its work.

Supposing, though, that the mortar mix is proven to be less durable than that specified, and that therefore the estimated economic life of the building would be reduced from 50 years to 45 years, the doctrine of substantial performance could still apply but the contract price would be reduced by the economic value of the difference of the shortened life span.

In our story, the doctrine of substantial performance cannot apply for two reasons: the breach was intentional and the pipe is inferior. The law will not enforce payment of the contract price, or any part of it, absent substantial performance or full performance.

Rather than dig up and replace the Cambodian pipe, Aardvark walks off the job while demanding payment for the work that has been performed.

23.4 Doctrine of Severability

The doctrine of severability is another doctrine that could permit a contractor to be paid for less than full performance. This doctrine can come into play when the services performed by a contractor are priced on a per unit basis, such as cubic yards of compaction or linear feet of trenching. Under certain conditions, a contractor may be paid for part performance of a contract where the contract price is "severable" and multiplying prices by units yields the value of the part performance.

23.5 Efficient Breach

Some advanced thinkers in the field of contract law have proposed a doctrine of "efficient breach": the doctrine has been accepted by few courts. This doctrine would permit a subcontractor to substitute "efficient" performance for the performance required by contract documents if the subcontractor thereby saves more than the owner loses. The subcontractor must compensate the owner for the loss but the wealth of the overall economy is increased by economic efficiency.

Under the doctrine of efficient breach, even the willful and surreptitious substitution of cheaper pipe could be considered a substitute for proper performance if the subcontractor could prove the economic efficiency of the breach.

23.6 Windfall to Owner?

It is possible that the owner will get a windfall from the subcontractor's breach. Let us suppose, for example, that the owner elects not to replace the water service and employs another contractor to finish the job at such a favorable price that the overall price is less than Aardvark's contract price. Should the windfall be paid to reduce the subcontractor's loss? The answer is "no." Unless one of the doctrines discussed above (substantial performance, severable contract, or efficient breach) applies, the only way the subcontractor is legally entitled to be paid is by full performance as specified by the contract documents.

24

Parc L'Cockaigne
Discovering the Cause of Construction Defects

"Myra, if that damn phone rings one more time, tell them I'm out playing tennis, or jogging."

"They won't believe me, Murray. Its raining. Besides, at 76, you're too old for that sorta thing. I'll tell them you're here and you'd be glad to talk to them."

"To hell with them."

"You've gotta talk to them, Murray. It's your job."

"I don't gotta do anything, at my age."

"Why didn't you say that when they voted you in as President of the Parc L'Cockaigne Condominium Homeowners Association?"

"I didn't think of it then." At the time, he felt honored to be chosen for the job. It was right down his alley, as he had been the administrative partner of Cygnus, Draco & Lyra, a large construction law firm in Los Angeles before he finally retired last year. Besides, it would give him something interesting to do in his spare time. He didn't realize how much of his spare time this non-paying job would consume.

Murray was keeping track of the calls that had come in that morning. Eleven calls and breakfast wasn't even over yet.

Myra continued, "They wanted someone with managerial capability and a keen sense of responsibility. They made a good choice."

"I didn't think it would be like this. Damn. There goes the phone again."

Myra answered it, said yes a couple of times, frowned, and handed it to Murray.

"Hello, this is Murray."

"This is Opal Pungle, in Unit 3G South. Our bedroom carpeting is already ruined. And the water's staining the draperies. I'm a nervous wreck. Can't something be done about this damn rainwater leaking? This is the third time it's happened. I've called you every time and nothing's been done.

Do I have to write you a letter? Or what? Hello. Hello, are you there? I don't hear anything."

"Yes, Opal, I hear you. I'm taking notes."

"Well, what're you gonna do, Murray? We're tired of waiting. We moved in here three months ago and we're always cleaning up and drying out. Our new carpeting and draperies are ruined. And we're getting water in our kitchen cabinets. The place is starting to smell like a root cellar. We wouldn'ta moved in if we knew it'd be like this. We paid a fortune for this condo. Hello, Murray, are you still there? Murray? Say something."

Murray finally got rid of Opal, but it wasn't easy. He didn't like putting her off. He felt guilty trying to calm down homeowners that were genuinely suffering from the distressful rainwater intrusion into their new condominium units. It wasn't his fault. His unit was leaking too. Whenever it rained, he and Myra had to put their beautiful bath towels on the living room floor to soak up the water next to the wall. Then they'd have to wring out the water into buckets and replace the towels. If they went out for shopping or anything, it'd be worse when they got home.

After a hectic unleisurely breakfast, Murray dialed Birdie Sparrow, one of the Senior Managers at Lambda Condo Management Company. She took care of Parc L'Cockaigne and two other similar condominium projects.

"Birdie, I've been getting calls all morning, 13 so far. This place is still leaking. You're going to have to come to grips with it."

"What can I do? I'm doing the best I can, Murray."

"Well, it doesn't seem to be enough."

"There's nothing I can do until it stops raining."

"Then what are you going to do?"

"I've been talking with Karl Lane. As soon as the rain stops, he and his crew will apply some sealants around the window trim."

"What good will that do? They did that before."

"Karl thinks he knows what to do this time. He's getting on top of the problem. He's going to use a different type of sealant. It costs $6.85 a tube. He says it's great stuff. What he used before was only $1.20 a tube. They'll be there as soon as the rain stops."

Karl was Lambda's carpenter, handy man, jack-of-all-trades, and general repair chief. He and his crew were able to handle most minor repair jobs, but if anything big comes along, Birdie takes bids and hires various plumbing, electrical, roofing, and painting contractors.

Although Murray spent most of his professional life involved in construction industry legal affairs, he really didn't know much about the actual construction and the proper application of construction methods and building materials. It always amazed him that common construction materials had so many limitations and qualifications and seemed to fail in such strange ways.

He also wondered how the carpenters and other artisans and craftspeople could be fully informed about the scientific, technical, and engineering characteristics of all of the materials they dealt with. How, he wondered, could Karl know that some sealant would work better than another? Would the expensive one be better than the cheaper one? Why? What's the expensive one made out of? How is it different from the cheaper one? Maybe the leaking has nothing to do with sealing. How does Karl know?

Why don't other peoples' buildings leak? Maybe they all leak and I just haven't heard about it yet.

* * *

Parc L'Cockagne was located on a secluded four-acre site in Santa Esmeralda Hills, California, an exclusive suburb of Los Angeles. The prestigious walled and gated high security project consisted of 20 buildings, each containing from 4 to 8 condominium units. Although there were 6 different models, they were similar in their general layout. The fronts all faced attractive landscaped courtyards while the rears faced common motor yards. The lower front portion of each unit had an entry and spacious living room with a luxurious master bedroom and bath suite on the second floor. The rear section was three stories, split leveled with the front two-story section. Over the two-car garage were the kitchen, laundry, and dining room. On the next level above were two bedrooms and a bath. The planning was very clever and efficient, with only a half flight of stairs separating each of the five levels.

Two tennis courts, a swimming pool, Jacuzzi, and sauna were provided for common use. Although the sales prices were well into the upper levels of the condominium market, the general appearance and the planning was exceptionally appealing, and the whole 120-unit project sold out in less than two months.

The general construction of Parc L'Cockaigne was Type V under the 1991 Uniform Building Code, wood frame construction. The exterior cladding was 5/8-inch thick redwood plywood with vertical grooves about 4 inches on center, marketed by one of the leading plywood manufacturers under the name Quality III. The 4-foot by 8-foot sheets were installed vertically with the vertical joints between adjacent sheets being disguised by a shiplap joint. The horizontal joints, such as at floor lines, were covered with a 1 by 4 redwood trim. The windows and doors were trimmed with 1 by 4 redwood trim at the jambs and 1 by 6 at the heads and sills. The outside corners of the buildings were trimmed with two 1 by 4 vertical redwood trim pieces attached to each other at right angles. The roof facias were redwood to match and the general visual effect was striking. The buildings sat gracefully in the lush landscaping. The red concrete tile roofs and red brick walks and patios complemented the warm redwood of the buildings and fences.

* * *

As it generally does in Southern California, the rainstorm moved on and the sky was once again bright, blue, and beautiful. While Murray and Myra were reading the morning paper and savoring their second cups of black coffee, they became aware of some noisy activity outside across the courtyard. It was Karl Lane and his crew, consisting of Erik and Ingmar. They had driven their pick-up truck into the courtyard and were busily unloading ladders and aluminum scaffolding planks. They set up the long ladders and a scaffolding plank against the exterior wall of Building G South. This gave them safe access to the second floor windows on the front of Opal Pungle's unit.

Murray went outside to talk to Karl and find out what was going on.

"Morning, Karl."

"Morning, Mister Nelson. That was some rain we had."

"Do you think you can stop the leaks this time?"

"Oh, sure. This new sealant's great stuff. Just hit the market. Sticks like crazy. Costs almost 7 bucks a tube. We can't use it sparingly, though. That'd be false economy. So it'll cost quite a bit to do the job. But it's supposed to be the best. We have 4 cartons of it. We'll get more if we need it."

"What's it made out of?"

"I'm not sure, Mister Nelson. I think it's some complex new chemical material that was developed in connection with NASA's latest space program. That's the way Teflon and Velcro were invented, y'know."

Murray went back to finish the morning's paper. Karl, Erik, and Ingmar spent the best part of two weeks, working 10 hours a day, to beat the next rain. They sealed the joints between the wood trim and the wood siding. The wood trim occurred around all the windows and doors, at the floor lines, and at the building corners. They only treated the units that had complained of leaking as it would have been wasteful to use such expensive sealant in places that didn't leak.

It was very comforting to the condo owners to see the efficient crew, neatly dressed in spotless white work suits and caps, moving from one unit to the other in orderly progression. The men were neat and clean and didn't leave any mess behind them. They worked diligently and turned down all offers of coffee or cokes, stopping only for a quick lunch in the shade of the sparkling white pick-up truck. The occupants of each completed condo breathed sighs of relief secure in the belief that they were now protected from the next onslaught of inclement weather.

* * *

During the first few months of occupancy, all maintenance requests had been referred to the developer, IJ Developers of Santa Monica. They still had personnel on the job and it wasn't too much trouble. But as soon as the sales office was closed, Iggy Jones, IJ's President, notified the Homeowners Association that they'd have to start maintaining the project themselves. IJ had repaired some minor leaking with sealants, but were gone by the time more extensive leaking had occurred. When they wouldn't return for any more repairs and adjustments, Lambda assumed the responsibility for all maintenance and repairs.

* * *

Murray received the monthly invoice from Lambda Condo Management Company and was shocked to see that the cost of Karl Lane's sealing operation came to $15,160. That was a little over $250 per unit repaired. At the next meeting of the Homeowners' Association, it took over an hour to hear from all on the 12-member board who wanted to be heard. Murray was getting sick of the bickering and complaining and finally called a halt to the debate. It was getting repetitious.

"This bill has to be paid. We might as well get used to it. The work's been done and this is what it cost."

Elmer, the owner of 4F North, philosophized, "Well, it'll be worth every cent of it if this is the end of the leaking. Every new project needs a breaking-in period when various minor defects are discovered and fixed. Maybe this is the end of it."

Fred, from 17A East, commented, "I sure hope so. We wouldn't want a bill like this very often. Will our maintenance reserve fund cover this?"

Keenan, of 8D West, protested, "This invoice shouldn't be paid out of the reserve fund at all. It should be a special assessment divided up and charged to the units that had the leaks. Why should I have to pay when my unit didn't even leak?"

Murray ruled, "Keenan, you're out of order."

Opal Pungle cleaned up and aired her unit and painstakingly dried the carpeting and jute padding with a hair dryer. She sent the draperies out to be properly cleaned, pressed, and rehung. It was expensive but they were beautiful. Myra Nelson washed and fluffed her new bath towels and put them back in the linen closet. Life was back to normal.

* * *

The Los Angeles area doesn't get much rain compared with the rainy spots of the world. An average of only 15 inches a year. In contrast, Hong Kong gets 85 inches and Singapore 95. Even New York gets 43 inches. But L.A.'s 15 inches is concentrated in the six months from November through April, 60% of it from January through March.

It was not too surprising, this being January, that it rained again. It rained off and on all night and most of the following day and then suddenly cleared up. It was a fair test of Karl's expensive sealing job. Unfortunately, the results were profoundly disappointing. Most of the newly sealed units leaked again, some not as severely as before. Many of the units that hadn't been repaired leaked for the first time. Others were as bad as ever or even worse.

Murray and Myra and all the others with water intrusion were not only surprised and frustrated, but were bitterly disillusioned. Not only were their homes wrecked, but apparently the $15,000 spent on sealing was down the tube. A total and abject waste. Murray talked it over with Myra.

Myra expressed her opinion forthrightly. "I don't think that Karl knows what he's doing. That sealant he was using must not be any good. It's probably the wrong kind."

"I've been thinking about it, Myra. I don't think it's a sealant problem. It must be something else."
"But, Murray, what could it be?"

"I haven't the faintest idea."

"How'll you find out?"

"There was a forensic architect I used to deal with when I was with Cygnus, Draco & Lyra. His name was Frank Grimm. He'd know what's happening here."

"Why don't you call him?"

Murray called one of his former associates at his old law firm and obtained Frank Grimm's number. He called Frank and asked him to come on out to see Parc L'Cockaigne.

* * *

Frank and Murray discussed the short but miserable history of the leaking problems and the repair attempts. Frank examined some of the defective buildings inside and out and couldn't really see what the root of the problem was.

He told Murray, "Buildings of this common type of construction don't usually leak so profusely, so there must be something different here."

"What do you think it is, Frank?"

"I don't know, but we'll find out."

"How will you do that?"

"I'll need some help in partially disassembling one of the buildings. If you could get Karl Lane and his two helpers lined up, we can spend a day looking into the problem."

"Okay, how about next Monday?"

"That's okay with me. How about 8 o'clock? We'll start on your unit. Okay?"

* * *

Frank Grimm, FAIA, took charge of the biopsy operation and directed Karl and his crew to erect their ladders and scaffolding planks on the two story side of Murray's building. He picked a section about 12 feet wide from one corner, containing one window on each floor. Then he asked them to remove the redwood trim at the second floor window at the west end. They carefully removed the redwood trim pieces at the head, jambs, and sill. Frank climbed the ladder and went across the plank to get a close look at the uncovered joints. He painstakingly photographed and sketched each of the joints. He then asked them to remove the building corner trim from ground to eaves. He examined and photographed the trim and the joint. He could already see the problem. Next he directed removal of the floorline trim from the corner about 12 feet eastward. He again examined and photographed the uncovered joint. Then they removed the window trim on the first floor so the plywood exterior cladding would be free for removal. The exterior plywood cladding from ground to eaves, 12 feet wide, was removed.

He had Karl and the crew lay out all the removed pieces on the lawn with the exterior side down facing the grass and the backside facing skyward. He photographed the backside of the cladding and also the naked building framing. The building without its exterior cladding looked strangely vulnerable with its bare studs, electrical conduits, and insulation exposed to view.

It was obvious that the asphalt saturated felt underlayment was improperly installed. It had been installed by being cut to fit the plywood and was stapled to the back of each sheet of plywood. There was no provision for overlapping at joints or integrating it with the window flashing. The plywood and the felt had been installed in one operation. The system, as it was installed, would leak at all plywood joints and at all window heads, jambs, and sills.

Just before lunch, Murray came out to see what was going on.

He inquired, "Well, Frank, what have you found?"

Frank replied, "The felt underlayment has been improperly installed. It's contrary to the building code and common sense. It's bizarre. I've never seen anything like it. I'm astounded that someone on the job, seeing the installation in progress, wouldn't realize that it was wrong, perhaps the job superintendent, a building inspector, or an experienced carpenter. It would be humorous if it weren't so tragic for the owners."

He described the functions of the parts laid out on the lawn, and further explained, "The 1991 Uniform Building Code, Section 1708, provides that all exterior wood siding must be installed over a weather-resistive barrier to protect the interior wall covering. The barrier must be equal to that provided for in UBC Standard No 17-1 for kraft waterproof building paper or asphalt-saturated rag felt. The barrier is commonly installed directly to the wood framing. The wood siding is then installed after the barrier is in place.

"The 30 inch wide felt, as used here, comes in long rolls and should be installed horizontally. It is lapped 2 inches over the sheet below, shingle style, and is overlapped 6 inches at vertical joints. It must be properly integrated with the window flashings, shingle style, so that any drop of water that somehow gets behind the wood cladding or wood trim will impinge on the felt and, impelled by gravity, will run down and harmlessly out at the bottom instead of on into the interior of the building. If it were to rain after the felt barrier is in place but before the wood siding is installed, the building would not leak."

"So, what do you recommend that we do now?"

"We must put your unit back together. Karl should go to a nearby building materials supplier and pick up a few rolls of 15 pound asphalt saturated felt so he can properly reinstall your siding and trim.

"Then arrangements must be made to repair all the rest of the units. It's necessary to remove all of the exterior siding and trim in the whole project and properly reinstall it over a weather-resistant barrier of kraft waterproof building paper or asphalt saturated felt.

"Some of the siding and redwood trim will be ruined in removal, so the whole project will have to be refinished to conceal it and any other minor damage done to the redwood in the reinstallation."

"It sounds expensive. What do you think it would cost?"

"I don't know for sure, but it must be in the range of $100,000 to $200,000. It'll have to be figured by contractors capable of doing this type of work. The main objective at this point would be to commence the work. The way it is now, it will continue to leak every time it rains. It must be done properly and quickly before any more interior damage is done."

* * *

Murray called an emergency meeting of the Board of the Homeowners Association for that evening at 8 o'clock. He asked Frank Grimm to attend and explain why the buildings were leaking.

As a recently retired construction lawyer, Murray couldn't help giving his analysis of their situation.

"There's no question that IJ Developers are responsible for the improper construction of our condos. They'll have to pay for properly reconstructing all of our buildings. They'll also have to pay for all of the repair work we've already had done as well as all of the interior damage. The work must be done immediately as our damages and inconvenience will only increase the next time it rains.

"I recommend that we hire Cygnus, Draco & Lyra, the construction law firm that I was with before retirement. They'll undoubtedly file a lawsuit against IJ Developers. It's my expectation that Iggy Jones doesn't know that the siding was installed in this fashion, so when he sees Frank Grimm's report and photographs he's not likely to fight our claim. I'd expect he has insurance to cover the costs. If not, he'll have to find the money somewhere else."

"Any questions?"

* * *

Points of Law 24

PARC L'COCKAIGNE
Condominium Defects Caused by Contractor Ignorance

24.1 Who Will Have to Pay?

Murray opines that IJ Developers will have to pay for the leaks and invites questions. Although his opinion is probably sound, there are still plenty of questions.

24.2 Insurance Coverage?

The first question is whether the loss might be covered by the condominium association's own property insurance policy. In addition, the individual condominium owners may have policies that should be reviewed. Most property insurance policies cover all risks of physical loss that are not explicitly excluded. So the determination of coverage consists of comparing the loss to the exclusions from coverage that are written into the policy.

24.3 Meaning of Policy Language

All policies are not worded the same. A rigorous analysis of coverage is required in every case.

Common exclusions that might apply here are losses caused by faulty design, faulty workmanship, and latent defects.

Since insurance policies are drafted by insurance companies and sold to consumers who are not experts, courts interpret ambiguity in the language of an insurance policy of favor of coverage.

Assume carrier says the leaks were caused by faulty workmanship and are therefore not covered. One court held that the term "workmanship" is ambiguous because it might refer either to a flawed product or a flawed process. For example, a handmade antique dueling pistol could be an example of fine workmanship and a cheap imitation toy cap pistol of poor workmanship. The term "workmanship" in this sense describes the product.

On the other hand, poor workmanship in the case of a carpenter, a boxer, or a linebacker would refer to performance, not product.

24.4 Cause of Defect

Here leakage is caused by the defective process that was employed to install the waterproof felt. Since ambiguities are construed in favor of coverage, that means that the policy excludes only losses caused by faulty products, and since neither the siding nor the felt is inherently defective the exclusion would not apply.

24.5 Efficient Cause

An intriguing feature of property insurance policies is that they exclude coverage for losses according to their "causes." Thus the lawyers and courts whose job it is to interpret insurance policies become involved in metaphysical speculations supportive of the old saw that there are only two learned professions: law and theology.

24.6 Esoteric Analysis of Causation

The history of any phenomenon includes a long chain of causation. The flutter of a bird's wing in China may start a chain of events that causes a blizzard in Des Moines.

The eastward rotation of the earth combined with the inertia of the atmosphere causes westerly winds in the northern hemisphere. The action of the sun caused moisture to evaporate from the Pacific Ocean. Jet stream dips south. Area of high pressure over the plains moves east. The northeastward movement of the Pacific plate creates the Santa Monica Mountains. The mountains deflect southwesterly moisture-bearing winds upward. Cold front invades Southern California causing moisture-laden air to rise, causing a reduction of atmospheric pressure which causes molecules to increase their distance from one another causing a decline in temperature. Water vapor condenses as it cools and becomes rain. Because of the eastward rotation of the earth the winds around an area of low atmospheric pressure move in a counter-clockwise direction, causing a south wind that drives rain against south-facing units. Mexico cedes California to the United States. Murray purchases Unit 3A South. Carpenter foreman devises a more efficient way to apply felt to plywood units before they are installed. Gravity causes rain water to flow downward. According to laws of physics water adheres to a wetted surface.

Since any of these phenomena can plausibly be argued to be a cause determining whether the loss is excluded from coverage becomes a fairly debatable issue. Suppose the policy contains exclusions for losses caused by rain, faulty design, faulty workmanship, and latent defect. If the court selects the negligence of the carpentry foreman as the efficient cause, then coverage is not excluded! The negligence of the carpentry foreman is neither design, a latent defect, rain, nor is it the type of faulty workmanship that excludes coverage since the term is ambiguous.

24.7 Strict Liability

Legal liability usually depends on some kind of fault such as negligence, crime, deceit, or breach of contract. Merchant builders who produce housing, in common with the manufacturers of other products, are subject to the doctrine of liability which makes them responsible, without fault, for defective construction. Therefore IJ Developers is liable to the condominium owners. If the losses are paid by property insurance carriers, they will be subrogated to the rights of the condominium owners (step into their shoes) and recoup their loss from the developer.

The developer in turn can recoup from the subcontractor who employed the negligent carpentry foreman. The developer and the subcontractor will tender the claims to their liability insurance carriers and the insurance carriers will defend them and pay any judgment entered against them unless the claims are settled out of court.

25

ABC Warehouse XII
A Corrupt Alliance - A Criminal Conspiracy

"Dammit, Carl, you're spending entirely too much on electrical maintenance."

Carl Daly looked uncomfortable as he was taking the flack from his boss, Allen Brady, President of ABC Warehouse Company. He was standing in front of Allen's desk with his weight first on one foot and then the other. Allen pointedly hadn't offered him a seat.

Carl feebly presented his case, "It's not my fault, Allen. This crappy electrical system is no damn good. Never has been. Always breaking down."

"Don't be ridiculous, Carl. This is a brand new building. I think Dwindle Electric is screwing us. Why don't you watch them closer? This is their third bill this month. Over eight hundred dollars. The last two totaled over a thousand. What the hell's going on? Can't you control those pirates?"

"They only do what has to be done, Allen. They're still cheaper than the last two electricians we had."

When the building was about a year and a half old, the fluorescent lighting fixtures in the warehouse started going out. Carl had one of his warehouse crew go up an 18 foot aluminum ladder to replace the tubes but that didn't seem to be the problem as the new tubes wouldn't light either. So Carl called Acme Electrical Maintenance Company to check into the problem. They found that some of the ballasts were defective and had to be replaced. On their first call, two fixtures were involved and the bill was $123.85. The following week four more fixtures conked out and the bill came to $232.50. Acme's foreman suggested that the repair bills could be less if Carl would buy a few cartons of ballasts and keep them in stock for use by the Acme electricians. Each carton would be $240 for 24 ballasts and Acme wouldn't have to charge a retail mark-up. They'd just charge for the labor and the service call. When Carl okayed the purchase of 5 cartons totaling $1200, Allen blew his cork. So, Carl fired Acme and started calling Live Wire Electric.

While the spare ballasts lasted, Live Wire's bills didn't look too bad, but as soon as the bills had to include ballasts as well as the labor, Allen came unglued again. So Carl fired Live Wire.

One of the bowlers on Carl's Friday night bowling team gave him the name of his son-in-law, Elgin Dwindle, who was just getting started in the electrical repair business. Dwindle Electric's bills started out fairly low but they were gradually edging upward.

* * *

For months, Teri Unger, Allen Brady's secretary, had been nagging him for a new copying machine, one that would enlarge and reduce, copy both sides in color, and would automatically collate the sheets. He said okay, finally, and the fantastic new top-of-the-line machine arrived. The salesman was present when the machine was delivered and uncrated.

Teri directed the operation, "Elmer, have them put it over here next to the shredder. That'll be nice and convenient to the work table."

"Okay, Teri. Anywhere you want. Where's the electrical outlet so we can plug it in?"

"Oh. Isn't there one there?"

"No. And we can't use the one the shredder's plugged into. We need a grounded outlet on a separate circuit."

"Well, what'll we do? Teri looked perplexed and disappointed.

Elmer replied, "There's nothing more we can do today, Teri. We need the outlet. It'll have to be a separate circuit on a 30-amp breaker at the panel. You'll have to call an electrician. Call me when the outlet's in and I'll come back to complete the installation and the demonstration."

After Elmer and his two assistants left, Teri went into Allen's office.

"We need an electrician, Mister Brady. They can't finish installing the copier without a new outlet."

"That oughtta be easy, Teri. Just call Dwindle Electric and tell them to get right on it."

"Okay, Mister Brady. I'll do it right now."

* * *

Elgin Dwindle, armed with a clipboard, a heavy-duty fluorescent flash light, and a tool belt full of screw drivers, wire cutters, electrical tape, and dozens of other tools, testers, and meters, showed up early the next morning. Teri explained what was needed and Elgin started looking around. He checked the main service switchgear and the subpanels.

About a half-hour later he reported back to Teri with the bad news, "The panel's full. We can't add another 30-amp circuit, according to the building code. It could be done if there's no building permit, but in that case you'd hafta get yourself another electrician. We're new in the business and don't wanta get caught doing work without a permit. To do this job right'll involve a lotta work and it'll cost mucho loot."

When Teri reported this new development to Allen, he reached for his phone and called Carl on the intercom. "Carl, drop everything and come into my office." He slammed the phone down.

"What's taking Carl so long, dammit?"

"You just called him, Mister Brady. I'm sure he'll be here soon. Can I get you a cup of coffee?"

"No, dammit."

Just then, Carl rushed in looking concerned and apprehensive. He wondered what he'd be reamed out for this time. "What's up, Allen?" He tried to lighten the moment by sounding casual and upbeat.

"That damn Elgin Dwindle was just here and he told Teri it's going to cost a fortune to put in an outlet for the new copier. What a robber. We gotta find a new electrician. Where do you find these crooks?"

"Dwindle's okay, Allen. It's this damn electrical system. Like I've been telling you, it's no damn good. Why don't you call Hyde Construction and ask them to look at it? After all, they built it."

So, Allen phoned Hyde Construction and talked to George Hyde. Hyde was friendly and sympathetic but passed the buck to the architect. "Judge & King designed the building, Allen. We only built what was in the plans and specs. No more, no less. You want to talk to Ivor."

Allen called the architect's office and asked for Ivor Judge. He explained the whole thing to Ivor who replied, "Actually, Allen, we didn't design the electrical system ourselves. It was designed by our electrical engineering consultant, Maynard Noble Engineering. I'd have to get them involved. I'll call Maynard and see what he has to say. I'll call you back as soon as I find out anything."

Ivor immediately called Maynard Noble, his electrical engineering consultant of many years standing. He explained the whole thing as Allen had described it to him. Allen's sketchy story sounded a little extreme and neither architect nor engineer felt that it could possibly be accurate. Allen was always going off the deep end.

Maynard said, "I'm not too familiar with the details of the ABC job, Ivor. Ken Kray, our senior project engineer who handled that job from start to finish isn't with us any more. Ken's in Saudi Arabia working on an enormous oil refinery project. Been gone over a year."

"Well, can you dig out your engineering notes, drawings, specifications, and your inspection reports and call me back when you're up to speed?"

"Sure, I'll call back within an hour."

* * *

Allen was getting the feeling that nothing was happening, so he called his attorney, Phil Quinn. He recited the whole story again to Phil. "What are we gonna do? We're getting screwed again. I know it'll cost me a bundle as usual. All I get from the builders and architects is a royal runaround." He sounded paranoid.

"Calm down, Allen. Wait till Ivor calls back. Nothing can be done before we know for sure what's happened and who's responsible. As soon as that's known we can decide what to do."

"Well, I can't wait forever. I wanta get that damn copier hooked up. And I don't wanta pay a fortune for it either."

* * *

Later, when Maynard checked back with Ivor, he sounded shaken and puzzled, "Ivor, something's strange here. We can't find any office engineering records for the ABC job. No files, no engineering design calculations, no inspection reports, no specifications, no drawings. Nothing. One of our junior engineers said that Kray took it all with him when he left."

"Can you get in touch with Kray?"

"I don't know. I'll try."

* * *

Ivor called Allen to report what Maynard revealed about the electrical engineering records, promised to keep him up to date, and would let him know when the missing records were found.

Allen immediately called Phil again. "Phil, nothing's happening and I'm getting the runaround. They can't find the electrical engineering records. Now I have to wait while they look for my stuff in Saudi Arabia. Do I have to put up with this Kabuki dance?"

"No, Allen. I think we ought to call Frank Grimm, the forensic architect, and have him look into the whole thing. We need facts. Then we'll know if someone's acting improperly."

"Okay, Phil, I remember Grimm. He helped us with the retaining wall problem."

* * *

When Phil called Frank Grimm, the answering machine message directed him to call Frank's mobile phone number. He called the mobile number and found some annoying interference on the line. "What's that noise, Frank? Maybe you could call me back on a better line?"

"It won't do any good, Phil, that sound is the wind. I'm on top of a six hundred foot telecommunications mast examining some substandard structural joints. I'll call you when I get back down to my car in about an hour."

Frank called Phil later and they made arrangements to meet at Allen's office.

* * *

The services of Frank Grimm, FAIA, Forensic Architect, were engaged to look into the electrical system of the ABC Warehouse. He was given carte blanche to examine all parts of the building and premises and to interview all persons involved in the design and construction of the electrical system. He was given copies of the design and construction contracts and all the relevant statements and invoices. He went to the building department and got copies of their pertinent records. He spent most of two weeks working on the assignment and announced that he would have his report ready in another week.

* * *

Allen and Phil met in Allen's office to receive Frank's report. He handed Allen and Phil each a neatly bound copy of his report. "I'll give you the highlights and then you can read the gory details after I leave. When I'm through I'll answer any questions you may have."

Frank's report exposed a most fantastic series of malpractices. Even Frank had to admit that this was one of the most bizarre sets of circumstances he had ever encountered in his 40 years of forensic practice.

In Frank's view, the electrical system started going downhill when Allen's brother-in-law, Sonny, and his Speedy Electric Company, started the electrical work, went broke, and then disappeared. That little fiasco cost Allen close to $55,000 to straighten out. Hyde Construction hired Faithful Electrical Corporation to finish the work of the electrical subcontract. Hyde had worked with Frank and Earnest Faithful for years and knew he could place his trust and reliance on them. They had never let him down.

When Faithful Electrical came on the job, their most senior field superintendent, Nick Fowler, was put in charge. He'd been with Faithful since he was 18 years old, having been hired by Frank and Earnest's father. Now Nick was nearing 65. This would be his last job before retirement.

Maynard Noble Engineering had assigned the field observation job to Kenneth Kray, a senior electrical engineer, who had done all of the engineering, supervised preparation of the drawings, and had written the specifications. No one in Noble's office knew the job any better than he did. He visited the jobsite twice a month and was a great help to Nick.

After Faithful had been on the job for several weeks, Ken Kray noticed that the main feeder conduits from the transformer vault were the wrong size. He pointed this out to Nick Fowler. "Look, Nick, these four subterranean conduits are only 2-1/2 inches I.D. The drawings show four conduits at 3 inches I.D."

Nick's genuine consternation was evident on his face. "Hell's bells. I didn't notice. And the truck yard paving is all in. It'd cost a fortune to change those conduits now. They're 115 feet long. Part of the warehouse floor would have to come up too. Frank and Earnest'll think I'm losing it. I'd hate like hell to hafta tell 'em I missed it."

"That would be embarrassing," Ken agreed.

Nick looked around, lowered his voice, and said, "Why don't we just leave it like it is? No one'll know the difference." He looked pleadingly at Ken.

Ken said, "Yeah, but the main feeder wires will have to be smaller. The capacity of the main panel will be reduced."

Nick argued, "There's plenty of spare capacity in the system. It'll never be missed."

Ken was busy jotting notes. "Well, I'll have to put it in my inspection report. There's no other way."

Nick pleaded, "What if you just didn't see it?"

Ken said, "I couldn't do that. I've already seen it." He lowered his voice and suggested, "Well, Nick, maybe if you slipped me 500 bucks I could think about overlooking it."

Nick replied quickly, "That's okay with me. I can give you $150 today and the rest the next time you're here on the job." Nick was relieved at this easy way out.

"Okay. Hand it over. Gimme the one-fifty. I'll drop by here tomorrow for the rest." Ken counted and pocketed the money.

Then Nick suddenly thought of the City inspection. "But what if the City electrical inspector notices it? What'll we do?"

"Just leave him to me," said Ken. He's my buddy. I've worked with him before. He owes me one. But it'll cost you another five bills."

Nick thought about it for a few seconds, swallowed, and said, "Okay, Ken. I can get it." He was relieved, but it was getting expensive.

A couple weeks later Ken was on the job touring the work with Nick, making notes, and answering questions. Just as he was leaving, he lowered his voice and said, "Nick, I'm going to need a coupla hundred more."

Nick whined, "I thought we had a deal."

"Yeah, but I've got expenses." Nick gave him $125 from his pocket. "That's all I have at the moment."

"That's all right, Nick. You can owe me. I trust you." Nick now sincerely regretted his deal with Ken.

Two weeks later, Ken asked Nick where they were going to buy the fluorescent lighting fixtures for the warehouse.

Nick replied, "We've placed the order with Amalgamated Wholesale Electric Supply."

"Well, cancel it. You can get a much better deal at Streamline Fluorescent Corporation."

"I can't do that, Ken."

"I'm sure you can, Nick. Just a couple of simple phone calls. I wouldn't want to have to put those service conduits in my inspection report."

"I'll see what I can do." Nick made the necessary phone calls.

When the Streamline Fluorescent fixtures started arriving, over 1200 of them, Nick opened a few cartons and saw that the fixtures were poorly made and shabbily finished. He opened a few fixtures and saw that the ballasts were made by Arrow Corporation, long out of business. His heart sank, but what could he do about it. He got them installed as quickly as possible before anyone got a good look at them. When they were 20 feet above the warehouse floor they looked all right.

The building was completed and survived all final inspections by the City and the architect. ABC Warehouse Company moved in, and had their grand opening party on June 1, 1990.

Frank Grimm took a breath and continued.

"Sonny Trumble, of Speedy Electric Company, who installed the substandard conduits, disappeared about two years ago and no one has heard of him since.

"Nick Fowler, profoundly depressed after the job ended, suffered a fatal heart attack in the sixth month of his retirement.

"At about the same time, Ken Kray met an untimely death in Saudi Arabia when mid-eastern terrorists incendiary-bombed the refinery where he was working."

Frank finished his narrative at this point. "Well, that's about the size of it. Any questions?"

Allen asked, "Who's gonna pay me for all I've been through?" He looked at Frank, then Phil.

Phil said, "The people who actually caused all the problems are home free. So now we must carefully examine the facts and analyze where the financial liability lies."

* * *

Points of Law 25

ABC WAREHOUSE XII
Skullduggery and Corruption

25.1 Who Is Liable?

Because of the skullduggery of the superintendent employed by the electrical subcontractor and corruption of the inspector employed by the consulting electrical engineer, the electrical system of the ABC Warehouse is under designed in capacity and the lighting fixtures are shoddy. What persons are potentially liable and under what legal theories?

25.2 Insurance Coverage

A disciplined review of potential financial responsibility for construction defects should always start with an examination of insurance coverage. Here, we will want to examine three types of insurance policies: 1) property insurance, exemplified by Allen's fire insurance policy; 2) commercial general liability insurance, exemplified by the liability policies carried by Hyde Construction Company, Speedy Electric and Faithful Electric; and 3) errors and omissions policies carried by the architect and the consulting electrical engineer.

The chances that the loss will be covered by Allen's own insurance policy are remote because such policies usually cover only losses that results from physical property damage. In rare cases, courts have found that the mere installation of defective materials (here, fixtures and ballasts) may constitute physical injury. Therefore, although the potential for coverage is remote, the property insurance policy should be nevertheless be carefully reviewed.

Standard forms of commercial liability policies cover only bodily injury and physical injury to tangible property. The liability policies of the contractors are therefore unlikely to provide coverage even though, if suit is filed, the liability carriers would provide a defense.

The coverage of errors and omissions policies of the architect and engineer, on the other hand, is not restricted to physical injury to tangible property. Therefore these policies would respond.

25.3 Performance Bond Coverage

An analysis of potential sources of recovery should include a determination of whether the contractors supplied performance bonds. We know that Allen's brother-in-law, the owner of Speedy Electric, was probably not "bondable." No bonding company would normally write a performance bond for such an inexperienced and financially weak contractor.

Hyde Construction and Faithful Electric, however, are legitimate contractors with bonding capacity and may well have furnished bond. As the name implies, a performance bond is a guarantee that the contractor will perform as required by the plans and specifications. If the contractor does not so perform, the bonding company must pay to fix the problem. But there is an important difference between insurance and a bond. An insurance company pays losses in exchange for premiums. A bonding company pays losses all right, but a bonded contractor who causes a loss is obliged to reimburse the bonding company. Therefore, even though they may have supplied performance bonds, Hyde and Faithful will have to absorb any covered loss themselves. The bond merely serves as a guarantee that they will do so.

25.4 Hyde Construction Company

As the general contractor, Hyde is required to build the project per plans and specs. Hyde Construction is not relieved of that responsibility by employing subcontractors. The prime contractor is legally responsible for construction defects caused by subcontractors.

25.5 Speedy Electric

As a subcontractor, Speedy Electric has no direct contractual relationship with ABC Warehouse Company. In many states, however, the contractual relationship would be supplied a legal theory called "third party beneficiary" which says that a contract which is made for the direct benefit of a third party is enforceable by that third party. Since the electrical subcontract was for the direct benefit of ABC Warehouse Company, the company would have a claim against Speedy Electric for breach of contract.

25.6 Faithful Electric

It was a field superintendent employed by Faithful Electric who was made aware of the substitution of 2-1/2" i.d. conduit for the required 3" conduit. The undersized conduit forced use of wire of lesser capacity than that required by the contract documents. The field superintendent is deceased. Faithful Electric, though, as all employers, is responsible for defective construction performed by its employees. This legal doctrine is called "respondiat superior": employers are responsible for the misconduct of their employees within the course and scope of their employment. Otherwise contractors would escape responsibility for their work, since virtually all work is performed by employees.

25.7 Judge & King

The field inspector of a consulting engineering firm failed to write up in his inspection report the undersized wiring. A consulting engineer bears the same relationship to an architect as a subcontractor to a prime contractor: an architect is legally responsible for the errors and omissions of its consulting engineer. Likewise, of course, the engineer is legally responsible for the errors and omissions of its field inspector.

25.8 The Impact of Corruption

The electrical superintendent corruptly offered, and the field inspector corruptly accepted, a bribe to conceal the under capacity installation. The same men connived at ordering shoddy substandard fixtures and ballasts. Here we deal not with negligence, but crime. The fact that the actions of their employees was intentional, concealed, and corrupt does not relieve the employers of liability. Corruption could, however, have an effect on insurance coverage. In many states, liability insurance policies are prohibited from covering intentional wrongdoing of the insured. Commercial liability policies exclude coverage for damages caused by intentional wrongdoing. The errors and omissions policies of the design professionals may also be undermined by the intentional nature of the wrongdoing. Substandard wiring and shoddy fixtures and ballasts were installed intentionally rather than negligently and the potential coverage of the liability policies is further called into question. Surety bonds, however, cover intentional wrongdoing as well as negligent errors.

25.9 Shoddy and Substandard

Is a contractor liable for intentionally installing substandard equipment and materials? The answer will depend on contract provisions. Most construction contracts either specify lighting fixtures by brand name, or cover them with a general provision that all materials and equipment must be first-class.

Focusing on Faithful Electric, let's assume that there is no such contract provision and the work was performed on a handshake. Interpretation of the contract would then depend on trade practice. The jury would decide whether Faithful Electric breached its contractual obligations based on expert testimony as to whether trade practice would sanction the installation of substandard ballasts under the circumstances of the case.

25.10 Contribution and Indemnity

Under a doctrine loosely labeled contribution and indemnity a party who has caused a loss may be required to contribute to, or indemnify against, the loss of another party "vicariously" liable for the same wrong. (The liability of a prime contractor for the defaults of a subcontractor and the liability of an employer for the misconduct of an employee are "vicarious" because in each case one party is liable for misconduct by somebody else.)

Here we have three wrongdoers: Speedy Electric for the substandard conduit, and the electrical superintendent and the field inspector for the substandard wiring and ballasts. The law would require the employees to indemnify their employers under the circumstances of this case. Moreover, the electrical subcontractors would have to indemnify the prime contractor, and the consulting electrical engineer would have to indemnify the architect.

25.11 Express Warranty

How about the companies that manufactured the substandard lighting fixtures and ballasts? If the fixture manufacturer gave a written warranty it would be required to indemnify Faithful Electric. If the manufacturer of the ballasts gave a written warranty it would be required to indemnify the fixture manufacturer.

25.12 Implied Warranty

Absent a written warranty, the law supplies an implied warranty that products are fit for their intended purpose and of merchantable quality. Since the fixtures and the ballasts do not fulfill these implied warranties, the manufacturers are liable. However, in most states, a claim for breach of warranty depends on "privity." This means that the only parties who can take advantage of warranties are those parties who dealt directly with the manufacturers. Thus the fixture manufacturer would indemnify Faithful Electric and the ballast manufacturer would indemnify the fixture manufacturer.

25.13 Strict Liability

The doctrine of strict liability avoids the privity requirement. This doctrine provides that a manufacturer of consumer goods is strictly liable to a consumer for defects. The doctrine of strict liability would give Allen a direct cause of action against both manufacturers.

25.14 The Economic Loss Defense

In many states, strict liability claims are defeated if the consumer's loss is "economic" in nature.

This brings us back to our old friend "property damage." In some states manufacturers are strictly liable for bodily injury and property damage, but not economic loss. "Property damage" may be defined, as in insurance policies, as physical injury to tangible property.

Our final analysis is therefore as follows. The installation of the fixtures and the performance of the ballasts did not cause physical injury to tangible property and the loss is therefore classified as economic and therefore not compensable by the manufacturers under the doctrine of strict liability. But if the ballasts dripped and damaged the fixtures, the warehouse or its contents – it could be another story.

26

ABC Warehouse XIII
Home Improvements

"Allen, I'm getting so sick of this house. I feel all cooped up! There's no room to move around. It's the same day after day. I'm stuck here all the time while you're out in the exciting business world."

"I'm sorry, Alice. One of these days we'll find a new house. As soon as we can afford it." He finished his coffee and got up from the breakfast table.

"When'll that be?" Alice snapped. She was discouraged, frustrated, and bored, bored, bored.

Allen was getting used to his wife's frequent complaining about their home. He was just leaving for his office as president of the ABC Warehouse Company. Gathering up his brief case and perfunctorily kissing his wife, he hastily departed.

Shortly after Allen left, Alice pulled herself together and started preparing for her weekly Tuesday afternoon bridge club. Four neighborhood women met for lunch and bridge in each other's homes. This week was Alice's turn. Rose Lambert lived next door to the west. Angela Warner and Heidi Grant lived in adjoining houses on the next street to the south, directly behind the Bradys and the Lamberts. Even though Angela and Heidi lived literally within a stone's throw from the Bradys and Lamberts, they couldn't walk because their back yards were separated by a 120 foot wide steep earth bank about 60 feet high. This mammoth filled-earth bank was constructed when the Morado Hills subdivision was built about 10 years ago. So Angela and Heidi, trading off each week, had to drive the long way around.

Alice found that planning the luncheon menu and late afternoon snack was becoming more and more difficult since an invidious competition had developed among the bridge hostesses.

When it all started a year and a half ago at Heidi's, it was an enjoyable simple lunch of a small salad and a few finger sandwiches after a few hands of bridge. When it was Rose's turn the following week she added coffee and home-baked cookies at the end of the game. Each week the fare got more elaborate and escalated in quality and variety until, finally, last week Angela had it catered and served by Chez Nous French Restaurant. It was absolutely magnificent.

Alice was keenly feeling the pressure to meet or possibly surpass the constantly rising standard. When Alice had reported her latest luncheon plans to Allen, he had put his foot down firmly. "No, Alice! No ridiculous catering. No stupid French waiters. Just make a nice lunch and serve coffee and a desert." Alice was concerned that she wouldn't measure up. What would her bridge friends think? Allen just didn't understand.

During the bridge session, Rose Lambert mentioned that she and Ray had a new wooden deck built near their family room and they'd later add a hot tub. "It was built by Garff. It was quite inexpensive, and it's simply beautiful. Ray and I can sit on it and enjoy a drink and the city view." She was reveling in their new up-market acquisition. The others concealed their envy with disingenuous expressions of admiration and acclaim.

"How positively wonderful," said Angela.

"How incredibly fabulous," added Heidi.

"How utterly fantastic," agreed Alice.

In the ensuing week, Heidi told her husband about the Lamberts' new deck and suggested that they should have one too, adjacent to their family room.

"How big is the Lamberts' deck?" George asked.

"Rose said it was 6 feet by 8 feet and is 8 inches above the ground. It seems small and skimpy to me. We should have a bigger one, don't you agree, George?"

"Okay with me, Dear. I'll ask Garff about it."

Garff was the neighborhood gardener who took care of most of the lawn mowing and tree pruning in the immediate vicinity. He was also a fairly adept handyman. He could replace a sink washer or a faulty light switch. A five-dollar tip to Garff sure beat paying a 45 dollar service call to a plumber or electrician. He was also fairly good at small carpentry and painting jobs. He charged only 12 dollars an hour with no deductions or mark-ups. He said he was an independent contractor.

Garff's real name was Garffield Otis Winthropp, but he liked to be called Garff. He was always easy to find because his old truck was generally parked on one of the nearby streets and he'd be found working busily nearby.

That Saturday George had a talk with Garff and they closed a deal for a wooden deck behind the Grants' family room. It would be at the foot of the hillside slope that rose 60-odd feet up to the Lamberts' and the Bradys' homes at the top of the bank.

Garff was making notes on a green pad of recycled paper on his clipboard. "How bigga deck d'ya want?"

"Well, how big is the Lamberts' deck?" asked George.

"6 by 8 feet by 8 inches high," Garff promptly replied.

"Ours would have to be bigger."

"I see. Okay, how about 7 by 9 feet?"

"Fine. How much will it cost?"

"You pay for all the materials and I get 12 dollars an hour, just like for gardening."

"What'll it cost altogether?"

"I don't know. It all depends on what the materials cost and how much time it takes."

"Well, dammit, how much did the Lamberts' deck cost?"

"73 dollars for materials and 60 dollars for labor."

"That's not so bad. Go ahead with it. When can you start?"

"Right away, Mister Grant. Gimme the money for the materials and I'll pick 'em up Monday morning." George gave him 80 dollars to get him started.

* * *

"What are you going to do about this house, Allen? Or are we going to buy a new one?" Alice was at it again. She'd reminded Allen about the Lamberts' and Grants' new decks at least 5 times a week for the past month. It was especially bad on Tuesday evenings after the bridge club. Allen knew he was getting dangerously close to the end of Alice's patience.

"Well, Alice, there's no realistic way we can buy a new house right now. We'll hafta wait'll business picks up at the warehouse. Maybe we could just fix up this place a little."

"A little? A little? What the hell are you talking about, Allen? It better be a helluva lot more than a little." There was a rising sharp edge to her voice.

Allen knew she wasn't kidding and resolved to do something heroic to get back into Alice's good graces. "Okay, Alice, I'll talk to Garff and see what we can do about building a wooden deck behind the carport so we can sit there evenings, have a drink, and enjoy the city view."

"No, Allen. That won't do it. Not behind the damn carport. It has to be opposite the living room. There has to be a hot tub and the deck has to be bigger than Grants'."

Allen knew when he was beat. "Okay, Alice, I'll talk to Garff soon."

A couple of days later Allen saw Garff's truck parked in front of Lamberts'. It was a 1971 light blue Chevy pick-up with an orange right front fender and a 2 x 12 plank for a tailgate. He had fashioned a pipe rack for carrying stuff overhead. The truck carried his 18-year old Mitsubishi gasoline engine lawnmower, neat coils of green garden hose, rakes, shovels, and assorted gardening tools. He also had an apple box full of tools including a hammer, a hatchet, an electric saw, some wrenches, a couple of screwdrivers, pliers, and a chisel.

Garff always wore khaki trousers and a matching short-sleeved shirt with "Garff" stitched over the pocket. He had a key ring attached to a belt loop with about 75 or 80 keys that didn't open anything. He also wore a hunting knife in a leather sheath on the belt. His personal equipment was completed with a 20 foot retractable measuring tape fastened to his belt. In his shirt pocket was a plastic pocket protector with 15 or 20 pencils, pens, and markers. He was ready for anything.

Allen had a discussion with Garff using the orange front fender as a desk. Garff had his clipboard out. They decided on a 15 foot wide wooden deck extending 7 feet out from the living room wall. It would be 16 inches off the ground to be level with the living room floor, so it would need handrails.

Garff, now becoming an authority on wooden decks, explained to Allen that porches and decks less than 18 inches above grade are exempt from the building code and do not require a building permit. He picked up this valuable information from a guy at the lumber yard.

Allen, after all his experience with the new warehouse construction and the concrete tilt-up addition, felt authoritative about dealing with contractors and architects. He asked Garff who would draw the plans for the deck.

"Hell's bells, Mister Brady. I don't need no plans. I'll just build it. We don't need no permit so we don't need no plans."

This sounded logical to Allen and would save a lot of time and money. Architects and contractors just complicate things. After all, it was just a simple 7 by 15 foot wooden deck. He then asked, "How much will this cost, Garff."

"Just the cost of the materials and 12 dollars an hour for my time."

Garff was speaking Allen's language. So Allen told Garff to go ahead.

Alice was now getting a little easier to live with. She now had something to look forward to.

* * *

Angela Warner laid down the law with Roger. She told him all about the Grants' and the Lamberts' new wooden decks. And the Bradys are building an even larger deck. "I want a swimming pool. We've got plenty of room for a pool and it would be great for the kids. I don't want a crummy wooden deck."

Roger soon realized that these demands weren't negotiable. Far from it. He knew Angela well enough to know when bargaining was out of the question. Angela was a sweet pliable girl before she started going to these damned bridge club luncheons. She had developed into an unrelenting, demanding monster. It wasn't long before the pool was under construction.

* * *

The Brady deck was finished quite promptly and was the main topic of conversation at the following bridge luncheon.

Rose Lambert, praising it effusively, said, "Alice, I think it's really wonderful, so restrained, and tasteful. But isn't it a little inconvenient? It's just outside the living room but you have to go out the kitchen door, through the carport, and walk back across the yard to get to it."

Heidi also heaped compliments, "Alice, how perfectly lovely and quaint. What a cozy little deck, and the flooring is beautiful with the mahogany stain and all. Although it isn't my taste. I'd prefer a lovely mauve Astroturf covering like we have on our deck."

Angela exclaimed, "Oh, it's so beautiful, Alice. Absolutely truly fabulous. Especially the cute little handrails. I just love the rich distressed antique finish. But aren't the railings a little low? Someone might fall off. People sue over every little thing these days."

When Allen came home from work that night, he was expecting a warm reception followed by an invitation to sit on the deck, have a drink, and enjoy the city view before dinner. But that's not how it was. Alice was morose, unhappy, and totally dissatisfied with the wonderful new deck.

She related the critical remarks of the bridge club, and Allen reassured her, "Just leave it to me, Alice. I'll talk to Garff and we'll take care of all the little shortcomings."

The very next morning he found Garff mowing the Lambert lawn and walked him back to the Brady house. "Garff, we gotta make some changes in the deck."

Garff was apprehensive. "Why? What's wrong with it, Mister Brady? It's just the way you wanted."

"Nothing wrong with your work, Garff. We just want to make some changes. It's too small. We wanta make it longer. 15 feet is too short. Add more deck to each side so it's the full width of the house, 60 feet. Okay?"

"Sure, Mister Brady. But I'll need some help. It's too big a job for me to do alone. I know a guy who can help me. It'll cost you another 10 bucks an hour."

"That's okay, Garff. We just wanta get going on it. Right away."

"We can start tomorrow. I'll pick up the materials first thing in the morning."

"Now wait, Garff. There are a coupla more changes we wanta make. First, we wanta take out that 6 foot wide living room window, remove the wall under the sill, and convert the window to a sliding door. D'ya think you can do it?"

"Sure thing, why not? I've never done anything like it before but I'll give it a try. Just common sense. What's to lose?"

"That's not all, Garff. The handrail is only 24 inches high and my wife thinks it's too low. Tear it out and build the new one 30 inches high. That oughtta be high enough. What d'you think, Garff?"

"Sounds okay to me, Mister Brady. Anything else ya want done?"

"No, Garff, that's about the size of it for now."

* * *

Allen was proud of himself for solving all of Alice's dissatisfactions with the deck so easily. Garff's charges were mounting up but this was still considerably less expensive than finding and buying a new house. And a lot less trouble since he'd also have to face up to the time and effort of selling the old house as well as paying a fat broker's commission.

A few days later, he was looking forward to arriving home after work, as Garff's new assignment was supposed to be finished. Maybe he and Alice would sit on the new deck, have a drink, and enjoy the city view before dinner.

When he arrived home, all was not tranquil. Alice and Garff were engaged in heated dialogue.

"I know the door's not perfect, Missus Brady," Garff was trying to explain, "But I'm sure I can fix it. All it needs is a coupla more screws, a little oil, and some more putty and paint."

"It'll take a helluva lot more than that, Garff. It looks downright crappy. And it doesn't slide properly or lock. And it rattles."

Allen walked out through the new living room door onto the deck and was astonished at the length of the deck. "Well, Alice, the deck should be large enough now. How d'you like it? Is it large enough?"

Alice moaned, "Oh, Allen. Now it looks too narrow. It's 60 feet long and only 7 feet wide. It's too long and skinny. It looks like the front porch of an old people's home. It's horrid and tacky. It'll never do."

Allen and Garff were exchanging sympathetic glances. Garff was glad he'd never married. Allen asked Garff, "How hard would it be to widen the deck out another 8 feet?"

"Not too hard, Mister Brady. But there's a slight problem. Right now the deck extends to the edge of the level part of the lot. Then the land starts sloping down the hill. The outer edge of the extended deck would be about 7 feet above the sloping bank."

Allen asked, "Is that a problem?"

"Hell, no. I'll just put in some precast concrete piers and 4 x 4 redwood posts to support the deck. Nothing to it! It'll be easy."

"Good. When can you do it?"

"I can start tomorrow morning."

"Great. Now there are a few other things I want done. I want Alice to be really proud of this deck. I want you to put in a gas line for a barbecue. I'm gonna buy the best one. We also want a sink in case I want to mix drinks out here. So, we'll need cold water and a drain. And I want a couple of electric outlets and some lighting. Okay?"

Garff was writing this all down on his clipboard. "I've got it all down, Mister Brady."

"Oh, I almost forgot. We need some piping and wiring for a hot tub."

Garff added it to the list.

"Okay, Garff. Do it." Allen was pleased that it was so simple. He was determined to please Alice, even if he had to spend a few extra bucks. Besides, he reasoned, it'd add to the resale value when we finally sell.

* * *

A few days later, Allen was sitting at his desk at the ABC Warehouse Company offices dreaming about the deck extension and how happy Alice would be. When it's done Alice will be the envy of the bridge club. However, his roseate reverie was harshly interrupted by a phone call from Alice.

"Allen. The deck is finished and it looks great, but..."

"But? But, what, Alice? But what?"

"It's much too hot out there, Allen. It's gotta have a roof. It'll be useless the way it is. Who'd be so stupid as to sit out there in the blazing sun?"

"Is Garff still there?"

"Yeah. He's working on the sliding door. It sticks and still won't lock. And it rattles."

"Put him on the line, Alice. Let me speak to him."

Garff welcomed the break. He wasn't making much progress on the sliding door. "Hello, Mister Brady."

Allen gave him instructions to build a roof over the center section of the deck opposite the living room. The roof should be 18 feet wide and run out 17 feet to 2 feet past the edge of the deck. "How does that sound to you, Garff?"

"Okay with me, Mister Brady. I'll start tomorrow."

"Do you know how to do it?"

"I'll figure it out, Mister Brady. Just common sense. Don't worry about it."

Garff had been doing so much work for the Bradys that Allen had authorized him to charge the materials to the ABC Warehouse Company account at Acme Wholesale Lumber and Supply Company. This was simpler than advancing cash for materials at each phase of the work. He gave Garff a check for labor every Friday. Garff would be there waiting each week for the check.

Allen was mightily disappointed when Alice didn't like the roof. The new deck roof was framed with light wooden members and covered with corrugated fiberglass-reinforced plastic sheeting. Garff had removed the house rainwater gutter so he could continue the house roof slope on out over the deck. He hadn't quite figured out how to handle the roof drainage, but he'd get back to that after he got the sliding door working right.

What ticked Alice off was that the heat of the sun came through the plastic sheeting and radiated onto the deck. It was like being in an oven. Also, the 6-foot sliding door looked way too small leading to the mammoth deck. Alice demanded that the door be enlarged before she'd allow the bridge club to come to her house again.

Allen wasn't too displeased about the door exchange, however, because it looked like an easier solution than ever fixing the defective 6-footer. He'd already paid Garff a small fortune in wages to fix it and it was still a long way from satisfactory.

So he brought up the subject with Garff. "Say, Garff, d'you think you could remove the 6-foot door and replace it with a 16-footer?"

Garff thought it over and looked at the wall, the floor, and the ceiling. He glanced at the old door and felt relieved that he could dump it and start over. "Sure thing, Mister Brady. No problem. Just common sense. I'll shore up the ceiling and roof, remove about 17 feet of wall, put in a new header and install the new door."

"Do you think you'll get the door right this time?"

"No sweat, Mister Brady, I learned a lot installing the first door. I know now what I shoulda done different."

"Okay, Garff. When can you start?"

"Right away."

* * *

So, the following Monday Garff and his helper, Pete, started getting the opening ready for the sliding door. They installed shoring inside the living room to support the ceiling and roof and started removing 17 feet of exterior wall.

After a few minutes of Garff's and Pete's uninhibited flailing away at the plaster wall, inside and out, Alice came running out screaming, "Stop! Stop it! There's dust all over the damn house. You guys gotta do something."

They stopped pounding on the walls and waited for the dust to settle. Alice insisted that they install a plastic dust barrier and this took the rest of the day. They had to acquire the plastic sheeting and light framing lumber and figure out how to build it. Alice made them move all the living and dining room furniture to the spare bedrooms. They resumed the demolition the next morning.

By noon they had the demolition completed and all the debris stacked on the deck floor. They installed a new wood header in the rough opening. The old header was a 4 x 6 so Garff reasoned that the longer span would require a much larger header, so he put in a 4 x 10. Just after lunch the new aluminum door frame was delivered and they installed it in the new opening right away. After the frame was in they installed the two 4-foot wide stationary glass panels and the two 4-foot wide sliding sections and it all worked perfectly. Garff was getting the hang of it. Then they started repairing and patching the interior and exterior plaster around the jambs and head.

By now it was late Friday afternoon and it was starting to rain. Garff had the door installation completed so the Bradys could use it over the weekend. He'd finish up the minor details and miscellaneous loose ends next week. Then, he and Pete started to remove the shoring holding up the ceiling and roof. They had both sliding door sections open for better access to the shoring in the living room. As soon as the shoring supports were removed the header sharply deflected and the doors would no longer slide. In fact, the superimposed load of the old and new roofs and ceiling cracked the glass in all four panels.

Garff told Alice that he'd leave the dust barrier in place over the weekend, as he couldn't close the sliding doors at the moment. "I'll fix it next week." He added, "As long as it's raining I can't work Saturday anyway so I think I'll go visit my sister in Arizona. I'll be back sometime next week."

While Garff and Pete were waiting around for Allen to come home so they could collect their wages, the City Building Inspector showed up unexpectedly, announcing, "I saw this construction from across the canyon." He walked around carefully examining the work, and, addressing Garff, said, "All this work will have to stop until plans are submitted and a building permit is issued. Much of what I've seen is in violation of the Building Code and will probably have to come out." He signed and handed an official-looking Red Tag to Garff.

Garff looked at it and replied, "You'll hafta talk to Mister Brady. It's his house."

* * *

It rained off and on all Friday night and for most of the day on Saturday. Then, on Saturday night it came down in bucketsful. Allen had never seen such a torrential rainfall during the ten years they had lived in the Morado Hills.

When he and Alice went to bed after the 11 o'clock news they had no inkling of what was in store for them within a few short hours.

* * *

The light persistent rainfall of Friday afternoon and all night, continuing into Saturday, thoroughly saturated the sixty foot high filled-earth bank that sloped down behind the Brady property.

Normally, the light rain would soak in around the roots of the ground cover and the nominal amount of surplus runoff would be intercepted by the two lateral concrete drainage channels that were spaced about 60 feet apart on the earth bank. The gigantic mass of sodden earth was poised, just waiting for a trigger.

The storm water from the Bradys' roof normally would be caught by the metal gutters at the front and rear eaves. It would then be conducted safely by metal downspouts into plastic subterranean piping leading to Tranquilo Terrace, the street in front of the house.

But Garff had radically changed the situation. The rainwater from the rear half of the house roof now drained off onto the new deck roof and then cascaded directly onto the sloping earth bank. The several hours of unremitting light rainfall running off the roof caused the bank to be supersaturated to a considerable depth. When the heavy torrential downpour arrived after midnight on Sunday morning it provided the impetus for the supersodden mudbank to pop out of the hill.

The westerly section of the mud mass flowed swiftly like water, wiping out the Grants' new deck and hot tub, and piled up against the rear wall of their home. When the depth of mud reached over 8 feet, it broke through the sliding glass door of the family room, filling the room to within 18 inches of the ceiling. The mud flowed on into the kitchen up to the height of the counters and out the kitchen door into the front entry hall.

Fortunately, the Grants weren't home when the catastrophe struck, as they were visiting Heidi's parents in Westlake for the weekend. The first floor rooms of their two story home were a total write-off.

The easterly segment of the mudflow filled the Warners' new swimming pool thereby suddenly displacing 30,000 gallons of water, most of which crashed into their dining room sliding glass door, shattering it, and on into the house saturating the carpeting and furniture in every room. The balance of the mudflow ran between the two houses and on into Fango Terrace. It was continuing to rain and the mud was so loose that it continued to flow, ooze, and seep for another day and a half. It ran down Fango Terrace from curb to curb and filled six storm drain catch basins for three blocks down the street.

When the Warners' dining room door crashed in they were rudely awakened and within seconds a two foot depth of water sloshed into their bedroom. They didn't know exactly what was happening but they quickly gathered up their twin girls and left in their car for the safety of a motel. They brought nothing with them except for the sodden nightclothes they were wearing. They were soaked to the skin and muddy, but alive. They didn't find out what really happened until the next day. The interiors and furnishings of every room of their home were water-logged and ruined.

When the filled-earth bank collapsed, it carried the Lamberts' entire back yard with it, including their new deck and hot tub. Their house was left perched precariously on the brink of the crater left by the sudden departure of the mudbank.

The Bradys knew when the bank cut loose as it dragged the new deck down with it, making a god-awful racket, collapsing the deck roof, and causing the house to shudder. It ruptured water pipes, gas piping, and electrical lines. The free flowing water pipes were adding to the mudbank problem until they were finally turned off by the Fire Department. The firemen also cut off the house gas line at the meter and deactivated the entire electrical system.

The Bradys spent the rest of the night huddled in blankets in the cold and darkness except for the dim light of a candle and a rapidly-dimming flashlight. They weren't aware of the full impact and extent of the catastrophe until daylight when they could see it all.

Shortly after daybreak Sunday morning, Allen telephoned Phil Quinn, ABC's lawyer, who was still in bed, to seek his advice. After explaining the whole thing to Phil, Allen pleaded, "I desperately need your advice, Phil. What'll I do now?"

Phil's succinct reply was, "Why didn't you ask my advice before you hired Garff?"

* * *

Points of Law 26

ABC WAREHOUSE XIII
Allen's Deck – The Independent Contractor

26.1 Employee vs. Independent Contractor

A moving van carrying office furniture runs a red light and totals a Mercedes. The furniture belongs to the Law Offices of Phil Quinn. Is Phil legally responsible for the damage to the Mercedes? It depends on whether the driver was Phil's *employee* or an *independent contractor*.

If the driver was supplied by a moving and storage company then the driver was an independent contractor and Phil is not liable. On the other hand, if the driver was Phil's office manager driving a rented truck, then Phil is liable. Under a doctrine of *agency law* employers are liable for torts of their employees committed within the course and scope of their employment.

The law doesn't evolve in a vacuum. The law reflects the social and economic goals and the ethical and moral precepts of a civilization. The social policy behind this rule is pretty obvious:

- the person who gets the profit from economic activity should be responsible for injuries caused by negligence,
- employers have the power to closely supervise employees, and
- by imposing liability on employers, the law will encourage them to train their employees to work safely.

On the ethical side, we feel that people should be careful not to hurt one another. We don't want those who profit from the operation of a business to push their employees to increase efficiency at the expense of safety. We draw the line, though, at the *independent contractor* level. An independent contractor is only responsible for the final result. If a person can't control the details of the way an independent contractor performs work that person should not be liable for the negligence of that independent contractor.

The earmarks of an independent contractor (as opposed to an employee) are:

- an independent contractor is employed to achieve a finished result and is not subject to control as to the means employed to achieve the result,
- an independent contractor performs its services in return for the a contract price rather than for salary or hourly wages,
- an independent contractor can't be fired or terminated without cause,
- an independent contractor owns its own tools, equipment, and implements of production.

26.2 Peculiar Risk

As so often happens in law, a line, once drawn, gets moved around. A bump in the line between *employee* and *independent contractor* is known as the doctrine of *peculiar risk*. This doctrine says that a person who is engaged in ultra hazardous activities should not be able to avoid liability by employing an independent contractor. The perfect example is blasting. Since blasting is an ultra hazardous activity liability can't be avoided by employing an independent contractor: the employer must take measures to make sure that the blaster operates safely.

Most construction operations qualify as *peculiar risk* and therefore project owners are usually liable for the negligence of their contractors and subcontractors.

26.3 Unlicensed Contractor

Contractor's license laws in most states provide that unlicensed contractors are treated as employees of the persons who hire them, and therefore the independent contractor defense is not available to a property owner who employs an unlicensed contractor.

26.4 Case Against Allen

The liability case against Allen stacks up like this:

- Garff was not an independent contractor but an employee. He was paid wages rather than a contract price for a finished project and he was subject to close supervision and control by Allen, even though Allen may not have exercised such close supervision and control.

- Even if Garff was an independent contractor, he was engaged in an ultra hazardous activity as shown by the facts of the case. Property and human life were at risk. Therefore Allen would be liable under the doctrine of peculiar risk.

- Since Garff wasn't a licensed contractor he is deemed to be an employee.

- Finally, Allen is liable under an old common law doctrine that imposes strict liability (liability without fault) on a landowner who allows dirt to escape from his property.

26.5 Liability of Alice

Allen's wife may also be liable. Granted, Alice did not employ Garff: Allen did. But in doing so, Allen acted as her agent to provide the deck she desired. Although the terms are different, the outcome is the same: just as an *employer* is liable for the acts of an *employee* performed in the course and scope of employment, a *principal* is liable for the acts of an *agent* performed within the course and scope of the agency.

27

Utopian Villas
Keeping Costs Under Control

"Look, Leo, we gotta do something about these sky high architectural fees. We can't pay this kinda money and stay in business." Iggy was explaining the reality of life as he saw it.

The normally calm architect, Leo King, FAIA, was listening politely but having considerable difficulty concealing his growing exasperation. He again, slowly and carefully, explained, "Architectural fees are based on the amount of time it takes to get the ideas, develop them into something useful, and then produce the documentation. It's that simple, Iggy. This is a big project and it'll take a lot of time, skill, and effort."

"I know all that, Leo. You're beginning to repeat yourself. But, there's no damn way we're going to pay almost $600,000 for architecture and engineering for this project. You've gotta recognize our problem as developers in the real world of fierce competition in a cut-throat market."

"But, Iggy, this project is over 20 million dollars. The fee is less than three percent of the construction cost."

Iggy continued pleading his case, "We paid top dollar for the land so now we're under the gun to produce the buildings at rock-bottom cost so we can still make a profit in this competitive market. Y'unnerstand?"

They were talking about the architectural and engineering fees for a luxury housing project for the north end of the San Fernando Valley. I/J Development Company had an option on a beautiful slightly sloping seven-acre site with street access on three sides on which they planned to develop a high security 150-unit condominium project.

Iggy was Ignatius Jones, the President, Owner, and Chief Executive Officer of I/J Developers. He also headed I/J Investments, I/J Property Managers, I/J Equipment Rentals, I/J Time-share Sales, and half a dozen other companies. This would be his third major condo project in as many years. He'd built the magnificent 120-unit Shangri-La Gardens in 1994 and the extravagant 120 unit Parc L'Cockaigne in 1995.

He'd never dealt with architects before and, regrettably, his past projects showed it. The design quality, though flashy for immediate sales appeal, was unimaginative and mediocre at best and there were far too many technical flaws, mostly unnoticeable to the casual observer. The buildings were prone to monumental leaking and other unexpected problems, and the homeowner associations were constantly griping about the astronomical maintenance costs. He'd had his fill of lawsuits filed by dissatisfied condo buyers and their rabble-rousing associations. So Iggy decided to have the new project designed by a competent experienced architectural firm.

His most serious competition, the impressive Edenbliss Townhomes Condominiums, developed by his brother-in-law, Irving Jason, had just hit the market and was selling out in record time. Iggy called Irving and got the name of the architects. He knew he'd have to pay the architects more than he was accustomed to but Irving said it was worth it.

He contacted Judge & King, AIA, and visited their office. He was highly impressed with their portfolio of completed work and was immediately convinced that they should design his project. So, Iggy got right down to the fundamentals. How much? He was genuinely shocked at the response. Actually, he was more than shocked; he was astounded, if not dumbstruck.

Iggy was profoundly discouraged. "Well, hell, Leo. We can't pay that kinda money. We gotta keep all our costs down. If our condos cost too much we can't sell 'em. Y'unnerstand?"

At the second meeting with Leo, Iggy was making extremely slow progress in getting the fee down to what he considered reasonable. He whined, wheedled, sweet-talked, coaxed, and cajoled. He started to walk out twice but returned when Leo failed to stop him.

Iggy finally said, "Look, Leo. We're getting nowhere. If you can't get the fee down to the real world, I'll just have to go back to my old way."

"What was that, Iggy?" Leo already knew but wanted to hear it again. Maybe, upon repetition, Iggy would realize how ridiculous it sounded.

"Well, I hired a coupla draftsmen to work in my office and a civil engineer to figure the retaining walls, beams, and shear panels and he signed the plans. It didn't cost an arm and a leg. Dammit, we gotta watch the costs."

"You know we can't compete with that, Iggy. You may have saved money but you got what you paid for. You paid a helluva lot more than our fee just correcting the errors."

Iggy wasn't listening. "It's getting late. I'll be back tomorrow. Same time okay? Maybe we can figure out how to squeeze the air outta your fee. Let's both think about it, Leo."

"See you tomorrow, Iggy. Same time. Four o'clock." After Iggy left, Leo went into his partner's office. Ivor Judge was packing his brief case to leave for home. He looked up and asked, "How'd you make out with Iggy?"

"We got nowhere. He doesn't realize how much time and effort goes into doing a proper job. All he can see is the cost. We may not be able to make a deal. I'm not too sure we want to work for him anyway. He could make life miserable."

"Well, even if he's a pain in the neck we could use a good job right now. We're running out of work."

"Yeah. I know. We're going to have to let some people go as soon as the Blakely job is done. That'll be the middle of next week."

* * *

When Iggy got back to his office in Santa Monica, Glen Harker, his field superintendent was sitting at a desk studying the racing news. He looked up. "Hiya, Chief. How'd you do with the architects?"

"Lousy! They still want a friggin' fortune. I'd be better off with the drafters and the civil engineer."

"You're right. Why do they want so much?" Glen was sitting around the office a lot these days since they were between projects in the field. He was hoping the new job would get under way so he'd have something to do.

"Leo showed me a thick book of their specifications and a set of construction drawings for a project similar to ours. There were dozens of drawings and pages of complicated details. I guess it all takes time and the cost sky rockets."

"But Chief, we don't need all those specifications. We can work all that stuff out with our subcontractors. We'll buy whatever's the cheapest available when it's time to buy anyway. And we sure as hell don't need all those complicated details. The subs and I know how to build buildings. Most of those specifications and details just run the construction cost up anyway. Why don't ya ask Leo what the fee would be if they left out all that unnecessary stuff?"

"Damn good idea, Glen. There's another thing. Leo wants to visit the job once a week during construction to see what we're doing, talk to the subs, write reports, and check shop drawings."

"That all costs money too. Tell him we don't need it. I don't need some smart aleck junior architect snooping around my job. The building inspector's bad enough. And I'll check the shop drawings. There probably won't be any shop drawings anyway. I always tell the subs to check their own."

As Glen left for home, Iggy had a renewed respect for his superintendent. Glen is a real construction man. Knows his stuff. Keeps his eye on the costs. Watches my interests.

* * *

At 4 o'clock the next day, Iggy, comfortably seated in Judge & King's conference room, laid it on the architect. "Leo, we're experienced builders and real estate developers. We know how to build buildings. We've built and sold dozens of them. And all our subs know what they're doing, too. We don't need a fat book of specifications. We'll work out the specs when we buy out the job with our subs and suppliers."

"Well, that could save some money." The architect made a note on his pad.

"Another thing, we don't need hundreds of useless details. My superintendent, Glen Harker, says he and our subs can work out their own details on the job. There's no point in wasting a lot of money on standard boilerplate details."

"Now, Iggy, that might be short-sighted. The building might not come out right. The construction details aren't just standard boilerplate. Many have to be specially designed so the members will transfer structural loads properly, provide weatherproof construction, control sound, and carry out the design intent."

"Well, dammit, that's what we want. No details. Also, we won't need any of your services during the construction period. We'll do it ourselves. That should save a bundle."

Leo was skeptical. "Okay, Iggy, if you're sure that's what you want. But if you should need our services during construction, you know we'll always be available at our standard hourly rates. I'll refigure the fee on that basis."

Iggy was firm. "Yes, that's what we want. Just enough drawings to describe the building and to get the building permit. No more. Keep it simple."

"Okay. I'll figure it out. I'll phone you tomorrow and if the fee is acceptable, I'll prepare a contract and we can get started. The middle of next week."

"Good! We'll talk tomorrow." Iggy rose and left. He was optimistic at last.

* * *

Leo figured out what drawings would be needed to get the building permit and how staff would be deployed on the project. He conferred with his engineering consultants. He computed the costs, added a small contingency factor, and determined the fee to be quoted. He ran it past Ivor during a quick lunch and phoned Iggy at two o'clock. Iggy beat him down another five percent on general principle and they finally agreed on a lump sum figure. Leo prepared an architectural services contract and messengered it to Iggy who promptly signed and returned it.

Now the project was off to an auspicious start. Iggy thought the fee was still way too high but felt that this was the best he could do with Judge & King. He'd just have to make it up somewhere else.

At the same time, Leo was extremely apprehensive of the outcome of the reduced scope of services. He hoped that he'd not later regret taking on this project.

The scheme, if built according to the Judge & King drawings, would be impressive. It was a 150-unit condominium project with parking for 330 cars. There were generous landscaped courtyards weaving through and around the clumps of three story buildings. There were enticing recreation areas and tennis courts and a beautiful swimming pool.

The building consisted of a subterranean parking garage of 152,000 square feet. The subterranean walls of the garage were of concrete block masonry and the top was a concrete slab supporting the condominium buildings and courtyards above. The residential structures were of wood frame construction aggregating 342,000 square feet. The exterior walls were cement plaster and the interiors were all gypsum wallboard. This would be Type V One-Hour construction as defined in the Uniform Building Code, 1994 Edition.

* * *

Judge & King commenced their assignment in early March 1995, and had the building permit in hand by the middle of September. Iggy and Glen had started taking bids from subcontractors and suppliers as soon as early preliminary drawings were available. They firmed up most of their costs during the building department's plan check period. By the time the building permit was issued I/J Developers were ready to break ground.

The construction period proceeded fairly smoothly and costs were held firmly under control. In fact, some of the costs dropped during the construction period since Iggy continued to take bids up to the last moment before each subcontractor would start work. Iggy was a consummate negotiator and he prided himself on his tight subcontracts. He was known among the subcontracting fraternity as an extremely careful spender. He usually exacted an additional discount from the subs when they discussed their final payment.

Glen Harker pushed the subs relentlessly on a tight schedule and the project was ready for sales by early October 1996. The market was receptive and by the end of November, they were sold out. Another outstanding I/J development.

Iggy spent most of December and January in Acapulco with his wife, celebrating the spectacular sellout. Glen Harker remained at the site office, which by then had contracted to a garage storage room, to take care of remaining odds and ends of punch list items and customer nit picking. Gradually the routine maintenance was transferred to the Valley Condominium Management Company hired by the Utopian Homeowners Association. By the middle of January, Glen had moved his operation to the I/J Developers office in Santa Monica and Utopian Villas was off his back. He felt free. Now he could spend more time in his scientific study of the daily racing form and tracking his investments at the local equestrian venues. He had already squandered his hefty end-of-job bonus, generous even by Iggy's standards. He was ready for a new project to start. He could use the money.

* * *

There was very little or no rain in November, December, and January but the rainy season got into full swing by mid-February, 1997. It had pelted rain all night and was still raining at daybreak.

Cindy Dawson got up early on February 14th to prepare an elaborate Valentine's Day breakfast to surprise her husband, Clyde. This would also be their six months wedding anniversary, a double celebration. When Cindy nimbly walked through the dining area on her way to the kitchen, she was uncomfortably aware of a squishy, sloshing effect underfoot. She dashed quickly back to the bedroom, shrieking and shouting, rousing Clyde from a sound sleep. "Clyde, get up! Come quickly! We're flooded!"

Clyde lurched out of bed, lost his balance, and caromed into the wall, knocking a picture onto the floor. Now fully aroused, he raced, barefoot, after Cindy out to the dining area. He ran on into the sloshy carpeting and stopped, then rapidly retreated to a dry area.

"What the hell's going on here?" he demanded. "What happened?"

"I don't know," she replied. "It was like this when I came down." She had gotten towels to dry their dripping feet.

"We better call the management company," suggested Cindy, "Maybe they'll know what it is."

While Clyde was phoning the condo manager, Cindy was looking around to assess the extent of water intrusion. She found that water seemed to be gathered next to the exterior walls of their Entry, Living Room, and Dining Room. All other areas appeared to be dry. However, the wet area seemed to be spreading. It was now closer to the Kitchen than it was 15 minutes before.

* * *

Mike Nadir, a Senior Property Manager at Valley Condominium Management Company, had his hands full that morning. He and his associates had received 17 phone calls from homeowners at Utopian Villas. He was in his car racing to the scene while the phone calls kept coming in. He was in constant touch with the office by cellular phone. His driving was dreadful as he was racing through the sparse traffic and concentrating more on the phone than on his driving. His mind was racing as well. What could be happening at Utopian Villas? They'd had numerous complaints about minor items of incomplete or sloppy construction but nothing big. This thing, whatever it was, sounded different, more ominous, more pervasive. The phone calls kept coming.

When Mike arrived he went directly to the Dawson Condo, a unit designated as Elm 102. The ten buildings were named after trees. He knew by the number that it was on the first floor. He looked at his list of complaints. So far, they were all on the first floor. Oak 105, Maple 101, Cedar 101, Hemlock 104, and Ponderosa 103 and 105.

He rang the doorbell of Elm 102 and was waved in by Clyde and Cindy.

"Come on in, Mike. Look at this mess," groaned Cindy.

"What's happening? Our new furniture and carpeting are ruined," added Clyde. He and Cindy had already moved all of the smaller pieces of furniture into their guest bedroom.

"What's going on, Mike?" Cindy was recalling how hard it had been to save the money for the down payment and, since then, the hardship of stiff monthly payments on the mortgage and furniture. Now their dreamworld was collapsing.

"I don't know!" Mike answered honestly, "But I'll find out. Don't worry. I'll be back to let you know." He dashed out to look at the other condos. After a couple of hours of examining similar damage in 23 condominium units, he went back to his office. He and his associates quickly organized help to dry out the units and protect the interiors. They had to roll back carpeting and remove draperies. They dried and moved furniture.

* * *

Hannibal Grayson, owner of Cedar 101, a retired Major General of Field Artillery, had no regular job to go to, so he had plenty of time to observe the clean-up procedure. He felt he owed it to his fellow condo owners to keep a close watch on the operation and to offer helpful suggestions whenever he deemed them appropriate. His was one of the first units to be dried out and secured, as the bullied workers expected this would shut him up. How wrong they were. Their deferential concession to his demands only confirmed his position as de facto commander of the troops. He limited their lunch breaks to 15 minutes and hourly rest breaks to 3 minutes and kept up the momentum of the operation. He wanted them to continue past midnight, after a 14-hour day but, sensing mutiny, he let them go after extracting promises of an 8 o'clock start the next morning.

After the clean-up crew left, the rain resumed, even more intensely than the night before. All the dried-out units were reflooded. The still-wet units got wetter. Mike Nadir hired more helpers for the next day's operation. He also rented wet vacuum cleaners to suck up the loose water, and large fans to dry out the interiors.

When General Grayson found that his unit and the others that were dried out yesterday were now reflooded, he accosted Mike Nadir. "What the hell are you and your cowboys doing out there? The friggin' building's still leaking. Worse than ever. Can't you fix it?"

"What we were doing yesterday was drying up the water and trying to save the interiors. We worked our butts off. But, we can't fix the building. We still don't know what the problem is."

"So, what the hell are you going to do about it?" the General pressed.

"I'm going to call the builders. They'll know exactly what to do."

"Well, why are you just standing there? Get your head in gear, man."

As Mike ran off, the General resumed his oversight of the continuing operation. He looked nifty in his camouflage fatigues and shiny dark brown combat boots.

* * *

Glen Harker arrived on the scene in less than an hour, followed shortly by a worried Iggy Jones. Mike ushered Glen to the disaster area and asked, "What's wrong, Glen? How'd the water get in? What can you do about it?" Mike showed him several of the flooded units and Glen could see what was happening.

Glen was staring at one of the painted exterior plaster walls at the point where it met the tile courtyard flooring. Something about the joint was causing this massive construction failure. But what in hell was it? He was trying to recall and visualize the exact sequence of installation of each minute part of the joint and the precise details of work interface. He considered what he knew of each trade's jurisdiction and practices and which subcontractor was responsible for each part of the construction assembly. He could plainly see that rainwater falling on the open courts and passageways between units was seeping under the exterior walls somehow and running on into the units. Which trade screwed up? He was trying to reason it out in his mind.

Who was responsible? The concrete contractor who formed and poured the garage ceiling slab? Maybe. The carpentry subcontractor who framed the exterior walls of the condo units? Maybe.

The sheet metal and caulking subcontractor who installed the sheet metal flashings? Maybe. The plastering subcontractor who installed the lathing and plastering on the exterior surfaces of the exterior walls of the condo units? Maybe. The drywall subcontractor who installed the gypsum wallboard on the interior surfaces of the exterior walls of the condo units? Maybe. The painting subcontractor who painted the exterior surfaces of the plaster exterior walls of the condo units? Maybe. The carpentry subcontractor who installed the thresholds at the entry doors of each condo unit? Maybe. The suppliers and manufacturers of the numerous materials used by the various subcontractors? Maybe. The tile subcontractor who installed the flooring on the open courts and passageways? Maybe.

Mike was standing there expecting Glen to suddenly spring into action and start issuing practical orders for the prompt and clever solution to all the problems. Then Mike could go around and tell all the homeowners that their problems were over. But Glen just continued to stand there staring at the offending joint. It seemed like his feet were glued to the floor and his brain was on hold.

Mike finally asked, "Well, Glen, what're you gonna do? We gotta move fast. These people are terminally disgruntled. Yesterday they were in shock and trying to help. But after last night's reflooding they're now getting mean, surly, and aggressive. Some are calling their lawyers. When I walk by, they scowl at me and make caustic remarks."

Just then, Iggy showed up. "What the hell's going on, Glen? Who screwed up?" Then he turned to the property manager and barked, "Get lost, Mike. Glen and I hafta figure out what to do." Mike reluctantly left them to rejoin his clean-up-dry-out crews. He moved with eyes downcast, intently avoiding inadvertent eye contact with any of the homeowners. Several had stayed home from work and were standing around in small groups. They looked menacing. The General walked from group to group and they seemed to be placing great store in whatever he was telling them. He had their undivided attention and respect. By then, all 50 first floor units were involved in the leaking. The second and third floor occupants were wondering if the problem would spread to the upper floors.

Iggy turned to Glen and demanded, "What the hell's happening here, Glen? This is a damn mess. Which sub is responsible? Whoever it is'll hafta pay for all this. We pay 'em a fortune and they short the job. What'd they leave out, Glen?"

"I don't know, Chief. I've never seen anything like this before. I don't know yet exactly what's wrong."

"Well, when'll you know? What can we do now?"

"The best thing would be to caulk the joint to keep out any more water. We can figure out what else to do later."

"Okay, Glenn, let's get organized to get the caulking done as soon as possible and then we can figure out who has to pay for it."

"Okay, Chief."

So Glenn ordered several cases of caulking compound and a few new caulking guns. He lined up two laborers and showed them how to load the caulking guns, just how much to cut off the plastic nozzle of the caulking cartridge, and got them started. He instructed them to use plenty of sealant so they wouldn't have to do the job twice. He then went back to the office to resume where he left off in selecting likely prospects for the afternoon races at Santa Anita.

* * *

It took a week for the two caulkers to seal all the joints, almost 4000 lineal feet in all. Homeowners would stop and look at what they were doing. The General checked several times a day, offering suggestions as to which units to do next. Most onlookers were somewhat shocked at the messy appearance of the caulking. Some felt that it didn't matter if it solved the problem. One of the caulkers got better as the week wore on and his later work was nowhere near as unsightly as the earlier work. The other worker never did really get the hang of it.

After the sealing operation was completed, the homeowners felt safe from further intrusion. The rain had stopped and the customary balmy spring weather returned. They started slowly to reinstall the old or new carpeting their units, rewallpapering where it had been damaged, sending furniture out for repairs and refinishing, reinstalling cleaned or new draperies and getting their lives back to normal.

The General had advised the homeowners to keep careful records of all their expenses and to send all their bills to the Valley Condominium Management Company. Some, not having the ready cash to do the work, either borrowed from their banks, friends, and relatives or piled on the credit card debt. Others continued to live in chaotic conditions until they would receive funds to work with.

Mike Nadir wrote a letter to Iggy to report that the homeowner demands had added up to over $180,000 in the first month and were still coming in. He estimated that they'd total at least $250,000 by the time they were all in.

Iggy asked Glen how much it had cost to do the caulking. "Not too bad, Chief. Only $2300."

"Have you figured out who's responsible?"

"Well, no, not exactly. But I'm working on it. I've never seen anything like this before."

"Well, dammit, Glenn. What're we gonna do? We can't just sit here waiting for something to happen, can we? Who's gonna pay for all this damage and the repair work?"

"Maybe our insurance will cover it," suggested Glen hopefully.

"Not a bad idea," replied Iggy. He scooped up the phone and got his insurance broker on the line. "Say, Linus. We gotta little problem here. We've had some leaking at Utopian Villas. Does our insurance cover it?"

Linus Moran asked, "What kind of leaking?"

"The usual kind. Rainwater leaking into the interiors of condominium units. The bills are getting kinda high."

"Well, Iggy, I'll have to check with the insurance company. In the meanwhile, send me all the information you have. Okay?"

"Yeah, Linus. We'll send it right away." Iggy was glad Glen thought of the insurance.

* * *

A few days later, Frank Grimm, FAIA, was sitting at his desk contemplating the interesting life he led as a forensic architect. Every week it was something new and unusual. By now he had seen just about every possible way for the architectural and engineering professions, aided and abetted by the construction industry, to achieve an unintended result. He had dug to the bottom of one costly fiasco after another.

His reverie was unceremoniously interrupted by the insistent harsh sound of his desk phone. It was Robby Scott, Assistant Claims Manager with Rampart Insurance Underwriters.

"Frank. Another leaking condominium. We want to know who's responsible. Who caused it? Was it the architect? Was it the developer/builder? Or was it lack of or inappropriate maintenance by the homeowners association?"

Robby briefly described the situation and promised to fax the names and addresses of all parties involved. He finished with, "Can you get right on it? No written report yet. Give me a preliminary rundown by phone as soon as you know anything. Okay?"

"Okay, Robby. I'll get right on it."

Frank's dormant fax sprung into action and started spitting out a long list of names and addresses of everyone involved with the design, construction, and operation of Utopian Villas.

* * *

Frank started his investigation by visiting Iggy and Glen at the I/J Developers offices in Santa Monica. He requested and they gave him a complete set of the construction documents prepared by the architects, Judge & King. He asked for a brief description of the rainwater intrusion problem. He took notes.

Iggy explained that all the units on the first floor had serious leaking but that after application of caulking, there had been no recurrence of rainwater intrusion. Frank asked if they had any idea what caused the flooding.

Iggy replied, "We don't know exactly, but we're pretty sure some of the subs shorted the job and left something out. Somebody screwed up. We paid them a fortune too. The ungrateful bastards. They're sending their kids through college on the money they cheat outa our jobs. We had a coupla new subs whose bids might've been too low. Maybe they did it. We shoulda canned their ass before they started."

Frank and Glen then went in their separate cars to look at the Utopian Villas. Glen showed Frank the mysterious joints and the caulking repair job. Frank took a few pictures, made a few notes, and started to leave. Glen asked, "Do you know what's causing the problem?"

Frank's laconic answer puzzled Glen. "Yep." He kept walking and Glen had to hurry to keep up.

"Well, then, who screwed up?"

Frank stopped and asked the breathless Glen, "Has it rained since the caulking was installed?"

"No, I don't think so."

"Well, the problem has to be properly repaired before it rains again. That caulking won't help very much in a persistent rain. The condos will be flooded again, maybe a little less than before."

Glen asked, "What do we hafta do?"

Frank replied, "I'll report to your insurance company." He hurried off to beat the late afternoon traffic back to his office. He rolled the Judge & King drawings out on his desk and examined them carefully, making appropriate notes. He then dialed Rampart Insurance and got Robby Scott on the line. "Robby, Frank Grimm here. I just saw the building and I'm sure I know what happened. I've seen a number of situations just like this one. The joint at the bottom of the wall separating the exterior and interior of the first floor condos is not properly designed and constructed and is not waterproof."

"So, who's responsible?"

"Iggy Jones and his construction superintendent think it's one or more of the subs who were cheating the job. They're not sure who's involved."

"Do you agree with their assessment?"

"No. Not at all. And it has nothing to do with maintenance, either. After all, it's a brand new building."

"So, who's responsible?"

"It looks like the architect. There's no question that it's a design failure."

Robby replied, "Good work. Keep all your notes and photos but don't start any reports yet. I'll get back to you."

"There's another thing, Robby. The caulking repair is inadequate. Doesn't solve the problem. It'll cost a lot of money to solve and repair the defect properly. Meanwhile, the next time it rains, there'll be a lot more interior flooding and damage."

"I'll get back to you, Frank."

It rained profusely that night and continued for the next few days thereafter.

* * *

Robby Scott called Linus Moran, Iggy's insurance broker, to report that Rampart Insurance had investigated the claim and would decline coverage as the problem was caused by the architect, not their insured, I/J Developers. He said this would be sent out in writing in the next few days. He recommended that I/J developers should make their claim against the architect. Maybe they have insurance coverage.

Linus asked, "How do you know it's the architect's fault?"

"We hired a forensic architect to look into the matter and that's his opinion."

"Can you give me his name? Maybe Iggy Jones would like to talk to him."

"Sure. It's Frank Grimm. You probably remember him. He worked with us on the Redwood fiasco in Bakersfield."

"Yeah. He really knows what he's talking about. Is it okay for I/J to talk to him?"

"Okay with me, but don't charge it to Rampart. We're finished with this case. Tell Iggy he'll have to pay Grimm's fee from now on."

* * *

Linus called Iggy to report the news. He explained that Rampart Insurance was now out of it. His claim would have to be against the architect. He suggested that it would be a good idea to turn the whole thing over to his attorneys.

So Iggy called Calvin Darnel of Darnel & Jimson. He explained it all to Calvin as he understood it and asked, "Whadda we do now, Calvin?"

"I'll be over to your office first thing tomorrow morning. I'll need to see all your documentation. Then we can decide what to do."

"What documentation d'ya wanna see?"

"Your architectural agreement, all your subcontracts, your insurance policies, and your correspondence files. I'll also want to see all your condo sales agreements and the homeowner claims as well as a summary of your repair costs."

"Then what?"

"I'll review the documentation and decide what to do."

When Calvin arrived the following morning, Iggy had all the requested documentation gathered and waiting in six heavy cardboard transfer cases. In the morning's mail was a new summary from Mike Nadir with claims resulting from the second round of rainwater inundation. The total now was well in excess of $500,000. According to Mike, most of the homeowners were now beyond reasoning with since none had yet been reimbursed for the original damage and now it had happened again. Many had sought legal help and the Homeowners Association had just hired Cygnus, Draco & Lyra, a large construction law firm in Los Angeles. The firm had been spectacularly successful recently in recovery of construction defect damages for several condominium homeowners associations.

Calvin told Iggy that he'd talked to Linus Moran, Robby Scott, and Frank Grimm over the phone and was fairly well convinced that the flooding problem stems from a design defect. "While we're waiting for a formal claim from the homeowners, I think it would be a good idea for us to have an informal discussion with the architects. We need a little more information before we finalize our strategy. I'd like you to line up a meeting with Leo King. I'll be there and I'll ask Frank Grimm to be present. You'll have to pay his fee. Okay?"

"Okay. I'll line up the meeting as soon as possible. How much d'you think Grimm will charge?"

"I don't know, but it'll be plenty. I'll ask him."

The meeting took place two days later at Judge & King's office. Leo King and Ivor Judge were both present, appearing nervous and concerned. Iggy introduced Calvin to the architects and to Glen Harker. Frank Grimm already knew the architects from previous projects.

Calvin started by addressing his remarks to Leo and Ivor. "As you've perhaps already heard, there has been unprecedented leaking at Utopian Villas. The homeowners have suffered untold hardship, inconvenience, and damages. I/J Development has spent a virtual fortune on repair efforts."

Leo replied, "Yeah, we heard about it."

Calvin continued, "I/J's insurance carrier has hired Frank Grimm to look into the situation. He has inspected the building and talked to Glen Harker. It's Frank's opinion that this is a design mistake." Then, turning to Frank, he asked, "Frank, why don't you take over at this point and explain the situation to us? Then we'll all understand the problem. Okay?"

Leo, defensively, interjected, "Now, wait a minute. You're not going to saddle us with Iggy's problems?"

Calvin, ignoring Leo, looked back to Frank and invited him to proceed, "Go ahead, Frank."

Frank Grimm explained, "I took a good look at the situation in the field." He started spreading colored photographs out on the table so all could see. "I could see pretty quickly what caused the problem. When living units are built over a reinforced concrete slab garage lid, the only effective way of waterproofing the joint where a wood-framed exterior wall meets the slab is to have a step in the slab at that point. Approximately 4-inches to 6-inches will do it. It's obvious that the Utopian Villas concrete slab has no such step. This is an improperly designed joint."

Leo was bursting to speak. Calvin nodded to him, saying, "The floor's yours, Leo."

Leo directed his remarks to Iggy. "Surely, Iggy, you must remember our pre-contract discussions. We reduced our architectural fee to rock bottom by eliminating all the construction details that you and Glen considered unnecessary. You said Glen and the subs know all about building buildings and didn't need any details. We also eliminated all the specifications because you'd write your own. Also, I'm sure you'll recall, you didn't hire us to visit the site during construction because you didn't want us harassing Glen Harker and the subs. In case you've forgotten these things, you can refresh your memory by rereading our agreement."

Iggy was looking at Glen and Glen was studiously examining an interesting spot on the ceiling.

Leo added, "When you check the agreement, Iggy, you'll also note the indemnification agreement where you agreed to indemnify Judge & King from any third party claims arising out of the services that were omitted."

Frank Grimm resumed the floor. "I haven't seen the architectural services agreement, but if all that Leo says is true, the offending joint was designed by Glen Harker and the subs. Now I realize why the drawings I reviewed had no more details than would be needed to obtain a building permit. Not having any construction period services, there's no way the architects could have known what was happening in the field."

Calvin suggested that the meeting be postponed to another time. "I haven't actually had the opportunity to review the architectural agreement. After I've done so, we can reconvene."

Leo said, "There'll be no more meetings on this or any other subject until our bill is brought up to date. I/J still owes us over $100,000."

Iggy said, "Don't worry about it, Leo. We can talk about it. Any time you want."

* * *

Points of Law 27

UTOPIAN VILLAS
Construction Documents for An Owner–Builder–Developer

27.1 Liability

There can be no doubt that I/J Developers is liable to the condominium owners for the damage to their carpets, furniture, and interior surfaces. Liability derives from the mere fact that I/J manufactured and sold the condominium units and, under the law, is strictly liable for damages caused by defective construction.

27.2 Strict Liability

The term *strict liability* reflects the doctrine that a mass manufacturer of housing units is liable for damages even in the absence of negligent or other wrongful conduct.

Some of the homeowners probably have their own insurance policies that will cover damage to their furniture and floor coverings, but those insurers will have the right to enforce claims against I/J Developers to recover payments made to cover such damages.

Iggy's best hope would be to transfer liability either to the architect, Judge & King, or to a subcontractor. Subcontractors, though, are not normally responsible for design. Their duty is to follow plans. They did!

27.3 Architect's Standard of Care

The duty of an architect is to prepare construction documents that meet the standard of professional care. The standard is established by average practice in the community where the job is located.

In this case, the standard must take into account the fact that the architect's client was an experienced developer who instructed the architect to omit details. Judge & King therefore has a plausible defense: the standard of care doesn't require an architect to supply details over the objection of the client!

Might Judge & King be liable to the condominium owners, as opposed to I/J Developers? This is a slightly different issue. Architects are often held liable for foreseeable damage to parties other than their own clients.

The standard of practice would be the same: established by average conduct of practitioners in the area. Does an architect owe a duty to potential condominium buyers to include, in its drawings, such details as the omitted curb even if the details are not wanted by the client?

It seems likely that the standard of practice would require architects, even over the objections of their clients, to supply details that are needed to protect life-safety. Since the omission of the curbs did not endanger life it is unlikely that Judge & King had an obligation to provide the detail.

27.4 Indemnity

Leo mentioned that I/J Developers agreed to indemnify Judge & King from any third party claims arising out of services that were omitted. This is an unusual provision for an agreement between owner and an architect, since owners usually rely on architects to provide all necessary services and won't agree to protect them from the consequences of omitting such services. In this unusual situation, however, the sophisticated owner decided to minimize the services performed by, and money paid to, the architect and was willing to indemnify the architect against claims that might arise out of the omission of the customary services. Therefore, if any third party should make a claim against Judge & King, Judge & King will tender the defense to I/J Developers, and I/J Developers will be obliged to provide Judge & King with a defense and to pay any judgment that might entered against Judge & King.

28

The Chief Estimator
Getting the Job by Hook or by Crook

"But, Uncle Con," pleaded Samantha. "You really need me. I want this job. I'm just right for it." She was perched expectantly on the edge of her chair, elbows on his desk.

Conan Fox, president of Commerce Constructors, was interviewing his 23-year old niece, Samantha South, his sister's youngest daughter. Her impressive C.V. was on the desk in front of him. It said she had two years of general business studies at North Texas Community College and had worked for two different general contracting firms in Fort Worth, first as a takeoff person and, later, as an estimator. She was now hoping to get a job in her uncle's firm as a construction estimator.

"We don't have much work right now, Sammy, and we don't really need an estimator."

"Maybe I could find some work for us to bid and then you'll need me," suggested Sam.

She was persistent and persuasive. Con was weakening. He wanted to hire her. If he didn't he'd never hear the end of it from his sister Veronica.

"Okay, Sammy. I'll give you a chance. You'll have a month's time to prove yourself and if work hasn't picked up by then I'll have to let you go. Agreed?"

"Oh, Uncle Con, you're a doll. You'll never regret it." She kissed him on the forehead.

He took her into the adjoining office and pointed out her work place. He introduced her to Al Booker, the bookkeeper and office manager. She disliked him from the start, as he was decidedly unfriendly. He knew Commerce didn't need any more help and resented the unfair competition of nepotistic favoritism.

Sam was grateful that Con hadn't delved too deeply into her C.V., as she had fattened it up a little here and there. She had never met some of her impressive references such as the Mayor of Fort Worth and one of the local District Court Judges. She hadn't actually completed the two-year course in general business studies, although she'd started it. She'd dropped out after one semester. Her grades were pretty good but she was hot to get into action. She couldn't wait. Besides, she'd learn it all on the job anyway.

Later that day she met Commerce's congenial field superintendent, Ramsey Bullock.

"Just call me Ram. Everyone else does. If there's anything I can do for you, Sammy, just ask. Anything at all." He was warm and friendly and she reciprocated. She had a friend.

* * *

The first week was uneventful and she spent most of her time getting acquainted, learning the ins and outs of the office, and getting her desk organized. She got business cards printed, Samantha South, Chief Estimator.

On Monday of the second week she spotted an announcement in the California Construction Daily of a warehouse addition that was going out to bid. She immediately left to visit the architect's office and, after ardent persuasion, talked her way onto the bid list. All it cost her was a friendly luncheon with one of the architects and a vague promise to keep in touch.

She brought three sets of the bidding documents back to the office and showed them to her Uncle Con. He had never before gotten onto a bid list so easily.

"Good work, Sammy. I knew you could do it. But the job isn't over yet. We've still got to produce the low bid."

It was a simple concrete block masonry building with a panelized wood roof and fire sprinklers. Fiercely determined to be low bidder, she waded right into the detailed take-off and the cost analysis. She telephoned several of Commerce's key subcontractors and sent out a number of faxes and postcards to solicit bids from others.

Titanic Masonry's estimator, Duane Elton, came in to study the drawings and specifications and make a takeoff. At noon, he invited Sam out to lunch. They went to a nearby coffee shop and he started working her over to discover the range of masonry bids. She had already received 3 bids ranging from $165,000 to $180,000.

"No, Duane, I can't tell you. It wouldn't be fair to the other bidders. We're a strictly ethical firm." She didn't sound too convincing so Duane decided to give it a shot. If she got too indignant, he could always say she misunderstood.

"Look, Sammy, this isn't a game. We're only in it for the money. If you'll tell me what your low bid is, I'll make it well worth your while."

"Why, what d'you mean, Duane?" She looked shocked and innocent.

"Well, I'll give you $500." Flat out. Now it was on the table.

"I'd think it was worth more than that," she countered.

"Well, how much, then?" Duane asked. He hadn't really expected it would be so simple.

"I think $1500 would do it."

Duane thought about it. She's playing hard ball. "Okay, Sammy, you've got a deal."

Sam figured she'd kill two birds with one stone. She needed a good low bid so she could win the job. So she told Duane he'd have to beat $155,000.

So, Titanic Masonry bid $154,780 and Commerce Constructors got the job. Sammy's stock went up 10 points in Con's mind. He congratulated himself for hiring her.

* * *

A few weeks later, she got Commerce onto a closed bid list for a significant job in Burbank. It was Liberty Center, a six-story office building over a four-level subterranean parking garage. The architect's estimate was a little over $16 million, the largest job Commerce had ever bid. With only five contractors on the list, Sam felt she could come up low bidder if she really put her mind to it.

Now that she had brought Commerce Constructors into the big time, she felt she should have another talk with her uncle. Not prone to beating around the bush, she got right to the point.

"Uncle Con, I should be getting more money. After all, I've brought in a good project that's now under construction and I'm sure we'll get Liberty Center." Con Fox was truly proud of his niece and agreed to a substantial increase.

After working on the Liberty job for a few days she told Con she needed help. So he agreed to her hiring an assistant. She placed an ad in the California Construction Daily, interviewed four applicants, and hired Elvis Grover, a recent graduate in Construction Science from Arizona State.

She and Elvis culled over the bidding documents thoroughly, analyzing every drawing and detail and combing out every paragraph of the specifications. With a steady stream of subcontractors' estimators visiting to examine the drawings and specifications, they had broad, in-depth bid coverage. She was sure they'd be successful if they had enough prices from subcontractors and suppliers. She drove all the proposals down to rock bottom.

She convinced Con that they had to have more bidding documents and he agreed to pay for three more sets. In the three-week bidding period, Sam had dozens of lunches and breakfast meetings with estimators, suppliers, and subcontractors. Some days she had two lunches. She felt certain she had a winning combination of low bids. Now it was absolutely essential for Commerce to be low bidder, as she knew the special understandings she had with her low bidders would be worthless if they didn't get the job.

* * *

The excitement was starting to build up at Environmental Impact Architects. Gregory Lamb, AIA was the partner in charge of the Liberty Center project. He was readying the conference room for the bid opening which was scheduled for 4 o'clock. He expected their clients to arrive at 3:45 so they could be ready for the contractors and the bids. The bidders had been invited to attend the opening.

Promptly at 3:45, Sylvester Stark arrived with his Business Manager, Terry Dane. Stark was president of El Dorado Investors, a real estate development firm that specialized in competitive quality office buildings. They would buy and develop a site, get it up and running, and sell it to a syndicate of investors, retaining an interest as well as the management. They'd been impressively successful, primarily because Sylvester knew the value of a buck and was a merciless negotiator. Terry, an experienced accountant, had developed a formidable understanding of real estate finance and the mathematics of leases and mortgages. Between them, very few economic opportunities fell through the cracks unexploited.

When the bids were opened, the low bidder was Omega Construction Corporation at $16,210,000. The high bidder was Prudential Builders at $18,950,950, clearly out of the ballpark. Sam's bid was $330,000 higher than the low bid. She was profoundly let down.

The architect was keenly disappointed that the low bid was $205,000 over the budget but tried to put the best face on it. "This is great. Only a little over 1% above the budget. Well within industry standards."

Sylvester was unimpressed. "You're forgetting, Lamb, that the figure you so glibly call the budget was our maximum. How do you propose we cut out over $200,000?"

Terry added, "We don't intend to pay a red cent over our maximum. $16,005,000." He slowly and clearly enunciated each digit in the number.

"And we definitely are not going to reduce the size of the building. That would reduce our rent roll," Sylvester further clarified.

"And we're certainly not going to agree to any reduction in quality. That would make it impossible to move the space at the rental rate we're aiming for," Terry added so the architect would completely understand their position.

The meeting broke up without any resolution of the problem. No one involved got any sleep that night. Particularly Samantha South and her Uncle Con.

* * *

But Sam was not just sleeplessly lying there agonizing over her future. Her mind was churning out ideas for productive action. By morning she knew what she'd do. When she came bouncing into the office Con and Elvis were moping around, as if preparing for a funeral. Al Booker took perverse pleasure in their misery. He didn't say it aloud, but his demeanor said, "I told you so." This proved what he already knew, that Sam was an incompetent intruder and a troublemaker.

Sam called El Dorado Investors and got Terry Dane on the line.

"Mister Dane, can I come to your office and talk to you? I have some information that could be of great value to you."

"What is it?" he brusquely asked.

"I'd rather not say over the phone."

"Then why should I talk to you?" He sounded impatient.

"It involves the general contract bids on Liberty Center. It could save you some serious money."

Money was the magic word. "Well, okay. Can you be here in 30 minutes? I have other appointments."

"Sure thing. I'll be there, Mister Dane."

Twenty minutes later she was sitting in Terry Dane's office in the Freedom Tower Building, one of El Dorado's investments.

"Okay, Miss South, what do you have?" He waited expectantly, with a sheet of paper and pen.

"After the bid opening yesterday we checked our bid and found a few peculiarities." He looked up.

"So, what's that to me?"

"Well, we could lower our bid considerably."

Terry was now listening intently. This was the kind of talk he liked to hear.

"How much?" he asked.

Sam opened her thick file folder, and thumbing through her voluminous notes and schedules, said, "First, Mister Dane, please call me Sam." Then, after a pause, "My friends call me Sammy." She smiled.

"Okay, Sammy, spill it. I haven't got all day." He was not a patient man.

"Well, Mister Dane, after adjusting for a couple of bidding anomalies, we could lower our price to $15,950,000."

Terry started making notes. He could readily see that this new price was $260,000 below Omega's bid and $55,000 below El Dorado's maximum price. He didn't look up immediately. He was thinking.

"Will you excuse me for a minute, Sammy? I'd like to talk to Mister Stark, our President. I'll be right back. Would you like to have a cup of coffee?"

"Yes, thank you. Black, no sugar." She smiled. She knew she had his attention.

A few minutes later he was back with Sylvester in tow. He introduced them, then asked Sammy, "Please tell him what you just told me."

Sam recited the same story and the developers listened intently. Finally, Sylvester asked, "Can Commerce Constructors get a bond?"

"I don't see why not?"

"How soon can you sign a contract?"

"As soon as you want."

"When can you start work on the site?"

"Any time you want." Sam was on a roll. This was proving to be much easier than she had thought it would be.

Sylvester brought the meeting to a close by saying, "Well, Sammy, I think we've got a deal. But I've got to clear up a few matters with the architects. We'll call you soon."

After she left, Sylvester called Gregory Lamb and made an appointment for 5 o'clock. "It's important. We've got some fantastic news."

* * *

At Environmental Impact Architects, Gregory Lamb was discussing the Liberty Center bidding results with one of his partners, Vance Wade.

"Sylvester and Terry are coming in to talk about something important. They're due in 5 minutes."

"What're they going to talk about?

"I don't know, Vance. The last time I saw them they were pretty ticked off about the bidding results. They want to cut over $200,000 out of the job but they don't want to take anything out of the building."

"That doesn't seem reasonable. Is it possible?"

The receptionist buzzed to summon Gregory to the conference room where Sylvester and Terry were waiting. They seemed unduly pleased and upbeat considering their morose departure after the bid opening yesterday.

Sylvester opened with, "Well, Greg, we've wonderful news. Commerce Constructors made an error in their bid of almost $600,000 and are now willing to do the job for $15,950,000."

This caught Greg unawares and he didn't immediately react.

Terry added, "And they can furnish a bond."

"And they can start right away," Sylvester enthused.

Greg, now recovered, answered, "Now, wait a minute. That would be unethical. You should limit your negotiations to the low bidder."

"Who says?" Terry asked.

"The Recommended Guide for Competitive Bidding Procedures and Contract Awards for Building Construction, AIA Document A501 and AGC Document 325(23). This is an agreed procedure issued jointly by the American Institute of Architects and the Associated General Contractors of America."

"What's that got to do with us?" asked Sylvester.

"Yeah," added Terry, "we never agreed to that. In fact I remember seeing something in the bidding instructions that said, 'The owner reserves the right to reject any and all bids.'"

"So, we reject all the bids except Commerce's revised bid," concluded Sylvester.

The architect hung in there, "Well, it's unethical. Our office cannot be involved with such a procedure. No future contractor will ever bid our work."

"Well, Greg, you won't be involved any further anyway. If you recall, the only services left in your contract are the checking of shop drawings and visiting the jobsite on a time basis whenever we call you. Isn't that right, Terry?"

"Yeah. That's right. I reviewed the architectural services contract earlier today and that's what it says."

Greg, persisting, pleaded, "Commerce Constructors never did anything like this before. Don't deal with them. It isn't ethical."

Sylvester made their position clear. "Look, Greg, we're not going to give up this bid. We can't afford to. We can't afford your ethics. We're going with Commerce."

Sylvester and Terry gathered up their papers and started out. Greg interrupted their departure to ask, "Did you receive our last invoice?"

Sylvester replied, "Terry takes care of that sort of thing," and kept on walking.

Terry answered "I think it's on my desk. I'll look at it tomorrow," and he hurried to catch up with Sylvester.

When Greg spoke with Vance Wade a few minutes later, he complained, "The bastards leave us to make peace with Omega, the legitimate low bidder, and they still owe us over $300,000 on our fee."

* * *

Sam had her work cut out for her. Now she had to figure out how to get $590,000 out of their cost breakdown. Her Uncle Con was furious with her. "How could you do such a thing? We'll be ruined."

"Don't worry, Uncle Con. I'll get the costs down. And we'll make some money. Just leave it to me."

Con was dubious but left the whole thing in Sam's hands. "Well, do whatever you have to, but it better not cost us anything."

Sam started getting her subs lined up. First, she called Seth Upton at Reliable Steel Fabricators and made an appointment for lunch. They met at Ken's Koffee Kafe for a sandwich.

"Seth, we've got to get all of the fluff out of your bid. We've got the job if we can get our price down. You have to help."

"Hell's bells, Sammy, you already squeezed the joy outta this one. And your $15,000 kickback isn't helping any."

"Dammit, I warned you before, Seth, don't ever refer to my advisory fee as a kickback. Now, let's talk sense about your bid. If we don't, there'll be no job. I know the only way we'll get the price down any more is by reducing the inefficiency, impracticality, and waste. Now, how're we gonna do it? You're the steel man. Be imaginative. You tell me."

"Well, we can use Guatemalan or Himalayan steel. We can reduce the sizes and shapes of a lot of the members. We can simplify the welding and other connections. We can do all sorts of creative things."

"Well, then, dammit, do it. Report back by Friday. That's all the time we've got." She rose without eating her sandwich and hurried out to her car. She had to get to another lunch with Fred Martin of Martin Plumbing. She also met with all of the other major subs in the next couple of days. She finally managed to get all the costs down a total of $720,000 without disturbing any of her advisory fees. In fact she raised the electrical advisory fee from $12,000 to $15,000.

She met with her Uncle Con to update him on progress. "I've got our costs down now and the subs are all ready to start." She showed him the breakdown and pointed out that she was able to raise Commerce's allowance for overhead and profit from $680,000 to $720,000."

Con was amazed. He wasn't surprised, therefore, when she asked for a wage increase.

"Okay, Sammy, you've earned it." He didn't say anything about the self-bestowed title of Chief Estimator that he'd seen on her business card. With such fantastic results she can call herself anything she wants.

"I've gotta get going, Uncle Con. I've got a meeting with Grant Kenter to get the surety bond lined up.

They met at Rolly's Folly, a new coffee shop near the insurance broker's office. When they were seated and had ordered, Grace went directly to the heart of the matter.

"Grant, we need a bond for a $15,950,000 contract. How much will it cost?"

"Well, Sam, I don't know for sure, but it's usually around one percent of the contract price, give or take a little, depending on the contractor's financial condition. I'll know for sure when I get a quote from the bonding company."

"And, I take it that you'll get a commission on that?"

"Yeah, Sam, but..."

"Well, if you want this business, you'll have to pay my advisory fee, which will come to 50% of your commission, payable up front. No fee, no bond, okay?"

"But, Sam, this is illegal. We're not allowed to share insurance brokerage and bond commissions with unlicensed brokers. I just can't do it."

"Look, Grant. You wouldn't really be sharing the commission with me. You'll be paying my advisory fee. Without my advice you can't write the business. You might as well get used to the idea. No fee, no bond. Tell me yes or no right now so I can start talking to other brokers." She got up to leave.

"Okay, Sam. You win. I'll do it." He immediately regretted it. He despised himself.

* * *

The construction contract was signed and Commerce Constructors immediately got the mass excavation underway. The estimating department was not so busy now, so Sam's assistant, Elvis, could stay on top of it with a minimum of her supervision. Sam turned her attention to expediting the job. Ramsey Bullock, the job superintendent had created the construction schedule and Sam was hustling the subs to meet it.

She'd been on the phone with various subs and suppliers all morning.

"No, dammit, Seth, you've gotta get those shop drawings in to the architect's office for checking. They're insisting on it. If you hold up the job, we'll backcharge you for liquidated damages." She hung up. She'd been calling all the subs required by the specifications to submit shop drawings. It was like pulling teeth to get them moving.

Two days later, Seth showed up with a large roll of Reliable Steel Fabricators shop drawings. Sam asked Elvis to deliver them along with the plumbing, electrical, and air conditioning shop drawings to Environmental Impact Architects.

She continued her telephone campaign to get all the shop drawings, samples, and material lists in. Her persistence was paying off as they all gradually came in during the next couple of weeks.

Sam and Elvis were helping Ram line up the subs and materials. Without their constant prodding, threats, and sometimes sweet talk, the schedule would never have been met. Now she was concerned about getting the shop drawings back from the architects. She called Greg Lamb several times a day. He repeatedly told her that the structural, mechanical, and electrical shop drawings had been sent on to the respective consulting engineers who were responsible for the specialized design.

Greg said, "We'll send them back to you as soon as we can." He was getting sick of her constant nagging.

"Let me know when they're ready and I'll send Elvis over to pick them up. Okay?"

A few days later the Reliable Steel shop drawings were back in Commerce's office. Sam unrolled them and examined the attached letter of transmittal. The drawings weren't approved. Far from it. They had so many red ink markings, it looked like the checker had bled all over them. The structural engineer and architect had spotted and rejected each and every deviation from the contract documents.

This concerned Sam, but, undaunted, she called Terry Dane to complain about the architect's unreasonable and uncooperative action. "Doesn't he realize this is a team effort? He has to do his part. He's a stubborn obstructionist."

She was surprised when Terry said, "You'll have to do what the architect says. We don't want any deviations from the contract documents."

"But, these were just minor adjustments to make the work more efficient and practical, and so we could control our costs," Sam explained.

"Sorry, Sammy, but we don't want any relaxation from the design requirements."

She persisted, "But, Terry, the architects are too idealistic. They don't understand the costs. The steel fabricators work with this stuff every day and they know the most practical ways to do things."

"Sammy, Reliable Steel is just going to have to follow the contract documents. After all, that's what they bid on."

"But, Terry..."

"No more buts, Sammy. We're holding you to the contract. That's what you bid on."

* * *

Now Sam was worried. She called Seth at Reliable Steel and asked him to come over and pick up the shop drawings.

"Can't you just have Elvis bring them over, Sammy? I'm kinda busy right now."

"No. We've got to discuss the shop drawings, Seth. There's a slight problem."

Seth arrived within the hour.

"Hi, Sammy, what's the problem?"

"The architect turned them down. You'll hafta follow the contract documents."

"Well, then, we'll hafta raise the price back up to our original bid."

"You know we can't do that, Seth."

"If the price isn't raised, then we can't do the job."

"I'm afraid you'll have to. You bid per plans and specs. All these creative reductions and deviations were your idea. Don't you remember? You're the steel man."

"No, Sammy. You can go to hell. We're not doing this job at that price. Go hustle someone else."

"If you walk out of this contract, Seth, my uncle will sue you."

"Drop dead." Seth grabbed the shop drawings and left, steaming.

During the next several days the same scene was re-enacted with virtually all of the subcontractors that had submitted shop drawings. Sammy knew that Commerce was not financially capable of finishing the Liberty Center job. She hadn't shared her problem with anyone else in the office, least of all with her Uncle Con. She was panic-stricken and fearful of facing him.

She was sitting at her desk, mind racing, and absent-mindedly half-reading California Construction Daily. She finally decided what to do.

* * *

When Con Fox arrived in his office the next morning, there was a sealed envelope leaning against his desk clock. "Uncle Con" was written on the face. Curious, he quickly tore it open. Hurriedly written was, "Uncle Con, Thanks for all you've done for me. I've left for Fort Worth to accept a better job offer. With love, Your fond Niece, Sammy."

* * *

The Liberty Center project unraveled quickly in the few days following Sam's unannounced departure. Word about the Reliable Steel fiasco spread to the other subs, who all immediately jumped ship, filed liens, and hired lawyers. The construction closed down immediately.

Sylvester and Terry conferred with their lawyers, who asked, among other things, to see the surety bond. It was only then they realized that no one had followed through on the bond. The first month's billing, which had been duly paid, had included a bond premium of $159,500.

Inquiry to Commerce Constructors revealed that no one there had seen the bond either. So, Con Fox asked Grant Kenter about it. Grant, after considerable stalling and sidestepping, amid profound embarrassment, finally admitted that the bond was never issued. He hadn't yet figured out how to hide the illegal commission kickback in his books and hadn't gotten around to requesting the bond from the surety.

* * *

Points of Law 28

THE CHIEF ESTIMATOR
The Crooked Estimator

28.1 Ethics of Competitive Bidding

Although law is based on ethical precepts, not all unethical conduct is illegal.

Under statutes and ordinances governing public contracting, the construction contracts of most public agencies are required to be awarded to the lowest responsible bidder. The purpose of competitive bidding laws is to prevent favoritism and corruption in the administration of public construction contracts.

There is no such rule governing private projects. Project owners who receive competitive bids usually reserve the right to reject any and all bids. Owners often reject the low bid because it exceeds their budget. Rejection of the low bid is usually followed by negotiations aimed at bringing the price of the project into line with the budget. Accepted ethical standards of the construction industry call for the owner to negotiate only with the low bidder. If the owner negotiates with the second low bidder, or an outside contractor who didn't participate in the bidding, the bidding procedure becomes a travesty in which the low bid is used as a starting point for negotiations with other contractors. This is naturally offensive to the low bidder, who would not have participated in the bidding at all had it known that its bid, even if low, was merely going to be used by the owner to determine an economic value of the project that could be used as a basis of trying to negotiate still lower prices from other contractors.

Bidding a job is expensive, time consuming and complicated. Legitimate contractors won't incur that expense if they know that, even if their bid is low, it may simply be used as a starting point for negotiations with another, favored contractor.

Therefore, Sylvester rightly remarks that he has legal freedom to negotiate with the second low bidder, Commerce Constructors, even over the protest of the architect that such conduct would be unethical and damaging to the architect's reputation.

28.2 Subcontractor Bids

A subcontractor bidding a job has a considerable interest in keeping the amount of its bid confidential. Once the amount of a subcontractor's bid becomes known, other subcontractors are tempted to beat the price by a few hundred dollars, perhaps without even incurring the expense of taking off and figuring the job. Here again, although there is no law against disclosing subcontractor bids, it is highly unethical. In our story Sammy is unconcerned about ethical considerations. But she crosses the line between unethical and criminal conduct when she takes bribes! Selling confidential information that belongs to an employer is a form of embezzlement.

28.3 Material Breach of Contract

Sammy made an under-the-table deal to permit Reliable Steel to cheapen the job in order to lower the price. Unfortunately for Commerce Constructors, the owner wasn't in on the deal, so there was a difference between the job that Commerce Constructors promised to perform and the quality of the work that was subcontracted to Reliable Steel. Since a material breach of contract excuses further performance, Reliable has the legal right to shut down the job and then recover from Commerce Constructors damages for breach of contract.

28.4 Reliable Steel's Choices

Reliable has two choices: it may either terminate performance and sue for breach of contract, or terminate performance and rescind the contract. The decision whether to stand on the contract or rescind it is mainly influenced by the different ways in which damages are calculated. If Reliable stands on the contract it will be entitled to receive the contract value of its part performance plus the profit it would have made by full performance. If Reliable should rescind the contract then it would be entitled to be paid the reasonable value of the work performed, but no unearned profit.

28.5 Commerce Constructors' Position

Sammy has put Commerce Constructors into a bad position. In order to fulfill their obligation to perform the contract, Commerce Constructors will have to either renegotiate subcontracts or sign with new subcontractors. This would turn the job into a half million dollar loser! If the company's net worth is less than $500,000, insolvency and probable bankruptcy will follow.

28.6 Sylvester's Dilemma

Assuming that the net worth of Commerce Constructors is far less than $500,000 (as seems likely), Commerce Constructors will go broke and leave Sylvester to pick up the pieces. In this case, the pieces will likely include $1,000,000 worth of mechanics lien claims for work done by subcontractors and unpaid by Commerce Constructors. In order to clear the title to the property, Sylvester will have to pay off these mechanics liens in addition to the money already paid to Commerce Constructors in some cases paying twice for the same work.

28.7 The Elusive Performance Bond

This loss could have been avoided by Sylvester if he and manager Terry hadn't let the bond fall through the cracks. An owner should not allow a contractor to start work on a bonded construction project until the owner has the performance bond in hand. When a bonded contractor defaults, the bonding company becomes responsible for performance of the contract, and therefore the insolvency or bankruptcy of Commerce Constructors, while annoying to Sylvester, would not have prevented the performance of the contract for the contract price.

A lot of time and money will be lost in rebidding the job and getting things going again. If the project is ever completed, it may easily cost 150% of the original budget.

28.8 Sylvester's Final Option

One option that Sylvester may have would be to let the property go back to the lender. Just to deal in round figures, let's assume the lot is worth $1,000,000 and, as improved by the projected office building, would be worth $8,000,000. To finance the construction, Sylvester took out an $8,000,000 construction loan which would have been enough to pay for the construction of the project (including architectural and engineering and permit fees) and construction loan interest.

Sylvester paid for the lot in part with a promissory note secured by a subordinated second trust deed for $900,000.

The excavation, grading, and foundation work that has been performed cost more than $1,000,000 but only increased the appraised value of the property by $200,000.

Subcontractors recorded mechanics liens worth $1,000,000 and the architect added another $300,000 mechanics lien.

Therefore the encumbrances against the property stack up as follows.

```
First trust deed ................................................................................. $8,000,000
Second trust deed ................................................................................. $900,000
Mechanics liens ................................................................................. $1,300,000
                                                            Total ............... $10,200,000
```

The appraised value of the completed project is $10,000,000. Therefore, Sylvester must face the fact that in order to finish the job he has a year and a half of hard work ahead of him only to end up with a $200,000 loss. Sylvester should also consider the increased interest on the construction loan caused by the delay. This will add another $300,000 or $400,000 to the overall expense of the project. He might do better to turn the project over to the bank and devote his time and attention to something else.

Assume that Sylvester, as most developers, has arranged his affairs so that his company can go into bankruptcy without affecting his personal assets. He will turn the project back to the bank and look for another deal.

28.9 The Poor Bank

The bank, no matter what it does, will lose a bundle. It has disbursed $1,500,000 from the construction loan account for architectural and engineering fees, excavation and grading work, and construction loan interest. It will take months to acquire the title to the property. It will have to decide whether to sell the partially improved property or build it out.

The bank will have to fight the mechanics lien claimants who will file stop notices, claiming undisbursed construction loan funds from the bank. The bank might have to pay out $1.3 million for mechanics liens and stop notices plus attorneys fees.

In most states, *anti deficiency* legislation will keep the bank from recouping anything from Sylvester or his company. As part of an overall settlement they might take an assignment of Sylvester's claims against Commerce Constructors, but chances are that that the claims will be against an insolvent company.

28.10 The End of Commerce Constructors

As for Conan, the owner of Commerce Constructors, he may escape personal liability by taking shelter behind the corporate veil! Conan has a good claim against Sammy (his niece) for embezzlement, but Sammy has left town and no kind of insurance will cover this claim. The big loser, then, would be the bank.

INDEX: Stories

Chapter Numbers

A

Abandonment of the Job, Contractor's, 11,12, 23
Accidental Fire, 8
Administration of Construction, 3, 12, 13, 14, 17, 20
Advisory Fee, 28
Agreement for Interior Design Services, AIA Document B171, 19
Air Conditioning Subcontractor, 3, 20, 21, 22, 23, 28
Allowances, Final Accounting, 3
American Arbitration Association, 5
Application and Certificate for Final Payment, AIA Document G702, 3
Arbitration Clause, 19
Arbitration, 1, 5, 14, 15, 21, 22
Architect, 1, 2, 3, 4, 6, 7, 10, 11, 12, 13, 14, 15, 16, 17, 19, 20, 21, 22, 23, 25, 27, 28
Architect, Forensic, 6, 9,13, 15, 22, 24, 25, 27
Architect, Repair, 13
Architect's Difficult Client, 16, 17
Architect's Lawyer, 15
Architect's Liability, 27
Architect's Ruling, 2, 14, 17, 21, 23
Architectural Agreement, AIA Document B141, 3, 16, 27
Architectural Fees, 27
Asphalt Paving Subcontractor, 20
Asphalt Saturated Felt Underlayment, 24
Assignment of Subcontractor, Owner's, 18

B

Backcharges, 3, 8, 19
Bankrupt Contractor, 8, 12
Bid Bond, 8
Bidding Phase, 16
Bidding, General Contract, 5, 8, 12, 16, 18
Biopsy Operation, 24
Boiler and Machinery Insurance, 17
Bond, Contractor's License, 6, 8, 11
Bond, Performance, 1, 8, 11, 12, 17, 21, 28
Bond, Sewer, 17
Bonded Subcontract, 8
Bribery and Corruption, 25
Builder's Risk Insurance, 13
Building Code, 10, 13, 16, 20, 24, 25, 26, 27
Building Department, 1, 3, 11, 20, 25
Building Inspector, 1, 7, 9, 10, 11, 12, 13, 25, 26
Building Permit, 1, 4, 6, 11, 12

C

Carpenters, 5, 7, 13, 24
Carpentry Subcontractor, 27
Carpet Laying Contractor, 19
Caulking, 15, 24, 27
Certificate of Insurance, 11, 12, 13, 17
Certificate of Substantial Completion, 3
Change Orders, 1, 3, 10, 14, 17, 18, 19, 21
Chief Estimator, 28
Clogged Roof Drain, 4
Closing Out the Job, 3
Color Selection, 19
Completion Date, 2, 5, 6, 10, 11, 12, 13, 17, 18, 19, 21
Concrete Block Masonry, 6
Concrete Cores, 1
Concrete Cylinders, 2
Concrete Floor Defects, 2
Concrete Subcontractor, 27
Concrete Subsubcontractor, 18
Concrete Tilt-Up Walls, 2, 12
Condominium Association, 9, 24
Condominium Defects, 9, 24, 27
Conflicts of Interest, 11
Consent of Surety, 3
Conspiracy, Criminal, 25
Construction Budget, 20, 28
Construction Contract, 1, 3, 5, 11, 12, 13, 14, 17, 20, 28
Construction Defect, 2, 6, 7, 9, 11, 12, 13, 15, 18, 19, 22, 23, 24, 25, 26, 27
Construction Documents for Owner-Builder-Developer, 27
Construction Documents Phase, 16
Construction Estimator, 2, 28
Construction Financing, 11, 13
Construction Phase, 17, 21
Construction Schedule, 2, 5, 8, 9, 10, 12, 17, 18, 19, 21, 27, 28
Contract Administration, 3, 12, 13, 14, 17, 20
Contract Documents, 3, 13, 28
Contract, Home Construction, 11, 13

Contract, No Extra, 1
Contractor Selection, Faulty, 12
Contractor, Carpet Laying, 19
Contractor, Electrical, 25
Contractor, General, 1, 2, 3, 4, 5, 6, 7, 8, 10, 11, 12, 13, 14, 15, 16, 17, 18, 19, 20, 21, 22, 23, 28
Contractor, Independent, 26
Contractor, Landscape, 6
Contractor, Painting, 6
Contractor, Separate, 19
Contractor's Abandonment of the Job, 11, 12, 23
Contractor's Difficult Customer, 17
Contractor's Final Billing, 2
Contractor's Lawyer, 5, 6, 8, 15, 18, 21, 27
Contractor's License Bond, 6, 8
Contractor's License, 8, 11, 21
Contractor's Progress Billing, 12, 16, 17, 18
Contractor's Request for Owner's Financial Information, 17
Cost Estimator, 28
Criminal Conspiracy, 25

D

Dampproofing, 9
Decorator, Interior, 19
Deed Restrictions, 16
Defective Concrete Floor, 2
Defects, Construction, 2, 6, 7, 9, 11, 12, 13, 15, 18, 19, 22, 23, 24, 25, 26, 27
Defects, Design, 7, 15, 27
Delamination of Glulam Beam, 7
Delay, Weather, 17
Design Defects, 7, 15, 27
Design Development Phase, 16
Developer, 15, 20, 21, 22
Draftsman, 11
Drapery Installer, 19
Drywall Crews, 5

E

Earthquake Damage, 7
Earthquake Insurance, 7
Earthslide, 6, 26
Electrical Contractor, 25
Electrical Design, Drawings, and Specifications, 25
Electrical Engineer, 22, 25
Electrical Foreman, 25
Electrical Maintenance, 25
Electrical Subcontract, 5
Electrical Subcontractor, 18, 25
Elevator Subcontractor, 20

Engineer, Electrical, 22, 25
Engineer, Mechanical, 10, 20, 22
Engineer, Soils, 1, 6, 13
Engineer, Structural, 1, 2, 4, 6, 7, 12, 13, 14, 20, 22, 27, 28
Engineering Geologist, 6
Engineering, Value, 20
Estimator, Cost, 28
Extra Work, 1, 3, 10, 14, 17, 18, 19, 21

F

Faulty Contractor Selection, 12
Final Payment, 2
Fire and Extended Coverage Insurance, 17
Fire Doors, 23
Fire Insurance, 8
Fire Safety, 20
Flashing Leaking, 7
Foolproof Construction Contract, 1
Footing Excavation, 1
Forensic Architect, 6, 7, 9, 13, 15, 22, 24, 25, 27
Forensic Architect's Report, 13, 22, 24, 25, 27
Furniture and Circulation Layouts, 19
Furniture Installer, 19

G

Gas Piping, 23
General Conditions, AIA Document A201, 1, 2, 3, 14, 17, 23
General Contract Bidding, 5, 8, 12, 16, 18
General Contractor, 1, 2, 3, 4, 5, 6, 7, 8, 10, 11, 12, 13, 14, 15, 16, 17, 18, 19, 20, 21, 22, 23, 28
Glazing Material Supplier, 15
Glazing Subcontractor, 15, 20
Glulam Beam Failure, 7
Ground Rules for Substitutions, 22

H

Handyman Unlicensed Contractor, 26
Hardwood Flooring Subcontractor, 11
Heating Subcontractor, 11
High Dome Strainer, 4
Home Addition, 11, 26
Home Improvement Contractor, 11
Homeowner, 11, 13, 26
Homeowner's Insurance, 11

I

Indemnification Agreement, 27
Independent Contractor, 26
Installer, Drapery, 19

Installer, Furniture, 19
Insurance Certificate, 3, 11, 12, 13, 17
Insurance, 3, 4, 6, 7, 8, 9, 11, 12, 13, 17, 20, 24, 27
Insurance, Boiler and Machinery, 17
Insurance, Builder's Risk, 13
Insurance, Earthquake, 7
Insurance, Fire and Extended Coverage, 17
Insurance, Fire, 8
Insurance, Homeowner's, 11
Insurance, Owner's Liability, 17
Insurance, Professional Liability, 7, 15
Interest on Past Due Payments to Contractor, ` 17
Interior Decorator, 19
Interior Design Services Agreement, AIA Document B171, 19

J

Job Close-Out, 3
Job Superintendent, 1, 2, 4, 5, 7, 9, 13, 14, 18, 19, 21, 23, 28
Jobsite Observation Report, 20, 23
Joint Control, 11
Joint Venture, 8

K

Keys, Masterkeys, and Keying Schedule, 3
Kickbacks, 28
Kraft Waterproof Building Paper, 24

L

Labor Relations, 5
Labor Union, 5
Land Survey, 1, 16
Landscape Contractor, 6
Late Completion of Construction, 5, 11, 12
Late Payment by Owner, 17
Latent Defect, 7
Leaking Roof, 7, 11
Leaking Walls, 9, 15, 24, 27
Leaking Windows, 15, 22, 24
Lender, 11, 13
License Bond, Contractor's, 6, 8, 11
Liquidated Damages, 2, 11, 12, 17, 18
Low Bidder, 5, 8, 9, 12, 14, 15, 16, 18, 20, 23, 27, 28

M

Maintenance Failure, 4
Masonry Subcontractor, 6, 28
Mechanic's Lien Claim, 3
Mechanic's Lien Release, 3
Mechanic's Lien, 8, 12, 13, 18

Mechanical Engineer, 10, 20, 22
Mediation, 1

N

Negotiated Contract, 1, 9, 11, 13, 18, 20, 28
No Extra Contract, 1
Notice of Completion, 3
Notice to Proceed with Construction, 12, 17

O

Operating Instructions, 3
Operating the Building, 22
Owner, 1, 2, 3, 4, 6, 7, 10, 12, 14, 16, 17, 18, 19, 23, 25,
Owner's Assignment of Subcontractor, 18
Owner's Late Payment to Architect, 16
Owner's Late Payment to Contractor, 17
Owner's Lawyer, 2, 3, 4, 6, 7, 9, 11, 12, 13, 14, 17, 20, 22, 24, 25, 26, 27, 28
Owner's Liability Insurance, 17
Owner's Program, 16
Owner's Protection, 11
Owner's Right to Reject Bids, 28
Owner-Architect Agreement, AIA Document B141, 3, 16
Owner-Builder-Developer, 9, 24, 27
Owner-Developer, 15, 20, 21, 22, 28
Owner-Nominated Subcontractor, 18

P

Paint Manufacturers, 23
Painting Contractor, 6
Painting Retaining Wall, 6
Painting Subcontractor, 19, 20, 22
Parapet Coping, 7
Patent Defect, 7
Performance Bond, 1, 8, 11, 12, 17, 21, 28
Plastering Subcontractor, 27
Plugged Roof Drains, 4
Plugged Weep Holes, 6
Plumbers, 6, 8, 9
Plumbing Subcontractor, 17, 22, 21, 23, 28
Plywood Cladding, 24
Pre-Construction Jobsite Meeting, 14, 18, 23
Pressure Regulator, 10
Professional Liability Insurance, 7, 15
Program, Owner's, 16
Punch List, 3, 13

R

Rainfall, 4, 6, 9, 11, 15, 17, 24, 26, 27
Recommended Guide for Competitive Bidding Procedures, AIA Document A501, 16, 18
Record Drawings and Specifications, 3
Reinforcing Steel, 1, 6
Release of Mechanic's Lien, 3
Repair Architect, 13
Responsible Managing Officer, 13, 21
Retainage, 11, 12, 15, 18
Retaining Wall Collapse, 6
Retaining Wall Design, 6
Retaining Wall Painting, 6
Retention, 11, 12, 15, 18
Roof Collapse, 4
Roof Debris, 4
Roof Drains, Clogged, 4
Roof Leaking, 7, 11
Roofing Materials, 23
Roofing Subcontractor, 8, 20, 21

S

Schedule of Values, 12, 16
Schedule, Construction, 2, 5, 8, 9, 10, 12, 17, 18
Schedule, Submittal, 17
Schematic Design Phase, 16
Sealants, 15, 24, 27
Security, 20
Separate Contractor, 19
Sewer Bond, 17
Sewer Subsubcontractor, 17
Sheet Metal Subcontractor, 7, 27
Shop Drawings, 14, 15, 18, 23, 28
Shoring, 7
Skylight Leaking, 22
Soil Pressure, 1
Soil Testing, 1
Soils Engineer, 1, 6, 13
Solar Energy Heating System, 20, 21
Steel Fabrication Subcontractor, 14, 20, 28
Structural Engineer, 1, 2, 4, 6, 7, 12, 13, 14, 20, 22, 27, 28
Structural Integrity, 20
Structural Steel Fabricator, 14
Subcontract Terminated, 18
Subcontract, Electrical, 5
Subcontract, Roofing, 8
Subcontractor Assignment by Owner, 18
Subcontractor List, 12, 17
Subcontractor, Air Conditioning, 3, 20, 21, 22, 23
Subcontractor, Asphalt Paving, 20
Subcontractor, Carpentry, 27
Subcontractor, Concrete, 27
Subcontractor, Electrical, 18, 20, 25
Subcontractor, Elevator, 20
Subcontractor, Glazing, 15, 20
Subcontractor, Hardwood Flooring, 11
Subcontractor, Heating, 11
Subcontractor, Masonry, 6, 28
Subcontractor, Painting, 19, 20, 22
Subcontractor, Plastering, 27
Subcontractor, Plumbing, 10, 17, 20, 21, 22, 23, 28
Subcontractor, Roofing, 8, 20, 21
Subcontractor, Sheet Metal, 7, 27
Subcontractor, Steel Fabrication, 14, 20, 28
Subcontractor, Tile, 27
Subcontractor, Waterproofing, 9, 20
Subcontractor, Window Wall, 15, 20
Subcontractor's Lawyer, 15, 18, 19
Submittal Schedule, 17
Substantial Performance, 2
Substitutions, 23
Subsubcontractor, Concrete, 18
Subsubcontractor, Sewer, 17
Subterranean Wall Leaking, 9
Superintendent, 1, 2, 4, 5, 7, 8, 9, 13, 14, 18, 19, 21, 23, 27, 28
Surety Bond, 1, 8, 11, 12, 13, 17, 21, 28
Swimming Pool, 26

T

T-Bar Ceiling System, 23
Tenants' Lawyers, 15, 22
Terminated Subcontract, 18
Test Hole Digging, 23
Testing Laboratory, 1
Testing, Soil, 1, 16
Tile Subcontractor, 27

U

UBC Type V Construction, 8, 24, 27
Underlayment, Asphalt Saturated Felt, 24
Unethical Bidding Practices, 6, 9, 12, 20, 28
Unlicensed Contractor, 26

V

Value Engineering, 20, 21, 22

W

Wall Leaking, 9, 15, 24, 27
Warranties, 3
Warranty Period, 3, 5, 7
Water Heater, 22, 23
Water Main Piping, 23
Water Pressure, 10
Waterproof Building Paper, 24
Waterproofing Subcontractor, 9
Waterproofing, 9
Weather Delay, 17
Weep Holes, 6
Winding Up the Job, 3
Window Leaking, 15, 22, 24
Window Wall Subcontractor, 15, 20
Window Wall, 15, 22
Wooden Deck, 26
Work Stoppage, 5

INDEX: Points of Law

Paragraph Numbers

A

Acceleration, 19.4
Additional Insureds 11.5, 12.9, 17.3
ADR (Alternative Dispute Resolution), 2.13
AGC Form Contracts, 14.1
Agency Law, 13.1
Agent, 13.1, 26.5
Agreement, Construction, 16.5
AIA Documents, 14.1, 14.2
Air Conditioning Subcontractor, 21.3, 21.4
All Risk Insurance, 4.2, 19.7
Allocation of Risk, 1.1
Alternative Dispute Resolution (ADR), 2.13
American Arbitration Association, 2.13, 3.2, 14.11
Anti Deficiency Legislation, 28.9
Application of Payments for Materials, 8.3, 8.4, 8.5
Arbitration Appeal, 2.13
Arbitration Award, 1.4, 2.13, 3.2, 19.9
Arbitration, 1.4, 2.13, 3.2, 14.11, 19.9, 20.3
Arbitrator, 1.3, 1.4, 2.13, 3.2
Architect's Consultants, 12.2, 14.13, 25.6
Architect's Decision, 2.13, 2.14, 14.11, 14.12
Architect's Estimate, 16.3, 17.1, 20.7, 20.9
Architect's Rulings, 3.2
Architect's Standard of Care, 27.3
Architects, 2.4, 11.8, 15.2, 20.9, 21.4, 21.6, 25.6, 27.3
Architectural Malpractice, 16.4
Attorney Fees, 8.7, 8.8, 14.13, 15.9, 15.11, 19.9, 21.2, 21.3, 21.6, 28.9
Award of Contract, 12.3, 18.1

B

Backcharges, 5.4, 14.13, 19.5
Bankruptcy, 8.2, 8.6, 8.7, 8.8, 12.14, 22.2, 28.5, 28.8
Bid Bond, 12.3
Bid Shopping, 16.7, 18.7
Bidding Process, 16.7, 28.1
Bodily Injury, 13.2, 25.3
Bond, Mechanics Lien, 3.4
Bonding Capacity, 8.1, 25.2
Bonds, 3.4, 8.1, 8.8, 11.6, 12.3, 12.5, 12.10, 12.14, 17.2, 21.4, 25.2, 28.7
Boundary Survey, 16.1
Breach of Contract, 2.1, 2.4, 2.5, 2.6, 7.1, 8.1, 10.2, 12.1, 12.11, 13.2, 15.3, 15.5, 16.6, 23.2, 24.7, 28.3
Breach of Contract, Economic, 2.7
Breach of Warranty, 22.2, 25.11
Bribery, 25.7, 28.2
Broad Form Liability Insurance, 9.6
Builders Risk Insurance, 17.5
Building Codes, 10.1, 11.10, 16.4, 20.1, 22.1
Building Department, 11.10
Building Inspector, 1.1, 10.1, 11.10, 20.1
Business Risk, 13.2

C

Cause of Defect, 24.4
CCD (Construction Change Directive), 14.7
Certificates of Insurance, 12.6, 12.9, 12.12, 17.2, 17.3, 17.4
CGL Insurance, 12.9, 15.8, 15.11, 19.7, 25.1, 25.7
Change Orders, 3.1, 21.2, 21.3, 21.4, 21.5, 21.6
Changes, 14.6
Claimant, Insurance, 4.3
Claims Asserted by Third Parties, 4.3
Claims Managers, 15.9
Claims, 3.2, 15.5
Clerical Error, 17.1
Collapse Coverage, 6.4
Commercial General Liability (CGL) Insurance, 12.9, 15.8, 15.11, 19.7, 25.1, 25.7
Comparative Fault, 6.6
Compensatory Damages, 11.3, 12.1
Competent Parties, 12.3, 14.6, 20.1
Competitive Bidding, 12.3, 16.7, 18.3, 28.1
Completion Date, 11.3
Concrete Subcontractors, 2.4
Conflict of Interest, 14.12
Consideration in a Contract, 12.3, 14.6
Construction Agreement, 16.5
Construction Change Directive (CCD), 14.7
Construction Contracts, 11.2, 12.3
Construction Defects, 15.1, 15.6, 15.11, 24
Construction Documents for Owner-Builder-Developer, 27
Construction Drawings, 14.2, 16.4, 16.5
Construction Lender, 21.4, 28.8, 28.9
Construction Loan, 17.6
Construction Schedule, 5.2, 5.6, 12.6
Construction Specifications, 16.5

Constructive Acceleration, 19.4
Consultants, Architect's, 12.2, 14.13
Contents Insurance, 4.2
Contract Award, 12.3
Contract Documents, 12.3, 14.1, 16.5, 16.7
Contract Drawings, 14.2
Contract Law, 9.1, 9.2, 12.11, 15.1, 15.3, 15.11, 18.1, 20.1
Contract Provisions, 2.10
Contractor Overbilling, 12.11
Contractor's Duties, 16.1
Contractor's License Bond, 11.9
Contractor's License, 8.1, 11.9, 20.1, 20.9, 21.5
Contribution and Indemnity, 6.8, 25.9
Cooperation, 5.2, 5.4
Coordinating Engineering Consultants, 12.2
Coordination of a Construction Project, 19.1
Corporations, 8.2, 28.10
Corruption, 25.7
Cracked Concrete Slab, 2
Credit Ratings, 8.1
Crime, 25.7, 28.2
Cross-Claims, 5.2

D

Damages for Breach of Contract, 2.2
Damages for Delay, 2.2
Damages, 18.5
Deed of Trust, 17.6
Default, 2.5
Defective Construction, 11.7
Defendant, 15.6, 15.11
Demand for Arbitration, 14.11
Developer, 20.4, 24.7, 27
Deviation from the Specifications, 23.1
Difficult Client for an Architect, 16
Dishonesty, 9.1
Dispute Resolution, 2.13, 14.11, 14.13
Doctrine of Contribution and Indemnity, 6.8, 25,9
Doctrine of Efficient Breach of Contract, 23.5
Doctrine of Equitable Contribution, 15.7, 15.10, 15.11
Doctrine of Illegality, 20.1
Doctrine of Implied Covenant of Cooperation, 5.2
Doctrine of Impossibility, 5.2, 17.7
Doctrine of Peculiar Risk, 26.2
Doctrine of Prevention of Performance, 2.12
Doctrine of Quantum Meruit, 1.3, 10.5
Doctrine of Respondiat Superior, 25.5
Doctrine of Severability, 23.4, 23.6
Doctrine of Strict Liability, 24.7, 25.12

Doctrine of Subrogation, 8.6, 19.8
Doctrine of Substantial Performance, 2.6, 2.8, 23.3, 23.6
Doctrine of Unclean Hands, 9.8
Document Submissions, 12.6
Drawings and Specifications, 2.3
Drawings, Construction, 14.2, 16.5
Drawings, Shop, 14.4, 15

E

Earth Movement Exclusion, 6.2, 6.3
Earthquake Exclusion, 4.2, 17.5
Economic Breach of Contract, 2.7
Economic Loss, 15.3, 25.3
Economic Waste, 2.6, 2.8
Efficient Breach of Contract, 23.5
Efficient Cause, 24.5
Elementary Sense of Justice, 1.1
Embezzlement, 28.2, 28.10
Employee vs Independent Contractor, 6.7, 26.1, 26.4
Engineers, 2.4
Equitable Contribution, 15.7, 15.10, 15.11
Equitable Remedy, 9.8
Error in Judgment, 17.1
Error, Clerical, 17.1
Errors and Omissions Insurance, 7.2, 15.8, 15.11, 25.1, 25.7
Ethics of Competitive Bidding, 18.1, 18.7, 28.1
Exception to the Exception, 20.3
Excuse of Performance, 2.6
Excused Breach of Contract, 12.11
Expert Testimony, 14.5, 25.8
Expert Witnesses, 2.4
Express Covenant, 5.5
Express Warranty, 25.10
Extra Work, 1, 1.3, 3.1, 3.2, 21.2, 21.3

F

Fast Track Projects, 12.3
Faulty Construction, 4.1, 24.3
Faulty Design Exclusion, 6.2
Faulty Design, 4.1, 6.2, 7.1, 24.3, 24.6
Faulty Inspection, 9.1, 9.4, 15.3
Faulty Maintenance, 4.1, 7.2
Faulty Materials, 4.1, 22.1
Faulty Workmanship Exclusion, 6.2, 24.6
Faulty Workmanship, 7.1
Final Payment, 3.4
Finality of Arbitration Award, 1.4
Financial Responsibility, 13.1

Fire Insurance, 4.2, 8.6
Fire Loss, 8.6
Flood Exclusion, 4.2, 17.5
Flow Down Clause, 14.8
Foolproof Construction Contract, 1
Foreclosure of Mechanics Lien, 12.13, 21.2, 21.3, 21.4
Forensic Architect, 15.8
Form Contracts, 14.1
Fraud, 9.4, 15.2, 15.10, 15.11, 20.6, 20.8, 22.2
Fraudulent Shop Drawings, 15

G

General Conditions, 16.5
General Conditions, AIA Document A201, 2.13, 3.2, 8.6, 14.2
General Contractor, 25.3
General Liability Insurance, 12.9, 15.8
Geotechnical Report, 16.2
Glulam Beam Failure, 7
Good Faith Settlement, 15.10

H

Habitability, 9.5
Home Improvement Contractor, 11
Homeowners Insurance, 4.4, 9.7, 13.3

I

Illegality, 20.1
Implied Contract, 18.3
Implied Covenant of Cooperation, 5.2, 5.4
Implied Covenant, 5.5, 10.2, 10.5
Implied Promises, 10.2
Implied Warranty of Habitability, 9.5
Implied Warranty, 25.4
Impossibility of Performance, 2.6, 5.2, 17.7
In Pari Delicto, 20.2
Indemnification, 25.9
Indemnity, 27.4
Independent Contractor, 6.7, 26.1
Insurance Adjusters, 15.9
Insurance Advisors, 17.4
Insurance Certificates, 12.6, 12.9, 12.12, 17.2, 17.3, 17.4
Insurance Claims Managers, 15.9
Insurance, 4.1, 4.2, 4.3, 4.4, 6.1, 7.2, 8.6, 9.6, 9.7, 11.5, 12.9, 12.12, 13.1, 13.2, 13.3, 15.8, 15.10, 15.11, 17.2, 17.5, 19.7, 24.2, 24.3, 24.7, 25.1, 25.2, 25.7, 27.2, 27.3, 28.10
Intentional Misconduct, 9, 9.3, 23.3, 25.7
Intentional Tort, 15.1, 15.2

Interest on Late Payments, 16.6
Interference by the Owner, 19.1, 19.2
Interpretation of Contracts, 2.3
Invitation to Bid, 18.3

J

Joint Control, 11.7
Joint Venture, 8, 8.1, 8.2
Judgment Error, 17.1
Jurisdictional Dispute, 5.1
Justice, Elementary Sense of, 1.1

L

Labor Relations, 5
Land Surveys, 16.1
Landslide Exclusion, 6.2, 6.3
Latent Defects, 6.2, 7.3, 24.3, 24.6
Law of Agency, 13.1
Law of Contracts, 9.1, 9.2, 12.11, 15.11, 18.1, 20.1
Law of Sales, 9.1, 9.2, 9.5
Law of Torts, 9.1, 9.2, 15.1, 18.1
Legal Counsel, 17.4
Legal Impossibility, 5.2, 5.3
Legal Object of Contract, 12.3, 14.6, 20.1
Legal Rate of Interest, 16.6
Legally Enforceable Contract, 12.3
Liability 6.1, 8.1, 8.2, 12.9, 13.1, 15.1, 15.11, 16.1, 17.3, 25, 25.1, 26.4, 26.5, 27.1, 28.10
Liability Coverage, 6.5
Liability Insurance, 4.2, 6.1, 6.8, 7.2, 9.3, 9.6, 11.5, 12.9, 12.12, 13.1, 13.2, 19.7, 25.7
Liability Insurance, Broad Form, 9.6
Liability Insurance, Standard Form, 9.6
License Bond, 11.9
Limited Partners, 8.2
Liquidated Damages, 11.3, 12.1, 18.4, 18.6
Low Bid, 9.1, 17.1, 18.3
Low Bidder, 8.1, 18.1, 28.1

M

Material Breach of Contract, 2.5, 12.11
Material Mistake, 16.7, 17.1
Materials Supplier, 8.3, 8.4, 8.5
Measure of Damages, 2.9, 12.1, 19.2
Mechanics Lien Bond, 3.4
Mechanics Lien, 3.3, 3.4, 8.3, 8.4, 8.5, 9.8, 11.4, 11.7, 12.5, 12.13, 12.14, 17.6, 18.8, 21.2, 21.3, 21.4, 28.6, 28.8, 28.9
Mediation, 2.13, 21.7
Mediator, 2.13, 21.7
Meeting of the Minds, 12.3

Merchant Builder, 20.4, 24.7
Misinformation Given by Owner, 10.1
Misrepresentation, 20.6, 22.2
Mistake, 17.1
Mortgage, 17.6
Multiple Parties, 15.6
Mutual Manifestation of Consent, 12.3, 14.6, 20.1

N

Named Peril Insurance, 19.7
NECA Form Contracts, 14.1
Negligence, 6.3, 7.1, 9.4, 13.2, 14.13, 15.1, 15.2, 15.5, 24.6, 24.7, 25.7
Negotiated Contract, 12.3, 12.4, 28.1
No Damages for Delay, 19.1
Notices and Procedures, 2.11
NSPE Form Contracts, 14.1

O

Ode to the AIA General Conditions, 14.2
Offer and Acceptance, 12.3, 17.1, 18.2
Offsets, 19.5
One-Year Warranty, 7.1
Opinion Testimony, 14.5
Or Equal, 23.2
Oral Contracts, 19.6
Overbilling by Contractor, 12.11
Owner Getting Something for Nothing, 1.1
Owner's Duty to Coordinate Separate Contractors, 19.1
Owner's Responsibilities, 16.1
Owner's Risk, 1.1
Owner-Builder, 20.4

P

Package Deal, 4.4
Partnerships, 8.2
Pass Through Clause, 14.10
Patent Defect, 7.7
Pay-if-Paid Clause, 21.3, 21.5
Payment Bond, 12.5, 12.10, 12.14
Payments for Materials, Application of, 8.3, 8.4, 8.5
Peculiar Risk, 26.2
Penalty Clause, 12.1, 18.4, 18.6
Perfect Construction, 2.1, 2.3
Performance Bond, 8.1, 11.6, 12.5, 12.10, 12.14, 17.2, 17.4, 25.2, 28.7
Performance of a contract, 5.2, 23.3, 23.6
Personal Liability, 8.2
Physical Damage, 15.3

Plugged Roof Drains, 4
Plumbing Subcontractor, 10.3, 13.3, 21.4
Pressure Regulator, 10
Prevention of Performance, 2.12
Prime Contractor, 2.4, 3.1, 15.2, 25.3
Principal, 13.1, 26.5
Priority, 17.6
Privity of Contract, 25.11
Progress Payments, 2.1, 2.2, 3.3, 12.8, 12.11
Promissory Estoppel, 18.3
Property Coverage, 6.2
Property Damage, 7.1, 12.12, 13.2, 25.13
Property Insurance, 4.2, 6.2, 7.2, 9.7, 12.12, 13.3, 19.7, 24, 25.1
Protecting the Interests of the Owner, 11.1

Q

Quantum Meruit, 1.3, 10.4, 10.5

R

Rain Exclusion, 24.6
Reading Plans, 14.5
Recission of Contract, 2.5, 28.4
Remedies for Breach of Contract, 2.2
Rental Value, 2.2, 11.3, 18.6, 20.7
Respondiat Superior, 25.5
Responsible Managing Officer, 13.1
Retaining Wall Collapse, 6
Retention, 2.1, 3.3, 11.4, 12.8
Risk, Allocation of, 1.1
Roofing Consultants, 4.5
Rough Sense of Justice, 10.5

S

Sales Law, 9.1, 9.2, 9.5
Schedule of Values, 12.6, 12.8
Scheduling, 5.2
Separate Contractors, 19.1
Settlement Agreement, 21.8
Settlement of Claims, 15.10
Severability, 23.4
Severable Contract, 23.4, 23.6
Shop Drawing Stamp, 14.4
Shop Drawings, 14, 14.4, 15, 15.2
Simple Justice, 3.1
Soil Report, 16.2
Soil Settlement Exclusion, 13.3
Soil Testing, 1.1
Sole Proprietorship, 8.2
Special Conditions, 16.5
Specifications, 16.5

413

Speculative Building, 20.4
Spike the Job, 12.6
Standard Form Contracts, 14.1
Standard Form Liability Insurance, , 9.6
Statute of Limitations, 7.1, 7.3
Straight Thinking vs Legal Thinking, 10.5
Strict Liability, 24.7, 25.12, 26.4, 27.2
Subcontract, 14.8, 14.10
Subcontractor List, 12.6, 12.7
Subcontractor, Air Conditioning, 21.3, 21.4
Subcontractor, Electrical, 25.4, 25.5
Subcontractor, Plumbing, 10.3, 13.3, 21.2, 21.4
Subcontractor's Duty, 27.2
Subcontractors Listing Laws, 18.7
Subcontractors, 16.7, 17.3, 25.3, 25.4, 25.5, 28.2
Subcontractors, Concrete, 2.4
Subrogation, 8.6, 19.8
Subrogation, Waiver of, 8.6
Substandard Bonds, 17.4
Substandard Maintenance, 4.5
Substantial Performance, 2.6, 2.8, 23.3, 23.6
Substitutions, 23
Surety Bond, 3.4, 8.1, 11.6, 12.3, 12.3, 12.5, 12.10, 12.14, 17.2, 25.2, 25.7, 28.7
Surveys, 16.1

T

Third Parties, 4.3, 27.4
Third Party Beneficiary, 15.4, 25.4
Three Party Insurance, 4.3
Tidal Wave Exclusion, 4.2
Timely Performance, 5.4, 18.6
Title Policy, 3.4
Topographic Survey, 16.1
Tort Law, 9.1, 9.2, 15.1, 15.3, 15.5, 15.11, 18.1
Torts vs Contracts, 9.3, 15.1
Total Cost Method, 19.2
Trade Practice, 2.4, 5.7, 25.8
Trust Deed, 17.6
Two Party Insurance, 4.3

U

Ultra Hazardous Activities, 26.2, 26.4
Unanticipated Underground Conditions, 1.2
Unclean Hands, 9.8
Unethical Bidding Practices, 18.1, 18.7, 28
Unexcused Breach of Contract, 12.11
Unexpected Events, 1.2
Unintentional Deviations from the Contract Requirements, 2.7, 2.8, 23.3
Unjust Enrichment, 1.2, 2.5, 9.8, 10.4

Unlicensed Contractor, 20.1, 26.3

V

Value Engineering, 12.4, 20.6, 20.7
Value of Extra Work, 1.3
Vicarious Liability, 25.9

W

Waiver of Subrogation, 8.6, 17.5
Warranty, 7.1
Water Intrusion, 4.1, 4.5, 9.1, 9.5, 9.7, 11.5
Windfall to Owner, 23.6
Withholding Information, 20.8
Worthless Claim, 22.3
Written Contracts, 12.3

Z

Zoning Codes, 16.4